Summer's last stand in the author's wild garden at Atlantic, Iowa. Masses of white snakeroot and aster in bloom in October. INSIDE COVER AND FLYLEAF. Masses of Pellett clover in bloom. BELOW — Memorial stone in Frank Pellett's gardens at Atlantic, Iowa.

THIS FIVE ACRE WOODLAND
SET ASIDE AND USED BY HIM
FOR MORE THAN 40 YEARS
IN THE STUDY OF NATURE
DEDICATED AS A
LIVING MEMORIAL TO
FRANK CHAPMAN PELLETT
NATURALIST – HUMANITARIAN
1879 – 1951
IOWA STATE HORTICULTURAL SOCIETY
AFFILIATED SOCIETIES AND FRIENDS
JULY 11, 1954

OTHER DADANT PUBLICATIONS

American Bee Journal

First Lessons in Beekeeping

The Hive and the Honey Bee

American Honey Plants

Together With Those Which Are of Special Value to the Beekeeper as Sources of Pollen

By

FRANK C. PELLETT

Field Editor American Bee Journal; Former State Apiarist of Iowa
Author Beginner's Bee Book; Practical Queen Rearing;
History of American Beekeeping;
A Living From Bees; etc.

Edited by

DADANT & SONS

FIFTH EDITION

DADANT & SONS • HAMILTON, ILLINOIS

Library of Congress Catalog Card Number 75-38346

ISBN Number 0-915698-01-3

PRINTED IN THE U.S.A. BY JOURNAL PRINTING COMPANY
CARTHAGE, ILLINOIS

Author's Dedication of the Fourth Edition

To the Memory of

MY INDULGENT PARENTS

who early encouraged me in my passion for the study of Nature, this book is affectionately dedicated

Frank C. Pellett's Creed

"The Universe is mine with all eternity to explore.

My limitations are only such as I myself shall make.

No one can injure me but myself.

The greatest calamity that can befall me is but temporary and in the light of the future will seem but a trifle.

I will therefore be serene, unruffled and content, knowing that if the thing which I desire is beyond me today it will come to me tomorrow."

Frank Chapman Pellett
1879 - 1951

In Memoriam

FRANK C. PELLETT
1879 - 1951

In the course of his life, Frank wrote thirteen books and co-authored several others, including his chapters in the 1946 and 1949 editions of "The Hive and the Honey Bee." His books include "Our Backdoor Neighbors," "Birds of the Wild," "Flowers of the Wild," "How to Attract Birds," "Success with Wild Flowers," "History of American Beekeeping," "Beginner's Bee Book," "The Romance of the Hive," "A Living with Bees," "Productive Beekeeping," "Practical Queen Rearing," and with his son, "Practical Tomato Culture."

The great monument, however, which he created and which will stand as both a challenge and a beacon is this book, "American Honey Plants." We are proud to be able to publish again this single, most complete work on honey plants.

We could write several pages devoted to the accomplishments of Frank Pellett, and to the honors that were poured upon him during his lifetime. But these would be cold facts that would do little to divulge the kindness, the warmth, and the enthusiasm of the man.

My first real memory of Frank and his family was as a small child of five years of age. Our family had just moved a short distance from the Pellett household, and I was first introduced to Frank and his wonderful family — his devoted wife Ada; sons Kent, Melvin and Fred; and his daughter Ruth (Mrs. William Barnard). Frank loved his family as he loved his work — and that love managed to spill over to all of his friends and neighbors. He approached life with enthusiasm and spread that same enthusiasm to his colleagues and to his friends.

So many memories, and all of them pleasant: Walking with Frank to town on a Sunday morning to buy the newspaper; looking through his wonderful butterfly collection; hearing the stories of the time he spent with American Indians, and seeing his collection of Indian artifacts; hearing of his work in the Honey Plant Test Garden at Atlantic, Iowa; his words of encouragement and advice as I grew up and entered the world of beekeeping.

Many of us here at Dadant & Sons remember Frank C. Pellett. He always had a kind word for everyone; all people were his friends. He was never too busy to help the other fellow and never resented, but rather applauded, the work done by someone else, whether or not it was at his instigation. We will always remember his enthusiasm, his philosophy of life, and his fullness of purpose which made him a great man.

G. H. Cale, Jr., Ph.D.
Dadant & Sons, Inc.

PREFACE TO FOURTH EDITION

When the first edition of American Honey Plants appeared in **1920**, there was no book in the English language dealing with the source of bee pasture. The literature dealing with the subject was scant and consisted of widely scattered references in bee magazines and a few brief bulletins. Each succeeding edition has been greatly enlarged and new subjects added. In addition to personal visits to most of the important beekeeping areas of the United States, Canada and Mexico, the author has corresponded with hundreds of beekeepers and has made a wide search of agricultural libraries in order to digest all important references to the subject.

After thirty years of effort the result is far from satisfactory since so much remains to be learned concerning the behavior of plants under different environmental conditions. It is a well known fact that plants which yield nectar freely in some areas yield little or nothing under other conditions. The quality of the honey from the same plants varies from year to year in the same location and varies even more in localities with different soils or climatic conditions. It is only possible to record the reports of different observers without explaining why they so often fail to agree.

An effort has been made to give due credit for the source of information with each plant discussed but there still remain dozens of observers who have given the author assistance in one way or another to whom he is indebted for much of the value which this book may possess.

The American Bee Journal maintains a honey plant test garden under the author's care. Hundreds of plants are under observation and much useful information has thus been secured.

Hamilton, Illinois FRANK CHAPMAN PELLETT

SOME PUBLICATIONS CONSULTED

Flora of Middle Western California—Willis Linn Jepson. Cunningham, Curtiss & Welch, San Francisco, 1911.

Manual of the Plants of Western Texas—John M. Coulter. Government Printing Office, Washington, 1891.

Gray's New Manual of Botany, seventh edition. American Book Company, Chicago, 1908.

Flora of New Mexico—E. O. Wooten and Paul C. Standley. Government Printing Office, Washington, 1915.

Flora of the Southeastern United States—John Kunkel Small, 1913.

Flora of the Northwest Coast—Charles V. Piper and R. Kent Beattie, 1915.

Flora of the Rocky Mountains and Adjacent Plains—P. A. Rydberg, 1917.

Select Extra-Tropical Plants—Ferd Von Mueller, Government Printer, Melbourne, Australia, 1895.

Origin of Cultivated Plants—Alphonse de Candolle. D. Appleton & Co., New York, 1919.

Plant Life of Alabama—Charles Mohr. State Geologist, Montgomery, Ala., 1901.

Weed Flora of Iowa—L. H. Pammel. Iowa Geological Survey, Des Moines, 1913.

Nouvelle Flore—Gaston Bonnier, Paris, 1916.

Manual of Poisonous Plants—L. H. Pammel, 1911.

Sturtevant's Notes on Edible Plants, 1919.

Honey Plants of California—M. C. Richter. Bul. 217, Agricultural Experiment Station, Berkeley, Cal., 1911.

Texas Honey Plants—C. E. Sanborn and E. E. Scholl. Bul. 102, Agricultural Experiment Station, College Station, Texas, 1908.

A, B, C and X, Y, Z of Bee Culture—Floral notes by John H. Lovell. A. I. Root Company, Medina, Ohio, 1919.

Flora of Southeastern Washington and Adjacent Idaho—C. V. Piper and R. Kent Beattie, 1914.

Flora of Northern United States and Canada—Britton and Brown. Charles Scribner's Sons, New York, 1913.

The Beekeeper's Directory—J. S. Harbison. H. H. Bancroft & Co., San Francisco, 1861.

Money in Bees in Australasia—Tarlton Rayment. Whitcomb & Tombs, Melbourne, 1918.

A Preliminary List of the Honey Producing Plants of Nebraska by

Charles E. Bessey. No. 40. Bulletin of the Agricultural Experiment Station of Nebraska. Vol. VII, Article IV. 1895.

Honey Plants of North America—John H. Lovell, 1926.

Profitable Honey Plants of Australasia—Tarlton Rayment, Whitcomb & Tombs, Melbourne.

Factors affecting the Usefulness of Honeybees in Pollination, Circular 650 U.S.D.A., by George H. Vansell, 1942.

Nectar and Pollen Plants of California, by G. H. Vansell, Bulletin 517, Cali. College of Agriculture. 1931. Revised 1941, by Vansell and Eckert.

Nectar and Pollen Plants of Oregon, Bulletin 412, Oregon Agr. Experiment Station, by H. A. Scullen and G. H. Vansell, 1942.

Honey Plants of Iowa, Bul. 7, Iowa Geological Survey, by L. H. Pammel and others, 1930.

Illinois Honey and Pollen Plants, by V. G. Milum, University of Illinois, 1943.

Honey and Pollen Plants of the United States, by E. Oertel, U.S.D.A. Agr. Circular 554, 1939.

Dependence of Agriculture on the Beekeeping Industry, U.S.D.A. Bureau of Entomology, E-584.

Valuable Plants Native to Texas, by H. B. Parks, Bulletin 551, Texas Agricultural Experiment Station, 1937.

Plantas Meliferas de Cuba, by Gonzalo S. Ordetx, Ministry of Agriculture, Havana, Cuba, March, 1944.

European Bee Plants and Their Pollen, by Rev. M. Yate Allen, Bee Kingdom League, Alexandria, Egypt, 1937.

Honey and Pollen of New South Wales, W. A. Goodacre, Sydney, Government Printer, 1938.

Plants and Beekeeping, by F. N. Howes, Faber and Faber, London, 1945.

AMERICAN HONEY PLANTS

The late Prof. A. J. Cook estimated that there are nearly eighteen hundred species of plants on which bees work in America. Most of these are minor sources, which the bees visit incidentally for the minute quantity of nectar that may be available, or for pollen. There are some plants rich in nectar which can never be important to the beekeeper because they are not sufficiently plentiful.

Honey production, as a business enterprise, is dependent upon a few species which yield nectar abundantly and which are sufficiently common to enable the bees to secure honey in large quantity. In order to make the most of his business, the beekeeper should have a thorough knowledge of the honey plants in all the country surrounding his apiaries. It often happens that a distance of but a few miles makes a great difference with the available honey sources. Many a man by moving an apiary a few miles has greatly increased the yield. It sometimes happens that the plant which is the main dependence will fail, and that by moving to some other source, a crop may be harvested. To know fully the honey plants of his region, their time of blooming and habit of nectar secretion under his particular conditions, is of fundamental importance to the man who would succeed as a beekeeper.

In many places the presence or absence of a single plant determines whether or not beekeeping is worth while. Over a large portion of the Middle West, the beekeepers depend almost entirely upon white clover for surplus, and in seasons when this plant fails they get no honey to sell.

Likewise, in many localities in the irrigated regions of the Rocky Mountain States, when alfalfa fails to yield, there is no surplus honey. Yet in all these sections, beekeeping would be impossible if there were no other plants. There are localities where tremendous honeyflows occur for a short period of time, where beekeeping is not practical because there is insufficient forage to support the bees the rest of the year. In such places beekeepers often take advantage of the flows by moving the bees away as soon as the plants cease to yield nectar, and returning them the following year at blooming time. This applies to some parts of the valley of the Appalachicola River in Florida. While the flow from Tupelo is sometimes remarkable, there is a shortage of pollen throughout the summer months.

The ideal situation for beekeeping is one where there are at least three plants which yield surplus honey in considerable quantity, and which bloom at different periods. Beside the main sources, there should be a great variety of minor plants yielding both pollen and honey throughout the season to support the bees between the main flows. In such a situation, there is seldom an entire failure of the honey crop; and, in good years, the beekeeper fares well, indeed.

There are many localities where the bees suffer seriously for lack of pollen at some seasons of the year. An available source of pollen is second only in importance to an abundant honeyflow. This being the case, the plants which are generally regarded as valuable for pollen, especially those blooming at seasons when pollen is not abundant generally, are included in this book.

HONEY PLANT REGIONS

Several attempts have been made to outline the principal regions of the United States. A careful examination of all these outlines brings out serious discrepancies. There are too many small regions within larger ones to permit of anything like accuracy with the present data and the present knowledge of the honey plants. In general, white clover may be said to be the principal honey plant of all the region from Nova Scotia west to eastern Dakota and south to Tennessee and Arkansas. Yet within that large area, there are many places where white clover is unimportant, and where other plants furnish the principal surplus. In much of Michigan white clover is of first importance, yet in the cut-over districts of the northern part of the State, raspberry, fireweed and milkweed furnish nearly all the honey that goes to market. It is good clover territory, and with the ultimate development of the region, clover will predominate.

In the irrigated regions of the Rocky Mountain States, alfalfa is the principal source of surplus, but sweet clover is rapidly crowding it for first place.

Basswood was once a very important source of honey over all the Northeastern States. The cutting of the basswood forests has gradually reduced the basswood area until there are now few localities, in Ohio, Indiana, Illinois or Iowa, where it is really an important honey source. In parts of Minnesota, Wisconsin, Michigan and Ontario, basswood is still sufficiently plentiful to yield large quantities of honey, but there it is being rapidly reduced.

In the cotton belt, where cotton would naturally be expected to be the principal source, the area would be divided into many small regions. Cotton may yield much honey in a locality where the soil is heavy and rich, while a few miles distant, where soils are light and sandy, there is little honey from cotton, although the plant is just as commonly cultivated. In the cotton region there would be a great many sub-divisions. In parts of Texas, mesquite is the principal source, in other catsclaw and huajillo (wa-he-ya), while in eastern Texas basswood yields heavily. Buckwheat is important principally in the region about the Great Lakes and south in the higher elevations to Virginia and Tennessee. Goldenrod is one of the most important sources of nectar in New England, while it is seldom of much value west of the Mississippi River, although growing abundantly.

In California and Florida there are several entirely different regions within the State. There is no one plant of major importance over all parts of either State. A large amount of work still remains to be done before the honey resources of America can be mapped out with anything like accuracy. Changing conditions are rapidly removing one plant and substituting another in many sections. When the author visited west Texas he was told by the beekeepers there that the clearing of the land and planting it to cultivated crops was rapidly curtailing the bee range, as no cultivated crops being planted were equal to the desert flora which was being removed.

In other sections, the planting of forage crops which are good sources of nectar, like alfalfa and sweet clover, is greatly increasing the available bee pasturage. In parts of California, the extensive growing of garden seeds is providing pasture sufficient for producing surplus honey of a kind seldom heard of in the markets a few years ago. Parsnip and celery honey are examples.

THE MINOR PLANTS

Although only a few dozen plants are important sources of surplus honey, there are hundreds of minor plants which are of value for the support they give the bees when no major plant is in bloom. The number and variety of these plants will largely determine the value of the locality, and whether it will be necessary for the beekeeper to resort to migratory beekeeping at times.

Blossoms of huisache (*Acacia farnesiana*).

Catnip is famous as a bee plant, yet it is doubtful whether a single pound of catnip honey was ever stored in America, unmixed with honey from other sources. If catnip could be grown in large fields like clover, it is probable that catnip honey would appear in the markets.

If the beekeeper is familiar with the minor plants, he will be able to locate outyards where the bees will be able to gather enough nectar from such sources to keep his colonies in the best possible condition for the surplus flows. It is a well-known fact that it is only the big colonies which produce large crops of surplus honey. A little nectar coming to the hive for some time in advance of the main flow, is the best possible stimulant for brood rearing. It often happens that

Golden wattle (*Acacia longifolia*).

bees will be poorly prepared for the harvest in one yard, while others only two or three miles distant will be in the best possible condition, because of the presence of some minor plants not within reach of the first.

A

ACACIA

The genus acacia composes a large group of trees and shrubs common to warm countries. They are particularly abundant in Australia which country claims 417 species. Tarlton Rayment lists them as the source of pollen only. Several of the Australian species are commonly grown in California and these apparently yield no honey.

Several species native to this country yield honey in great abundance and two, the huajillo, (*Acacia berlandieri*) and the catsclaw, (*Acacia greggii*) are famous as the source of large crops of Texas honey. (See Huajillo and Catsclaw.)

The sweet acacia in Texas commonly called Huisache, is found along the Gulf Coast in Alabama and as far east as South Carolina and westward to California. This species apparently yields little if any nectar and is visited by the bees in large numbers for pollen in early spring.

Catsclaw is reported as yielding in April in Texas but in Arizona as coming much later.

Among Australian species cultivated in this country may be mentioned the black wattle, (*Acacia decurrens mollis*), the Sydney golden wattle, (*Acacia longifolia*) and Acacia melanoxylon. All are reported as the source of abundant pollen. L. H. Bailey in his Cyclopaedia of Horticulture lists 67 species grown as ornamentals in this country.

Native species of Acacia are more abundant southward in Mexico where many different kinds are known to occur. Some are found in Central America and southward to Brazil in South America.

ACER, see Maple.
ADAM'S NEEDLE, see Yucca.
AESCULUS, see Buckeye.
AGARITA, see Barberry.
AGAVE, see Century Plant.

ALABAMA—Honey Sources of.

The total area of Alabama is slightly less than 52,000 square miles. The altitude ranges from sea level at the Gulf Coast to 2,407 feet at the top of Cheawha Mountain near the line of Clay and Talladega Counties, in the northeastern part.

Due to the proximity to the Gulf of Mexico, Alabama has a mild climate with long summer and short and mild winter. Extremes of temperature are rare, freezing weather seldom lasting for more than 48 hours. The winter and spring months are often accompanied by long periods of cloudy and rainy weather. The average annual rainfall is slightly above fifty inches which is distributed over the year in such manner as to provide ample moisture for crops at all seasons. Along the Gulf Coast it is more than sixty inches which is heavier than in any other portion of the United States except the North Pacific Coast. July is the wettest month in the year.

Frost rarely occurs in Alabama between the first of April and the

Acacia or black wattle (*Acacia decurrens mollis*).

end of October and pastures remain green throughout most of the year.

The long season together with a large variety of plants which produce nectar and pollen make it a favorable location for the breeding of bees. The sale of live bees and queens is of special importance in Alabama where many beekeepers specialize in this field.

There is a district in Alabama where sweet clover is the principal source of surplus honey. In this region good crops are the rule. In addition, rattan, tulip-poplar, black gum, hawthorne, field peas, privet, locust, redbud, cotton, bitterweed, asters and occasionally white clover, yield honey. There is the usual spring stimulation from fruit blossoms and willows in Alabama and a large number of minor sources which add something to the total yield, but which alone are unimportant.

The yellow jasmine is common over much of the state, particularly in the coast pine belt. Holly is represented by at least ten species, some of which are found throughout the state. Gallberry is most important. The tulip tree is found in the mountain region and is of most importance in the Tennessee Valley. The China-berry is commonly grown as an ornamental and blooms in March. Several species of magnolia occur in Alabama. Some are found in the mountain region while others are best adapted to the coast region and central prairie belt. Sourwood is found from the mountain region to the central plain. Ti-ti is an important source of surplus in the coast region.

ALASKA—Honey Sources of.

By the accounts given in Bancroft's History of Alaska and in translations made by Rev. George Kotteometinoff from the records of the Orthodox Russo-Greek Church at Sitka, the honeybee was first introduced into Alaska in 1809 by a monk named Cherepenin. These bees came from the Department of Kazan, in Siberia, and were brought that honey might be added to the scanty food supply of the pioneer teachers of the Faith as well as to supply the candles for the church services. By decree of Church, only beeswax candles can be used, and it is recorded that at Sitka, in 1816, no services could be held for six months because the supply of wax ran out. As early as 1819 apiculture was taught in the church school and was continued up to 1894. It would appear that the bees never flourished and seldom swarmed. There are a number of records of new importations to take the place of dead colonies. Very early a white clover was introduced to help out the honey supply. About 1830 bees were taken from Sitka to Fort Ross in California. As late as 1905 there were about 30 colonies at the Russian school at Sitka. These bees were in straw skeps and were kept on shelves under the eaves of the house. In winter they were kept within the same projecting eaves. In 1906 the Experiment Farm at Sitka made an unsuccessful attempt to keep bees in Langstroth hives. It is not probable that beekeeping will ever be a commercial project in

Alaska. References to beekeeping at Sitka by Dr. Sheldon Jackson are to be found in the Report on Education in Alaska, Bureau of Education. Prof. C. C. Georgeson, in the reports on work done at the Experiment Station in Alaska also mentions beekeeping. Bees were observed collecting nectar and pollen from plants given below during the years 1905 to 1912. It should be observed that a majority of these plants have pendulous flowers. In a climate such as at Sitka, where the normal precipitation is 120 inches, only pendulous flowers could protect the nectar:

Willow (*Salix speciosa*).
Crab Apple (*Pyrus rivularis*).
Salmon Berry (*Rubus spectabilis*).
Salmon Berry (*Rubus nutkanus*).
Cloud Berry (*Rubus Chamaemorus*).
Nahgoon Berry (*Rubus stellatin*).
Wild Red Raspberry (*Rubus strigosus*).
Blue Berries (*Vaccinium uliginosum*).
Blue Berries (*Vaccinium ovalifolium*).
Blue Berries (*Vaccinium vitis Idaea*).
Seaside Portulaca (*Claytonia sp.*)
White Clover (*Trifolium sp.*)
Wild Tansey (*Achillea borealis*).
Yellow Water Lily (*Nymphaea advena*).
Water Smart Weed (*Polygonum sp.*)
Elder (*Sambucus racemosa*).
Cow Parsnip (*Heracleum lanatum*).

—H. B. Parks.

ALBERTA—Honey Sources of.

Alberta appears to divide itself into three natural beekeeping regions. In the far north along the Peace River and in the Grande Prairie district, fireweed is the principal source of surplus. Yields of as high as 100 pounds from a single colony have been secured at Beaverlodge. With the settlement of this region fireweed will decline in importance while the extension of such forage crops as alfalfa and sweet clover will furnish more dependable pasturage.

In the park region of the central part of the province in the vicinity of Edmonton, fragrant giant hyssop is an important source of nectar, probably from one third to one half the crop coming from that plant. Fireweed and snowberry are also important in this region.

In the southern part of the province there is a large area of dry range country where beekeeping is not dependable, although some sweet clover is grown and bees within reach of it store some surplus.

In the irrigated country in the vicinity of Lethbridge, large yields of honey are secured from alfalfa and sweet clover and good crops are the rule. Larger average yields are gathered in the irrigated section than in any other part of the province. All the region which can be irrigated is promising beekeeping territory.

The bees find early pollen from the pasque flower, (*Anemone patens*) which is common on the prairies, and from the poplars which are widely distributed in all parts of the province except the dry sections. The willows also yield both pollen and nectar for spring pasture, as do the dandelions which are becoming common in the older settled sections and rapidly increasing in importance. The box elder, commonly known as Manitoba maple, is a well known source of both pollen and nectar. White and alsike clover are present in some neighborhoods and will become more important with the development of agriculture. Wild raspberry is common in some bush country areas and aster and goldenrod are generally present over the province, yielding some honey in late summer and autumn. Canada thistle yields well in neighborhoods where it has been permitted to establish itself.

The June-berry or service-berry is known as Saskatoon in western Canada and is generally recognized as of some value to the bees. Caragana is widely planted as an ornamental and as a shelterbelt about farmsteads. It yields freely and is the source of surplus.

ALDER (Alnus).

The alders are a group of shrubs or trees common from New England and Canada west to Michigan and south to Texas. The bark is sometimes used for tanning and as a dyestuff, and to some extent in medicine. The blossoms appear early in spring, and are the source of an abundant supply of pollen at a season when it is often much needed by the bees.

Several species of alder are also found on the Pacific Coast, blooming January to March, and in some localities they are important as sources of pollen. The white alder has been reported as a source of honey, but this is probably honeydew.

Pollen-bearing blossoms of the alder.

In American Bee Journal, (p. 239, Oct. 1861) appeared an account of a late harvest from alder after frost had killed vegetation generally, which caused trouble to the bees in winter.

ALFALFA (Medicago sativa).

Alfalfa is the most important honey plant west of the Missouri river. It is also the most valuable forage plant in the same region.

Alfalfa has become widely distributed because it is a valuable forage plant and through the interest of farmers. Numerous varieties of alfalfa are now offered in the markets but most of these are special strains of the common alfalfa, (*Medicago sativa*).

Alfalfa apparently came originally from south central Asia and has been carried to most of the important agricultural regions of the temperate parts of the globe. As a result of natural adaptation to the regions in which it has been grown, strains which manifest peculiar characters have appeared. Thus it becomes important to secure seed of a strain adapted to the peculiar conditions of the section where it is to be planted.

Grimm Alfalfa

In the north central regions of Iowa, Minnesota, the Dakotas, and Nebraska, hardiness is a most essential characteristic. Winter killing is a serious problem to the farmer. Thousands of acres of alfalfa are killed every severe winter.

In 1857 a German immigrant named Wendelin Grimm settled in Carver County, Minnesota. Grimm brought with him a few pounds of alfalfa seed which he sowed on his new farm. Some of the plants survived the severe winters and from these the seed was saved for increased acreage. Through a long series of years the seed was selected from this importation until the strain became well known throughout the Middle West by the name of the man who brought it from Germany. It is not a separate species but simply a strain which has been selected over many generations for hardiness.

There is a tendency of the taproot to branch, which probably accounts in part for its superior hardiness. Wherever winter killing is serious Grimm alfalfa is to be preferred, but it yields less forage in the warmer and more humid regions than does the common alfalfa.

Turkestan Alfalfa

In 1908, Prof. N. E. Hansen visited Turkestan as a plant explorer for the United States Department of Agriculture. From regions with light rainfall and severe temperatures he selected alfalfa for trial in this country. These first importations gave so much promise in the cold and dry regions of the northern plains that much enthusiasm developed for Turkestan alfalfa.

This demand had the very natural result of setting seed dealers in search of an abundant supply. As a result seed from all parts of Turkestan came to this country and was sold to the trade. The most of this seed proved to be inferior to the common strains already in use here.

If it were possible to depend upon seed sold under this name as being of the quality of the original Hansen importation, the Turkestan

Blossoms of alfalfa and yellow sweet clover.

alfalfa would be valuable, but due to the uncertainty of origin one is hardly safe to depend upon this variety as offered in the markets.

Baltic Alfalfa

The Baltic alfalfa is very similar to Grimm and received its name from having been grown for a series of years in the vicinity of Baltic, South Dakota. It is reported as resistant to winter killing and to succeed under similar conditions with Grimm. Not much has been heard of it of late and it is probable that the Grimm, which it so closely resembles, is largely replacing it.

Cossack Alfalfa

The Cossack alfalfa appears to be a hybrid between the common alfalfa and the yellow flowered Siberian alfalfa. It was brought from Siberia by Prof. Hansen in 1907. It has proved to be quite hardy and has received a large amount of advertising in the plains region. While the writer has tried Cossack alfalfa alongside of Grimm on his ranch in northern Nebraska, there was nothing to indicate any superiority to that variety. Both made a favorable showing and were about equally satisfactory.

Peruvian Alfalfa

In the valleys of southern Arizona and California the Peruvian alfalfa is commonly grown. As its name indicates, it came to this country from Peru in South America. It has winter killed very generally in all northern states.

In the region where it succeeds it is very productive and yields one or two more cuttings each year than the common forms. It is grown under irrigation and in a mild climate; hence its distribution is limited to a comparatively small area. It becomes very woody when sown thinly.

Arabian Alfalfa

Arabian alfalfa is another of the varieties suited to mild climates. It is not cold resistant, and is suited especially to the southern states. It is not drought resistant.

While it makes a good stand and is quite productive for about two years, the plant is short lived and by the third year the stand will begin to thin. Since it does not seed freely, it is rather difficult to propagate profitably.

Yellow Flowered Alfalfa

The yellow flowered alfalfas from Siberia belong to a different species, (*Medicago falcata*). Unlike the common alfalfa, there is a much branched root system instead of the one taproot usually recognized in the alfalfa plant. The crowns are set very low, which provides some additional protection.

The first importation was by Prof. N. E. Hansen in 1897. Since that date numerous other importations have been made and the plants widely tested by the U. S. Department of Agriculture and at the South Dakota experiment station.

This species has been found in the wild state in England, Norway, Sweden, Austria, France, Russia and other countries. To secure the hardiest strains the explorers went to Siberia, where it is exposed to both cold and drought.

Since it can be depended upon to make only one crop in a season, it is nowhere generally cultivated. To date it is of interest mainly as an object for selection and hybridization. Prof. Hansen considers this species especially valuable for bees.

Wild Alfalfas

In addition to the above list of commonly cultivated forms, there are a number of wild alfalfas to be found in Asia. Perhaps some of these may yet prove worthy of cultivation and distribution in other parts of the world. Among the twenty-one species to be found in the far East may be mentioned *Medicago arborea,* which grows to a height exceeding ten feet. Since it gets too woody to be palatable, it is hardly as well suited to cultivation as the common alfalfa, but it should be of great interest to the beekeeper for use in waste places. It is said to flourish in hot and dry situations where few other plants will live. It has been grown to a limited extent in California as an ornamental plant. It produces large yellow flowers in abundance.

From two to five cuttings of alfalfa are secured each season, depending upon the available moisture and the length of the growing period. It thrives best in the irrigated regions with its roots in the rich soil supplied with abundant, but not excessive, moisture.

The honey from alfalfa varies in color and quality in different localities. In Colorado and Idaho it is of very light color and with a spicy, mild flavor of excellent quality. In the Imperial Valley of California it is much darker in color and of poorer quality. Alfalfa honey granulates readily, but is generally regarded as a high quality of honey. The tendency to early granulation makes it more desirable to market in the extracted state than in the sections.

The yield varies greatly, according to season, but the heaviest yields come when there is a vigorous growth of the plant. Where grown without irrigation a much greater variation in yield can be expected. In Nebraska and Kansas, alfalfa often yields good crops of nectar without irrigation, while east of the Missouri River it is seldom of much value to the beekeeper. The author had a small field of alfalfa in western Iowa for several years. Only one season did the bees pay much attention to it. That season was very wet in the early part of the summer, thus promoting a vigorous plant growth. Later the weather turned very hot and dry. Conditions similar to those of the irrigated sections of the west were approximated. The roots of the plants were supplied with an abundance of moisture, while the air was hot and dry. Under such conditions alfalfa is at its best as a honey producer.

It seldom yields to any extent in humid climates. Given sufficient moisture at the roots, the hotter and dryer the atmosphere, the better seems to be the yield of nectar. The conditions which most favor nectar secretion are also favorable to seed production.

When weather conditions are just right there is an occasional crop of honey from alfalfa in the humid states. Such crops are reported from Wisconsin, New York and other states east of the Missouri, but, as stated, the flow is not dependable as is the case westward. The beekeeper noting the crop under exceptional conditions is likely to be misled as to the value of the plant under average weather conditions eastward.

Harry K. Hill, of Willows, California, states (Western Honeybee, April, 1913) that he gets three distinct shades of pure alfalfa honey in the same year. The honey from first extracting is much darker than later extractings.

In many localities it is the practice to grow a crop of sweet clover in advance of seeding to alfalfa. In localities where the plant is grown for seed a much longer blooming period results and the beekeeper profits accordingly. Where cut for hay it is usually done just when it is reaching its time of greatest nectar secretion.

ALLTHORN (Koeberlinia spinosa).

Allthorn is a very curious green shrub apparently composed entirely of thorns and without leaves. The leaves are very minute and are absent during much of the year. Near the ends of the sharp thorns are borne clusters of small flowers of a pale white color in early spring. It is common to the valley of the Rio Grande River from ElPaso, Texas, to the coast.

Allthorn, also called Junco and Crown of Thorns, is confined to dry desert areas where it is said to yield nectar freely. Since the region in which it is found is not extensive and largely uncultivated it is only locally important to beekeepers. It flowers in early spring and is said to yield heavily for short periods.

ALFILARIA, see Pin Clover.
ALGAROBA, see Mesquite.
ALLIGATOR PEAR, see Red Bay, also Avocado.
ALLIUM, see Onion.

ALMOND (Prunus amygdalus).

The cultivated almond is closely related to the peach and is native to Southern Europe, where it has been grown for centuries for its nuts.

The tree is extremely early in its blooming period and more tender than the peach. It is grown in large orchards in parts of California, and to some extent in Arizona. The tree will not thrive on wet lands.

Bees gather both honey and pollen from the flowers and in the almond belt of California it is of great value for early brood-rearing. There it blooms in February and beekeepers sometimes move their bees from a distance to the almond orchards to build them up for the orange flow.

Ralph de Ong of California, writing in Western Honey Bee, (May 1918) states that honey from almonds and peaches is bitter in flavor and for this reason should not be added to any marketable honey but used only for brood rearing.

In California Experiment Station bulletin 346, Tufts and Philip state:—

> "Pollinizing agencies, such as the honeybee, are necessary to a good crop of fruit. One colony of honeybees should be provided for each acre of orchard."

ALSIKE CLOVER (Trifolium hybridum). Hybrid or Swedish Clover.

Alsike clover is one of the very best honey plants of America. The beekeeper who lives within reach of large fields of this crop is fortu-

Almond orchard.

nate, for there is no better honey, and under favorable conditions the crops harvested from alsike are such as to give little ground for dissatisfaction. Some beekeepers have estimated that alsike will produce 500 pounds of honey per acre in a good season.—American Bee Journal, page 409, 1886.

Alsike thrives on clay soil, or lands inclined to be wet, where the other clovers do not succeed. It is sown very generally in a meadow mixture with timothy or red-top. In localities where grown for seed there is a long period of bloom, which is greatly to the advantage of the beekeeper. It is good for either pasture or hay, and although by itself alsike does not yield as many tons of hay per acre as red clover, when mixed with red clover the two together make more and better hay than red clover does alone.

Alsike is intermediate in size between white and red clover. The blossom looks like that of white clover, except it has a pinkish tinge of color not found in the white clover. The stem is upright and branched and on land with sufficient moisture reaches a height of two feet or more.

While alsike will grow nearly everywhere that red clover will grow, it thrives best in the northern part of the country. Minnesota, Wisconsin, Michigan, Ontario and New York all report alsike as especially valuable to the beekeeper.

The honey is white in color, mild in flavor and is regarded as one of the best for table use. At times the yield is very heavy. In American Bee Journal, Nov. 2, 1899, are given several instances of large yields from this source. In one case a single colony of bees gathered 72 pounds in four days, or 18 pounds per day. Another report was of 251 pounds in 21 days or 12 pounds per day from alsike.

In number two of the first volume of the "Review," Editor Hutchinson states that 10 colonies of bees gathered 300 pounds of extracted honey from alsike, with only two acres within reach. This, of course, takes no account of the honey consumed by the bees, but indicates that the yield is good for the acreage within reach. In the following number of the same journal an Ontario beekeeper reports that he had not known a failure from alsike in eight years.

See also Clover.

ALTHEA, see Hollyhock.
AMERICAN ALOE, see Century Plant.
AMERICAN CRAB APPLE, see Crab Apple.
AMERICAN IVY, see Virginia Creeper.
AMORPHA, see False Indigo.

AMPELOPSIS (Pepper-vine or Snowvine).

There are two species of *Ampelopsis* which are important honey plants in the south. The pepper vine or pepperidge, *Ampelopsis arborea,* is found in swamps and along streams from Virginia to southern Missouri and south to Texas and Florida. It is also found in the West Indies and Mexico. It is a vigorous climbing vine with tendrils few in number, or sometimes wanting. It is reported as the source of considerable honey of fair quality in Georgia and Alabama. Jes Dalton reported that in Louisiana it yields nectar freely and that the blooming period lasts nearly two months. He states that rain does not interfere with the flow as is the case with so many plants, and that the bees will be working it freely within two hours following a shower. Doctor Oertel describes the honey as thin and reddish.

In Georgia it is called snowvine or crossvine, and F. M. Baldwin reports (Dixie Beekeeper, May, 1920) that it yields an average of fifty pounds per colony in the swamps along the Altamaha. He describes the honey as amber in color and of mild flavor. Wilder reports as high as 80 pounds per colony from this source.

Ampelopsis cordata is found over much the same range of territory and is known by similar common names in various localities. It is found frequently climbing over bushes in woods and on river banks in the southeastern states. It has a broad ovate or heart-shaped leaf and the blooming period is earlier than in the foregoing species. (See also cow-itch).

This species is also found in the Mississippi River bottoms as far north as Warsaw, Illinois, and is the source of some honey in the Dadant apiaries in mid-summer.

AMSINCKIA.

Prof. George H. Vansell of Davis, California, seems to have been the first to call public attention to the fact that Amsinckia is impor-

Amsinckia in bloom.

tant to the bees. In a letter to the author he states that in California there are six or more species and that they are of decided value to the bees for nectar.

The author in company with Prof. Vansell found *A. lycopsioides*, to be common along the roads and in margins of fields about Davis, Cal. It was blooming in late March. This species is locally called fireweed because it irritates the skin of men working in the harvest. It is also called leather-breeches and wooly-breeches. It has also become established in waste ground in Massachusetts and Connecticut. In the east it blooms from May to July. In California the blooming period is said to be over by May.

Vansell states that the bees work the various species all through the blooming period. Apparently, where abundant it is of considerable importance for spring stimulation.

Rydberg lists seven species of Amsinckia as found in the Rocky Mountain region.

ANAQUA (Ehretia elliptica).

The anaqua is a common tree with small white flowers in open panicles and oval leaves, which is common from New Braunfels, Texas south to the lower Rio Grande. It is commonly reported as a valuable source of nectar by Texas beekeepers. Also called Knockaway.

ANDROMEDA.

"Andromeda (a scraggy shrub of the heath family) blooms in the central northwestern part of Florida for about four weeks in March and early April; yielding but little, three years out of four. The honey, too, is reddish yellow, thick and pungent, not very valuable as a surplus honey plant."—E. G. Baldwin, Gleanings, March 15, 1911.

Xolisma ferruginea is found on pine lands from South Carolina to Florida. It is an evergreen shrub or small tree with leathery oval leaves. In some localities it is known as "Wicky," although that name is usually applied to the Laurels.

F. M. Baldwin (Dixie Beekeeper, Feb., 1922) states that wicky is common on poor sandy spots near Deland, Sanford and Orlando, Florida, and that when climatic conditions are favorable yields large quantities of surplus honey at the time the orange is blooming. He states that it is little known to the beekeepers and that the honey is often mixed with orange and credited to that source. He credits it as one of the best sources of nectar in that region.

ANGLEPOD, see Bluevine.

ANISE (Pimpinella anisum).

Anise is an herb long cultivated for use in medicine and cookery. It comes from the region Egypt to Greece and is well known in American herb gardens.

According to L. R. Miller of Culver City, California, it yields a very light colored honey with a mild and elusive flavor. He reports large acreage on waste land near him and states that he always secures surplus honey without moving his bees.

APOCYNUM, see Dogbane.

ANTELOPE BRUSH (Purshia tridentata).

Antelope brush is a low much branched shrub common to the prairies of the Rocky Mountains, the Pacific northwest and dry hills of the southwest. It is found from British Columbia and Washington to New Mexico, Montana and California.

In northeastern Oregon it is known as buckbrush. The fragrant

Apple orchard in bloom.

yellow flowers are the source of both pollen and nectar, according to J. Skovbo, of Hermiston. He states that the blossoms come with fruit bloom and that the bees work it as freely as apple. He further states that the honey is dark and strong.

APPLE (Malus).

America's best and most widely used fruit is the apple. It is a native of Asia, but as a cultivated fruit is grown in most of the temperate regions of the world. Hundreds of varieties have been developed by plant breeders until there are few horticulturists who are familiar with them all. They range in size from the small cultivated crab-apples not more than an inch in diameter to the big Wolf River, often five or more inches in diameter. Some varieties are hard and sour and suitable for little else than making cider, while others are of the finest quality.

There is no more beautiful sight than an orchard in bloom in spring. The blossoms secrete nectar freely and in favorable weather the bees fairly swarm over them. The honey is light amber in color and of good quality. The trees bloom so early that it is of greatest value to stimulate spring brood-rearing, though strong colonies easily store surplus when the weather is suitable for the bees to fly freely during the period of bloom. Large orchardists often offer inducements to beekeepers to locate near their orchards for the better pollination of the fruit blossoms which results from the presence of large numbers of bees. In orchard districts there is frequently complaint on the part of the beekeepers that the bees are killed by the application of poisonous spray while the trees are in bloom. When American beekeepers learn to winter their colonies in such a manner as to maintain a reasonable strength in early spring, surplus honey in quantity may be expected from the orchard districts in favorable seasons. The weather is often too wet or too cold for the bees to fly during apple blossom, and this condition the apiarist can never overcome.

The amount of nectar available from apple blossoms can be judged from the following quotation from Gleanings, page 389, 1883:

> "My best hive gave me 36 pounds in ten days; four or five of them very bad weather. A friend of mine who has had a large experience says he has known them to gather 10 to 15 pounds per day, which, considering the quantity of bees in a hive at this season of the year, seems to me an immense yield."

Vansell has found sugar concentration in apple nectar to run as high as 57%.

APRICOT (Prunus armeniaca).

The apricot is a well known cultivated fruit, somewhat intermediate between the plum and peach. It blooms very early in spring and is valuable as a source of early nectar and pollen. It is grown in large acreage in some portions of California, where it is regarded as valuable, by the beekeepers.

ARBUTUS (Epigaea repens). Trailing Arbutus or Ground Laurel.

A trailing plant with evergreen leaves. The rose colored flowers, in small clusters, appear in early spring. They are fragrant and attractive to the bees. There is an occasional report to the effect that arbutus is valuable to the bees as a source of early nectar.

ARCTOSTAPHYLOS, **see Manzanita.**

ARIZONA BUCKTHORN, **see Gum Elastic.**

ARIZONA—Honey Sources of.

Arizona is a desert state with scant precipitation, high temperature and low humidity. In the northern portion of the state there is an area of mountains which rise to elevations from 3,500 to 14,500 feet where, because of altitude, more moisture is available and large forests occur. At the lower elevations there is little agriculture except by irrigation.

In the southern part of the state spring and summer temperatures rise to 110 to 125 degrees Fahr. In the mountain and northern plateau regions they seldom exceed 85 to 95 degrees. Bright skies prevail throughout the year and the percentage of sunshine is very high. In the south portion the winters are very mild and bees can fly nearly every day in the year. Because of this fact there is a serious loss of bees from useless exertion at seasons when no honey is to be gathered. The winter problem is one of conservation of bees and stores rather than of protection from cold.

The rainfall varies greatly according to the elevation above sea level. In the southwestern section about Yuma the average is less than four inches annually and increases with altitude to something like 24 inches in the mountain regions. The greater part of the rainfall occurs in the months of July, August and September.

Large numbers of bees are kept in the irrigated valleys where good crops of honey are secured from Alfalfa, sour clover, cotton and the garden crops. Large areas are devoted to the cultivation of such truck crops as melons. Considerable citrus fruit is also grown, especially in the vicinity of Phoenix and Yuma. Some honey is secured from wild radish and mustard which come up in the fields, as well as from the eucalyptus, pepper-trees and athel trees grown for shade and wind-breaks.

The desert flora yields much honey, although with somewhat uncertain variations. Mesquite, catsclaw, screw-bean and paloverde yield surplus in large quantities. Sahuaro or giant cactus and ironwood are two important sources of surplus honey little known in the United States outside of Arizona. Arrow-wood or cachinilla and baccharis are of considerable importance as sources of stimulation along the streams and ditch banks. Squawbush (Lycium) is a winter bloomer valued highly for stimulative purposes, common to the desert sections of the southern portion of the state. Jackass clover or Wizlizenia is common in alkaline places where the soil has been stirred, and phacelia and filaree are also important when moisture is available. The rayless goldenrod (*Isocoma*) is widely distributed over the deserts and is commonly called rabbit brush. Clematis and sunflower are to be found where conditions are favorable.

In the northern region at higher elevations, a very different flora is to be found. Sumac and ceanothus, cleome and horehound which are well known sources of nectar in other sections are common. Lemon weed, (*Psoralea lanceolata*) common on lower slopes of mountains, is reported as a valuable source of nectar. Mountain mahogany (*Cercocarpus*) is reported to yield heavily. Several species of eriogonum are found in northern Arizona and rhamnus and manzanita occur in the forests. Several species of rabbit brush or chrysothamnus are common in the lower woodland regions. Wild cucumber, Virginia creeper and poison ivy are common along the streams in the higher elevations. Oregon grape (*Berberis*) is very common over the yellow pine and upper woodland areas, growing in dry warm situations of south slopes of canyons. The yellow barberry (*B. Fremonti*) locally called agarita is common at edges of forests at six to seven thousand foot elevations.

ARKANSAS—Honey Sources of.

The upland soils of Arkansas are mostly thin and deficient in humus and over large areas are sour, requiring applications of lime for success with legumes. The mountainous areas are timbered with pine, oak and hickory and offer little forage for the bees.

Along the streams and in the valleys are belts of good farming lands, where staple crops are profitably grown. The best sections for beekeeping are along the streams and in the swampy districts where a much greater variety of plants are available.

The climate of Arkansas is mild with short and mild winters and long warm summers with abundant rainfall. While there are limited areas of good bee pasture, the state, as a whole, is not of the best for honey production.

White clover is the source of some surplus in Arkansas, although it is too far south for such heavy yields as occur in the northern part of the range of this plant. Sweet clover is valuable in some localities. Tupelo, holly, blackgum, redbud, locust, tulip-poplar, blackberry, heartsease and asters are important sources of nectar. Cotton yields in some sections of the State. Fruit bloom is valuable in spring, and where colonies are sufficiently strong some surplus may be expected.

In Bulletin No. 170 of the Experiments Station, "Beekeeping in Arkansas," by W. J. Baerg, the following additional sources are mentioned: White maple is said to be the earliest source of honey and pollen in spring, with elm also valuable for pollen. Dandelions furnish early nectar and pollen and shadbush or service-berry is listed as the source of a limited amount of nectar in woodlands. Persimmon is valuable over most of the state, rattan vine of especial importance in the central section, with hairy vetch cultivated to some extent as a forage crop. Thoroughwort and Spanish needles furnish much surplus late in the season.

ARNICA WEED (Amblyolepis setigera).

Arnica Weed is an annual weed with alternate leaves and showy head common to Texas prairies. The flowers are yellow and appear in late winter, the plant blooming from Christmas to March or April. The honey is reported as dark in color and of poor quality. Not an important source of honey. Several other plants with yellow flowers which bloom at this season are known as arnica weed.

In Alberta gumweed is known by this name. (See Gumweed.)

ARROW-HEAD (Sagittaria).

There are several species of arrow-head which are common water plants named from the shape of the leaves. They are to be found in marshes, ponds and still water in all parts of North America from Nova Scotia to Florida and west to the Pacific Coast; also found in Europe, Asia and South America.

Arrow-wood on banks of irrigating ditch in Arizona.

The Chinese arrow-head is sold in the markets of China and Japan, where the starchy bulbs are used as food. It is said to be cultivated in the vicinity of San Francisco to supply the Chinese market there. The American Indians also used some of the native varieties for food, which was prepared by boiling or roasting the roots. In California *S. latifolia* is known as tule-potato and is probably the source of some of the so-called tule honey.

The flowers of most varieties are white and the blooming period is long. Some bloom from June to September and are worthy of cultivation as ornamentals. In swampy regions they sometimes occur in large areas and in such situations should be of interest to the beekeeper. Jepson states that 100 acres of pure growth of this plant occurs just below the San Joaquin bridge, near Banta, California, where the tubers are fed to hogs.

Little is known as to relative importance of the various species, but H. B. Parks states that several species are much visited by bees and that they add much to the honey crop in swampy regions of Texas. They are undoubtedly equally valuable in other places where abundant.

ARROW-WOOD (Berthelotia sericea) Cachinilla.

Arrow-wood, in New Mexico commonly called by the Spanish name cachinilla, is a small shrub growing in dense patches in wet places along streams and irrigation ditches from the Rio Grande valley of west Texas through Arizona and southern California. It grows to a height of from four to six feet with straight stems, formerly used for arrows by the Indians, thus giving rise to the common name.

In the Rio Grande Valley near ElPaso, in the Salt and Colorado River Valleys in Arizona and the Imperial Valley of California, the writer found this little shrub regarded favorably by the beekeepers as a minor source of honey. It is reported as sometimes yielding surplus, the honey being dark and of good flavor.

It is found from Colorado and Utah to Texas and California, but is far more common along the southern boundary.

ARTICHOKE, see Sunflower.
ASCLEPIAS, see Milkweed.

ASH (Fraxinus).

There are more than twenty species of ash trees common to various sections of America. Some are well known timber trees, furnishing lumber for furniture and for interior finishing. The flowers are small, inconspicuous and of a greenish color. Their principal value to the beekeeper is as a source of pollen, although Richter lists the Oregon ash (*Fraxinus oregona*) as a source of honey also.

E. G. LeStourgeon considers that in Texas ash is the source of "propolis of a green color and very viscid consistency." He writes that apiarists in localities where ash abounds complain of excessive propolizing.

ASH LEAVED MAPLE, see Box Elder.

ASPARAGUS.

The garden asparagus is an introduced plant widely cultivated. It is very attractive to the bees and yields pollen plentifully.

There are a few places where the plant is grown for commercial purposes where the bees get some honey from it. The amount and quality of honey from asparagus is mentioned in Gleanings, page 139, 1883, as follows:

> "Within a radius of three miles there were thirty or forty acres of asparagus grown; and as it generally blossomed during a dearth of honey I had a good opportunity to judge its value as a honey-producing plant. My bees worked on it from early morn until dark and gathered from one-half to 2 pounds per day. The honey is light in color and has a peculiar acid taste, somewhat resembling unripe clover honey. It was very thin and thickened slowly."

In a private letter to the author, Davis Brothers of Courtland, California, wrote as follows:—

> "In this immediate locality is the center of the world's asparagus production. There are thousands of acres in this Sacramento River delta district.
>
> We find asparagus to be somewhat temperamental, yielding nector heavily only when weather conditions are just right. Shade temperatures of 100 degrees or more combined with low humidity, will injure asparagus blossoms and blossom buds beyond recovery for that season. And that is not an unusual happening at the time asparagus is in blossom here, June 15 to August 1st. However, when favorable weather does continue throughout the blossom period, the yield is excellent of a mild extra light amber honey. This happens on an average of one year in three or four.

ASPERULA BEDSTRAW (Asperula galioides).

Asperula bedstraw is an old world plant that has become established in several localities in the Northeastern States. Asbery Singleton reports that it has been introduced into Manitoba and that it is spreading slowly over the prairies where it is worked eagerly by the bees in June.

ASPEN (Populus). Poplar or Cottonwood.

There are several species of poplars. It is a widely distributed group, some species being found in most all sections of the country. They are important for pollen, though some honeydew is reported from them also. They thrive especially well on the low lands along streams and are the most common trees of the plains region from Dakota south to Oklahoma and western Texas. Much propolis apparently comes from these trees. This group should not be confused with the tulip-poplar, which see.

ASTER.

The aster family is very widely distributed, being common in Europe, Asia and South Africa, as well as America. There are more than 200 recognized species, of which at least 125 are found in the United States. They are extremely common in the Eastern and Southern States, although some kinds are to be found in every State in the Union, and from Canada to Mexico. Every American beekeeper may be sure that his bees are within reach of at least one species of aster, and, in most localities, there are several species. Some species produce nectar much more abundantly than others, and it is probable that the flow from all kinds is more or less affected by soil or climatic conditions. So few beekeepers differentiate between the species that it is very difficult to secure satisfactory information regarding their comparative value.

White or frost aster (*Aster ericoides*).

Asters are very seldom mentioned as sources of nectar in the southwest. Yet twenty-one species are listed as occurring in New Mexico. They seem to be of importance principally in the Eastern States. There are numerous reports of honey from asters in the Southeastern States of Georgia, Alabama and Mississippi, the amount of surplus increasing northward.

In most localities, the aster honey is mixed with that from goldenrod, and the two sources are usually spoken of together. Like the asters, they are of wide distribution, and, like them, they seem to produce nectar more abundantly in the moist climate of the Eastern States. Both bloom late in autumn, the crop often being cut short by frost.

According to Lovell, the asters are never common enough to yield a surplus in Maine, and the honey is always mixed with goldenrod.

As to the quality of the honey, there are many conflicting reports. Many reports are to the effect that the quality is poor and not suitable for table use. The fact that the honey is seldom unmixed with that of other fall flowers, may be responsible for this impression. C. P. Dadant had one year, in Illinois, a crop of about six barrels which was almost pure aster. This honey was secured late in the season, after other plants had ceased to yield, and was almost white, and of very fine quality.

There are numerous reports that a strong odor is apparent in the apiary when asters are yielding. We quote some of these:

> "We had a fall flow from wild asters that filled the hives with honey for wintering, and gave a few gallons of extracted honey. The honey is of good color and weight, but rather strong for table use. It also granulates very quickly. When the bees are gathering this honey the hives give off a rank and somewhat sickening odor, which can be detected for quite a distance away. * * * This odor disappears as the honey ripens and the flow ceases, but the strong taste never entirely disappears. It is as strong as basswood and not nearly so pleasant."—D. E. Andrews, Bloomington, Ind., page 98, American Bee Journal, 1907.
>
> "The odor is not unpleasant, but is very noticeable when the bees are bringing much of it in, and it can be distinguished a considerable distance from the hives. The amount of 'smell' is such a good criterion as to the amount of honey that one can tell the quantity he is getting from these indications alone."—W. H. Reed, Herrodsburg, Ky., page 228, Gleanings, 1911.
>
> "In the Shenandoah Valley, in Virginia, where I lived for fourteen years, there were many acres of white aster. There were several years when the bloom was in sheets, affording a good yield of surplus. The honey was very light amber, of fine quality, and was considered next to white clover. At such times a strong odor, which was distinctly sour, could be noticed."—Burdet Hassett. Page 257, Gleanings, 1911.

Much has been written concerning the danger of aster honey for winter stores. So many reports of disastrous results from wintering on aster honey have been published, that it is generally understood not to be safe for winter stores. Not only does aster honey contain gums

Purple stemmed aster (*Aster puniceus*).

which are indigestible to the bees, but the plants bloom so late that the honey may not be properly ripened. The fact that the honey granulates readily also probably accounts for the trouble in some cases.

In some localities, asters seem to be a dependable source of surplus, while in others they yield in appreciable quantity only in rare seasons. Kentucky seems to be in the heart of the territory where asters are important. The following are typical reports:

"We have never failed to get a good crop of surplus honey, and plenty left for the bees, from aster for more than twenty years, till this year."—H. C. Clemons, Boyd, Ky. Page 90, Gleanings, 1909.

"In this section the asters are invaluable as fall forage for bees. Let the season be cold or hot, we are certain to have a continuous blooms from early in September until a really hard frost occurs. My Italian bees have never failed to secure enough honey from aster to carry them through the winter, even when there was hardly a pound of honey in the hives at the end of August—Daniel M. Worthington, Elkridge, Md., American Bee Journal, page 125, 1869.

Large-leaved aster (*Aster macrophyllus*).

"Blue aster (*Aster azureus*), known among farmers as blue devil, or stickweed, in my judgment, is one of the best we have, from the fact that it produces honey in the fall of the year. It is usually in full bloom until about the middle of October, and if the weather is warm enough for the bees to fly, they get plenty of honey to winter on from this flower."—West Virginia. Page 869, American Bee Journal, 1906.

It is probable that most of the species are of more or less value for honey under favorable conditions. The writer has seen bees working on arrow-leaf (*Aster sagittifolius*) on sunny days in Cass County, Iowa, the first week in November, after everything else had been killed by frost. This species occurs in dry, open woods, from New Brunswick to Ontario, and west to Dakota, and from New York to the Ohio valley, and along the mountains to Georgia and Alabama.

Generally speaking, the small-flowered species with willow-shaped leaves, are best for honey. *Aster tradescanti* is probably first in the list as a source of surplus. It is found from Ontario to Saskatchewan, and throughout the States east of the Mississippi, south to the Gulf States. *Aster salicifolius* is probably one of the best in Iowa and Illinois, being common on low ground.

In a private letter, F. W. L. Sladen writes concerning the asters in Canada, as follows:

"I have this year had confirmation that *aster cordifolius* is a useful source of surplus honey in favorable seasons in the Gatineau valley in September. During a period of very fine weather between September 11 and 22, a crop of 12,000 pounds of honey, principally from this source, and from the late flowering species of goldenrod, was obtained by Joseph Martineau, at Montcerf, Quebec, from 300 colonies. The honey was light amber color, and a pleasant flavor, and not unwholesome for wintering, not granulating in the combs. (See Experimental Farms report 1914-15, page 996). Other valuable species of aster in Canada for honey production are *A. lateriflorus* (Maritime provinces to Ontario); *Aster umbellatus* (Maritime provinces to Eastern Manitoba), and *Aster puniceus*, (Maritime provinces to Ontario)."—Ottawa, October 2, 1917.

Arrow-leaved aster (*Aster sagittifolius*).

Aster puniceus, the purple-stemmed aster, is found from Nova Scotia to the Rocky Mountains and south to Northern Alabama. It is one of the most attractive of the asters, growing on wet land and in the borders of swamps. Lovell writes that in Maine he has seen the bees on this species in large numbers on September 17.

The white field aster, or frost flower (*Aster vimineus*), is common from Eastern Canada to Minnesota, and south to Arkansas and Florida. It grows in dry, open fields, along roadsides, and in waste places. It is a late bloomer, belonging to the group of field asters which are important for nectar. Some other species, however, yield more freely.

The swamp aster (*Aster acuminatus*) occurs on wet land, but as far as available information goes is less valuable for honey.

The large-leaved *Aster macrophyllus*, is a northern species, found in open woodlands. Graenicher observed ninety-five species of insects on the flowers of this species in Wisconsin, which indicates nectar in abundance in that State.

Several other species are known to produce nectar freely, *A. multiflorus, A. lateriflorus, A. dumosus, A. paniculatus* and *A. vimineus* being reported from various localities. *A. ericoides* is reported as valuable in Missouri.

"There is an abundance of *Aster ericoides* now in full bloom. The bees are working on it more vigorously than they have on white clover or any other bloom."—George E. Wilkins, Wright County, Mo. Page 699, American Bee Journal, 1904.

ASTRAGALUS, see Loco Weed. Also milk vetch.

ATHEL TREE (Tamarix aphylla).

The Athel tree is extensively planted as a windbreak in the hot and dry regions of Arizona and southern California. It was first planted by Prof. J. J. Thornber of the University of Arizona who secured specimens from Algeria, in 1909.

It is now regarded as the most important windbreak for hot and dry irrigated regions and is commonly found in the Salt River and Colorado River Valleys. It is a fast growing tree, sometimes growing as much as twenty feet in a year. It will stand but little frost, hence its range is limited to the warmer regions.

Swamp aster (*Aster acuminatus*).

Apparently it is a valuable source of nectar and its general planting will add much to the available bee pasture. There are numerous reports of honey from this source in Arizona and the Imperial Valley of California.

R. M. Kellogg of Glendale, California describes the honey as light yellow with a disagreeable flavor and granulating with a coarse grain.

AVOCADO, (Persea).

The avocado or alligator pear is cultivated commercially in Florida and California for its fruit which has a tender, juicy flesh enclosing a hard seed. It is native to the American tropics and is of comparatively recent introduction to orchard cultivation. Seedling trees grow to a height of fifty feet or more, but the budded ones are usually dwarfed. The small greenish flowers are produced in abundance, but apparently the nectar secretion is less abundant than in most fruit trees.

Concerning avocado for bees Roy K. Bishop writes in American Bee Journal, (October 1925):—

"Bees gather pollen from the avocado. The quantity of honey if any is small. Bees visit the blossoms quite freely at times, but never with the vigor and hum displayed in case of other fruit bloom such as apricots, apples and plums."

In a letter to the author dated June 2, 1925, Mr. T. Ralph Robinson of the Bureau of Plant Industry writes as follows:—

> "This spring when I was at Homestead (Florida) I was served with honey said to be made from avocado bloom, and where avocado orchards are numerous as at Homestead, there is doubtless an opportunity for producing considerable honey. While the avocado flower is small and fragile a full bearing tree might have million such flowers. Each flower secretes nectar on both first and second opening of the flower from a different set of nectaries each day. Pollen is shedding for a few hours only and that during the second opening."

In the report of the California Avocado Association for 1922-23 there is a paper entitled "Avocado Pollination and Bees" by Orange L. Clark dealing with studies of the pollination of the avocado by bees in the orchard of the Point Loma Theosophical Homestead from which I quote as follows:—

> "During the last blooming season there were 17 hives of bees in the orchard. Because of drought here, as elsewhere in southern California, there was less outside bee pasture than usual and bees worked very much more abundantly on avocado blossoms here than during any previous season, and produced an abundant honey harvest."

Ordetx reports that in Cuba the secretion of nectar in avocado is greatly affected by atmospheric conditions and when these are favorable nectar is produced freely permitting the bees to gather an abundant honey surplus.

AZALEA (Rhododendron).

The azaleas are closely related to the mountain laurel and are likewise sometimes reported as poisonous. (See Laurel; also Poisonous Honey). The flame-colored azalea is common in the mountains of the Eastern States from Pennsylvania to Georgia. It has a profusion of showy, flame-colored blossoms coming just when the leaves appear.

The Science News-letter under date of April 23, 1927 mentions the journey of Dr. K. Krause into northern Asia to investigate the source of poisonous honey mentioned by Xenophen.

> "According to Dr. Krause's description, based on personal experiments with the poisonous sweet it causes a giddiness and sometimes a brief loss of consciousness, followed by a short period of malaise, as though one had been on a spree."
>
> "Where the bees get the toxic nectar is still an unsettled question, but suspicion settles most strongly on two species of *rhododendron* abundant in the region whose foliage is known to be poisonous to cattle."

Although there are several species of azalea or rhododendron common from eastern Canada to Georgia, there are few reports available which indicate that they are anywhere regarded as important by the beekeepers.

B

BACCHARIS.

Mule fat (*Baccharis viminea*) is a loosely branching shrub, very leafy, growing six to seven feet in height. It is willow-like in appearance and strangers sometimes mistake it for a willow. It is common along streams, according to Jepson, in the Sacramento and Napa Valleys and south to Southern California. The Spanish name, Guatemote, is sometimes locally corrupted to water-motor. There are several species of *Baccharis* occuring in river valleys and along streams in New Mexico, Arizona and Texas and south to Central and South America.

Blossoms of guatemote.

T. O. Andrews, of Corona, California, reports mule fat as a valuable source of nectar in his locality. He states that it blooms nearly all summer, but most profusely in August and September. The nectar is not abundant and such honey as is gathered is used for brood rearing or mixed with that from other plants. Very little surplus is stored from it. Its principal value lies in the fact that it blooms when little nectar is available.

Coleman lists chapparal broom (*B. pilularis*) as the source of much honey for about a week when first it comes in bloom. (Western Honeybee, Jan., 1922.) Alma Hasslbauer (Beekeeper's Item, Dec. 1921), states that there are three or more species in Texas which are commonly confused with willow. She credits some heavy yields of amber honey to this source. She also credits Dr. Chas. T. Vorhies with the statement that desert broom. (*B. sarothroides*) is the source of fall honey in Arizona.

H. B. Parks, in Gleanings (Feb. 1922), states that along the Gulf Coast Baccharis grows by the acre. In fact there are places where thousands of acres are completely covered with this plant. He credits the large average yields in this region to this source. The Mexicans call it Yerba Dulce.

Along the Gulf Coast of the Southeastern States, the Sea Myrtle (*B. halimifolia*) is of considerable importance. J. G. Puett, of Moultrie, Georgia, reported that his bees worked it freely toward the close of the season and in some years it furnished a good flow. This species is found in low grounds near the coast from New York to Florida and west to Texas.

The author found several species of baccharis which were highly valued by the beekeepers of Texas, New Mexico, Arizona and southern California. In the vicinity of Phoenix, Arizona, *B. glutinosa,* is known

Desert broom in winter.

as guatemote or seep willow and is reported as the source of surplus. Desert broom, already mentioned, grows less commonly, being found in smaller clumps.

BACHELORS BUTTONS, see Chicory, also centaurea

BALLOON VINE (Cardiospermum halicacabum).

The balloon vine is a herbaceous climber with alternate leaves and clusters of small white flowers, followed by a three-celled inflated pod. This species is common along streams in south Texas. *C. molle,* a Mexican species, is also found to some extent in the mountains west of the Pecos River, according to Coulter.

Balloon vine is reported as the source of considerable honey in Texas. Scholl lists it as a fair yielder, but plants not abundant. Other reports indicate that it is the source of considerable surplus along the Gulf Coast of that State.

The balloon vine is commonly cultivated in many places and where it occurs in a wild state is probably an escape. It is now found quite commonly in many southern locations, including South Carolina, Florida and Alabama. It is also found in places in the north from New Jersey and Washington, D. C. to Iowa and Kansas.

According to H. B. Parks, (Gleanings, Dec. 1936) most of the balloonvine honey gathered in the Rio Grande Valley comes from the perennial *C. corindum.* Found in the same locality the bees also work freely on the related plant *Serjania incisa.*

BALSAM CLOVER (Trigonella caerulea).

The balsam clover, often called blue flowered sweet clover, is an annual common to some European countries where it is grown for forage. It has failed to find general favor in this country although numerous trials have been given. More than 100 samples of seed were distributed by the American Bee Journal in 1935 to all parts of the country but most reports were disappointing. In a few cases where the climate was cool and moist, favorable conditions seemed to prevail.

On the experimental grounds of the Agricultural College of Saskatchewan it had been previously tried with some promise. One report from Pueblo, Colorado told of a six weeks period of bloom with the bees visiting it freely.

BALSAM APPLE,
see Wild Cucumber.

BANANA (Musa sapientum).

Since the banana plant is little grown in the United States, it is seldom mentioned as a honey plant, yet it secretes nectar very abundantly, and in countries where bananas are grown on a large scale it must be important to the beekeeper. We are showing herewith two illustrations, one of the plant in fruit and one showing the opening of the bloom.

Blossoms of the banana.

The banana.

The following description of the possibilities of this plant is reprinted from page 83 of the American Bee Journal for 1880, and was written by a correspondent in Clifton Springs, Florida:

"Recently noticing bees working upon blossoms, I concluded to examine them. To my surprise, I found that each blossom had a sack on its under side, which contained several drops of nectar of the consistency and sweetness of thin syrup. This sack gradually opens, allowing the contents to escape, unless appropriated by some insect. The blossom hangs in a position that rain cannot enter to dilute or wash out the nectar. Procuring a teaspoon, I emptied into it the contents of a dozen blossoms, which filled it full. Each stalk, on good land, will produce a head having a hundred hands or divisions of blossoms, and each hand averages six blossoms, giving 600 blossoms to the stalk. Estimating 100 teaspoonfuls to the pint (88 of the one used filled a pint measure) we have 50 spoonfuls, or half a pint to the stalk. Planted in checks 8x8 feet, there will be 680 plants to the acre, yielding, according to the above estimate, 42½ gallons of nectar. But usually more than one stalk in a hill blossoms and matures fruit annually. The blossoms used were below those that produce fruit, which later, I am told, are much richer in honey.

BAPTISIA, see Indigo-weed.

BARBERRY (Berberis).

The common barberry, introduced from Europe, has become naturalized in the Northern States and occurs in thickets and woodlands from the Atlantic Coast to Iowa and southward. It blooms in May and June and is well known as an attractive plant to the bees, though seldom sufficiently abundant to be important. It is much cultivated for ornament.

Triple-leaved barberry, or agarita.

In Texas, the triple-leaved barberry (*Berberis trifoliata*) is known as agarita, the Mexican name, and as wild currant. It is common over a large part of the southern portion of the State and is an important source of early nectar. It is a shrub 4 to 6 feet high, with stiff leaves and bright yellow flowers that grow in dense clusters along the stem. The red berries ripen in May and are often called currants. They are used for jelly or sauce, as well as for barberry wine. The plants are to be found in thickets, in open woodlands and along roadsides, in fence corners and other waste places where the seeds have been scattered by birds.

> "As a honey plant it is one of much value to me. It blooms here early in February, and the bloom continues for several weeks, some bushes blooming later than others. The pollen yield is abundant, bright yellow in color. It is the second bloomer of the year on my list of Texas honey plants, coming after mistletoe and before our main fruit bloom."—Louis Scholl, Gleanings, Feb., 1907.

Honey from agarita is reported as light amber in color and of good flavor. Blooming so early in the spring, few colonies of bees are in condition to gather surplus from this source although a full super per colony is sometimes stored by the middle of March. Although the bloom is sometimes killed by late frost, agarita yields nectar so freely that it is well worth while for the beekeeper to plan his practice to get his bees to storing strength ahead of this flow. Too many beekeepers utilize the agarita flow for building up the colonies, much as fruit bloom is utilized in most localities.

Unlike most of the shrubs common to the region where it grows, agarita, apparently is not greatly influenced in its blooming period by rainfall. It has a rather prolonged season of bloom in early spring from January to April. At times the bees gather the juice from the ripe berries and store it in the combs. When this happens pink spots appear in the combs of white honey.

Richter reports the California barberry (*Berberis pinnata*) yields surplus in Monterey County after manzanita has bloomed. The honey is amber in color. The plant is rather common from Berkeley Hills south to Monterey, blooming in March and April, according to Jepson. (See also Oregon grape).

There are many different barberries cultivated as ornamentals. These come from various old world countries and are distributed by the nurseries. Some, if not all, are visited by the bees for nectar.

BARLEY (Hordeum vulgare).

Barley, the well known field crop somewhat resembles wheat to which it is closely related. The plant is nectarless and yields no honey but the bees do at times store the sap from freshly cut green stubble. The following from a private letter received from Duncan Chalmers of Edmonton, Alberta is unusual:—

"About ten years ago I noticed a small quantity of pink honey in my extracting combs. This was in late autumn after the main crop had been extracted, and a super of empty combs had been given each colony to catch the fall pick-up.

"Two years later after the main honey season was over and the bees had been quiet for a time they broke out into a riot and everything indicated a fresh source of nectar. Field bees were all going in one direction and returning bees were carrying heavy loads. The bees were collecting sap from fresh cut barley stubble. That season the usual time for seeding barley was so wet that it had to be postponed for several weeks. The autumn was free from frost and the grain matured sufficiently to make stock feed. The base of the straw was still green and in the morning for several days there would be a large drop of sweet sap at top of each stubble.

"There were about twenty acres in the field and my 75 colonies were working on it. I never experienced anything like it before. The flow lasted for about three days. The result was about 25 pounds of sap stored by each colony. The fresh sap was almost water white and as evaporation went on it gradually took a pink shade."

BARNABY'S THISTLE, see Star Thistle.

BASIL or MOUNTAIN MINT (Pycnanthemum virginianum).

Basil is common from New England to the Missouri River and southward. There are numerous reports from beekeepers to the effect that bees work upon it very eagerly from morning till night. Surplus is reported from several Illinois counties.

There are about eighteen species of mountain mint known to Amer-

Surplus honey is secured from the mountain mint which blooms profusely for a period of several weeks.

ica. Small lists seventeen in his "Flora of the Southeastern United States." They are especially common in this region, although some are found from Maine to Florida and west to Arkansas and Texas. There is one species, *P. californica,* common to the Sierra Nevada and Coast ranges of mountains of California. The narrow-leaved mountain mint

Basswood tree in bloom.

(*P. flexuosum*) is reported by Missouri beekeepers as valuable in that state.

Probably all the species are of more or less value to the bees and where abundant may become important.

H. B. Parks has observed that bees working on the blossoms of these plants are often caught by ambush bugs and flower spiders.

P. pilosum in our test garden has proved especially attractive to the bees as well as a promising source of essential oil. Tests from small measured areas by Prof. Arthur Schwarting indicate a high yield of volatile oil from which either menthol or thymol can be obtained. This species is found from Pennsylvania to Iowa and Arkansas. (Am. Bee Journal, April 1947).

BASSWOOD (Tilia americana).

The basswood, also known as linden, whitewood, and sometimes as limetree, is one of the best known sources of honey in the Eastern States. There are other species closely related which also produce

Blossom and leaf of the basswood.

nectar, and which, perhaps, would not be distinguished by the casual observer. The natural range of the basswood is from Canada to Florida and west to Nebraska and Texas. It is also grown as a shade tree in other Western States and is mentioned by Richter in the bulletin on "Honey Plants of California," as an introduced species of value.

The tree thrives on rich lands, and in the cooler regions of the country reaches a large size. The wood is soft and white and much in demand for making sections, separators and other bee supplies requiring a soft wood cut in thin sheets. For such purposes basswood has no superior. The wood is also sought for use in the manufacture of furniture, packing boxes, etc., as well as for paper making.

The blooming period is short, seldom yielding to exceed ten days or two weeks, and often for a much shorter period. The honeyflows from basswood are irregular and only to be depended upon about two or three years out of every five. A heavy flow from this source occurs only occasionally, but when it does come it is worth waiting for, for enormous yields are sometimes secured. The honey is white in color, with rather a strong flavor, but is usually regarded as high quality. Good basswood locations are no longer plentiful, as the cutting of the forests over the entire country has resulted in a large reduction of this along with other trees.

The European linden (*Tilia europaea*) has been planted in many places and, according to Sladen, is the source of surplus honey, at Charlottetown, Prince Edward Island.

BASTARD INDIGO, see False Indigo.
BASTARD PENNYROYAL, see Blue Curls.
BAY, see Magnolia, also Red Bay.
BAYBERRY, see Wax Myrtle.
BAY-LAUREL, see California Laurel.
BAY-TREE, see Magnolia and California Laurel.
BEACH APPLE, see Ice Plant.

BEAN (Phaseolus).

There are two varieties of the garden bean which are the source of nectar in quantity. Surplus honey from beans is seldom reported except in Southern California. Ventura County is said to produce as high as 72 per cent of the lima beans raised in the entire United States; 800,000 sacks is the reported output for 1910. The honey from lima beans is almost water white in color and of fine flavor, according to M. H. Mendleson, of Ventura, California, who produced many tons of this honey. The black-eyed beans yield a dark amber honey, but of good quality. A peculiarity of honey from lima beans is that it will sometimes sweat and ferment when left too long on the hives of weak colonies, near the coast. This, according to Mendleson, seldom happens with strong colonies, nor does it happen in any case in the interior. Honey from beans granulates very readily.

The honeyflow from beans is regarded as very dependable. Occasionally the blossoms are blasted by hot east winds, but not frequently. In Ventura County the average from this source is reported as 50 pounds per colony per year, with as high as 140 pounds in an exceptional season. The beans bloom through a long season, beginning in July and continuing till September, with the principal crop of nectar harvested in July. The bloom is prolonged by irrigation.

The principal counties of California where beans are important are: Ventura, Orange, Santa Barbara, Los Angeles and San Diego.

From Delaware comes the report that Lima Beans are grown on a considerable scale for canning in some neighborhoods in that state.

Where no dusting is done for control of insect pests it is reported that a nice crop of white honey is secured.

Lima beans are grown commercially in New Jersey and Maryland also. Several growers who have a large acreage maintain bees for the purpose of insuring pollination, even though it is generally thought that this plant is self fertile and needs no such assistance. Several letters were received in answer to enquiry which stated that the yield was larger and the beans more uniform where bees were present.

Mr. H. Brown of Cape May Court House, New Jersey, writes as follows:

> "In the bean field where I had my 150 hives of bees the production per acre was 20 percent more than on the other field where there wasn't any bees. Whether this increase in production was due to the bees or to the managers farming ability, I cannot state."

In the June, 1943 issue of American Bee Journal appeared an article by J. M. Amos, (Delaware) giving results of a carefully checked experiment to determine the value of the bees. From it we quote:

> "Observations indicate that nectar secretion by lima beans is greatest when the plants first come into bloom and remains intense for about a week following which it tapers off considerably. A succession of plantings is necessary to provide a good honeyflow.
>
> "At the prevailing prices in 1942 of $70 per ton for green lima beans and $20 per ton for white ones (assuming that the beans graded 60% green and 40% white), the increased production from the use of bees in pollination would be valued at $11.36 per acre."

The wild bean, (*Phaseolus polystachyus*), blooms profusely in late summer and attracts bees in large numbers. It is probably nowhere sufficiently abundant to be important.

(See velvet bean).

BEARBERRY, see Manzanita, also Kinnikinnik.
BEAVER TREE, see Magnolia.

BEE BALM (Melissa officinalis).

A sweet perennial herb, cultivated in gardens, from Southern Europe and North Africa. Sometimes escaped; flowers yellow or whitish, several in each auxiliary cluster. Plant erect and branching, with broad opposite leaves. Attractive to the bees, but not sufficiently abundant to be important.

BEECH (Fagus grandifolia).

The beech is a large tree common to Eastern America. It is known from Nova Scotia to Ontario and southward, sometimes in extensive forests. Its principal value to the beekeeper is as a source of pollen, though honeydew is sometimes secured from the leaves.

BEE SAGE (Hyptis emoryi).

The bee sage is a shrub which grows in compact clumps with numerous slender stems. It is common to valleys in the Mohave and Colorado deserts of California and south into Mexico.

Harry Cross of ElCentro, California reports it as a valuable source of light stimulation during the winter months from October until April with the bees most active on it in March and April when it appears to be a good producer of nectar.

BEEWEED, see Rocky Mountain Bee Plant.
BEGGARTICK, see Spanish Needle.

BELLFLOWER (Campanula) Canterbury Bell.

The genus campanula is a very large one with dozens of different species. They are widely distributed in Europe, Asia and North America. At least sixty different species are cultivated in this country as garden flowers. Although many of them are attractive to the bees they are rarely mentioned in lists of important honey plants. This may be due to the fact that they are usually present in relatively small numbers.

Edgar Scholes of Victoria, British Columbia mentions Canterbury bells as among the best garden honey plants, placing it above the catmints. He writes to the author to tell of an English monastery which is a large honey producer where a large border of Canterbury bells is planted for the bees every year.

In his garden he finds lavender to be the only plant which is more attractive to the bees.

The black bindweed is a common weed.

BINDWEED
(Polygonum convolvulus).

The black bindweed is a common weed throughout the Eastern States. It is of European origin, introduced in this country. It belongs to a family of plants which produce honey in quantity, but, is itself, of little value. The first indication that reached the author that this plant was sometimes of value to the bees, was the receipt of a specimen, by mail, with the information that the bees were working upon it. Careful watch was then kept for several years, before they were seen to seek it in his locality. Since that time there have been a few occasions when the bees have sought it freely and when it seemed to yield some nectar. It is doubtful whether it is ever of much importance as a source of honey.

The seed has been widely distributed with grain seeds and is very troublesome in fields of small grain. The vine closely resembles the wild morning glory, to the casual observer, but the blossom is small and inconspicuous, followed with a seed somewhat like buckwheat, hence it is often called wild buckwheat.

BIRD OF PARADISE TREE (Poinciana gillesii).

The Bird of Paradise Tree is commonly cultivated in southern New Mexico and adjacent regions and has escaped and is occasionally found wild there. The terminal racemes of large yellow flowers are very attractive to the bees and the author found numerous reports of this attraction from beekeepers in the Rio Grande Valley of New Mexico.

BIRDSFOOT TREFOIL (Lotus corniculatus).

Birdsfoot Trefoil is a new forage crop which is attracting much interest. It was called to public attention as a pasture crop by Prof. D. B. Johnstone-Wallace who first noticed it growing in a pasture near Claverack, New York in 1934.

Seed was planted in the American Bee Journal test garden in 1938. Growth was vigorous and from the start it appeared to be very drouth resistant. After nine years it has shown itself able to withstand dry weather from which other plants suffered severely. There is every indication that the plant will adapt itself readily to a wide variety of soils and its use will probably become quite general.

Rev. Yate Allen recognizes it as an important honey plant in his book, "European Bee Plants" and reports that much of the honey going to market in England as "White Clover" is found to contain pollen grains from this trefoil.

With us the bees work it freely at times. In some cases they have visited the trefoil flowers even more freely than those of Alpha sweet clover in nearby plots. New York beekeepers report surplus honey of good quality from this source. Surplus is reported from Oregon also.

Numerous species of Lotus are native to this country, especially in the west. Fourteen species are recorded for Arizona and several more for California. But little is heard of the native species as honey plants except in California where one known as wild alfalfa is commonly recognized. It is probable that more honey is gathered from the others than is generally known.

According to McKee and Schoth, (Cir. 625 USDA) Lotus corniculatus is practically self sterile and almost entirely dependent upon insect visitors for seed formation.

See Wild Alfalfa.

Birdsfoot trefoil has proved drouth resistant.

BIRD'S NEST PLANT, see Carrot.
BIRTHWORT, see Trillium.
BISCUIT ROOT, see Cogswellia.

BITTERSWEET (Celastrus scandens). Waxwork.

A well-known climbing shrub common in woodlands. The orange-colored pods displaying the scarlet seeds are often gathered for winter bouquets. The flowers are small and greenish in raceme-like clusters at the termination of the branches.

"The bees work freely on bittersweet."—Miss Mitchell, Keokuk, Iowa.

BITTERWEED (Helenium tenuifolium).

The author's first experience with bitterweed honey was in south Missouri in 1904 or 1905. There had been a good flow from white clover, followed by a dearth for a time, and the unfinished sections were filled out with bitterweed honey. The sections looked very nice and a northern beekeeper who had recently settled near the town of Salem, innocently sold his honey to the townspeople. The next time he came to town there were numerous persons looking for him, and he found it necessary to take back most of the honey he had marketed on his previous visit. The honey from this source is so bitter that a very little of it will spoil a fine crop of the best white honey. A few cells are sufficient to make a whole section absolutely unpalatable.

The range of the plant is given as from Arkansas and Texas to North Carolina. It probably does not appear to any extent north of Tennessee.

Chas. Mohr says of it (Plant Life of Alabama, page 54):

> "The bitterweed, originally from the sunny plains west of the Mississippi River south of the Arkansas Valley, was first observed in Mobile in 1866. It has spread along the embankments of the railroads to the mouth of the Ohio River, literally covering in many places the waste and uncultivated grounds, and reaching out along byroads and borders of fields and woodlands. In its northward spread it has largely taken the place of the mayweed (*Anthemis cotula*), a European weed of early introduction."

Regarding honey from this source J. J. Wilder says (American Bee Journal, Vol. 54, page 410):

> "It is truly a nectar-laden plant. Though it does not grow in great fields as yet, bees will store from 30 to 35 pounds of surplus from it. Its flowers are of a deep yellow; the honey, light yellow, heavy body, soon granulates when extracted. It is bitter; in fact it is about as offensive to the palate as quinine. In most sections of the South the cotton plant begins yielding two or three weeks before the bitterweed, and if it were not for the well-established fact that bees do not desert a honey plant for another as long as it yields well, nearly all the summer and fall honey would be unfit for market on account of the bitterweed. In sections where the cotton does not yield much, the honey is all bitter, and a small amount of it will ruin a tank of good honey. Bitterweed is also a great pollen plant, furnishing abundance of bright yellow pollen throughout its blooming period. Even the stems and foliage of this plant are intensely bitter, and no animals eat it."

Pammel cites a quotation which states that it has been reported as fatal to horses and mules in several of the Gulf States. It is said to contain a narcotic poison and to be the cause of bitter milk.

A relative of this plant, the northern sneezewood (*Helenium autumnale*) is also a good honey plant, and probably less bitter than the southern or narrow-leaved sneezeweed just described. Neither, however, can be said to be a desirable addition to the honey-producing flora, because of spoiling good honey from mixing with it. The northern sneezeweed is found in various localities from Connecticut to the Dakotas and southward. It is also found in places in the Rocky Mountain States.

The bitter honey seems to be as good as any for brood-rearing and, where present, the beekeeper should use care to avoid mixing it with his marketable produce, and use it for feeding the bees. The bitterness is said to come from the pollen grains present in the honey, which improves greatly with age, as the pollen grains settle to the bottom of the container.

Carl M. Teasley writes from Mississippi in American Bee Journal, August 1940, as follows:—

> "I have never known of a crop failure from this source. It always blooms during a dry time of the year which is ideal for nectar secretion. The bloom usually lasts for two months and the end of the flow arrives just before aster and goldenrod begin to yield. Bitterweed gives colonies an early fall stimulation and a start on winter stores. Taking my own bees as an example, I know that a surplus to the amount of forty pounds per colony can be had. Most beekeepers in the South extract the honey from bitterweed and feed it back late in the fall so that the bees use it for stores."

BLACK ALDER, see Holly.

BLACKBERRY (Rubus).

The blackberries, dewberries and raspberries are closely related plants, all of which are good honey sources. The blackberry is especially well known in the Southeastern States, where it thrives in fence corners and moist woodland borders. In north Georgia it is one of the principal sources of surplus honey. Farther north the nectar yield is apparently not as good, and in some localities the bees apparently do not get much honey from this source. Lovell states that in New England there is very little nectar available from either wild or cultivated blackberries. Richter states that the Himalaya variety of blackberry yields some honey in Yuba County, California. John W. Cash reports an average of 25 pounds per colony of surplus at Bogart, Georgia. The honey is amber, very thick, and does not granulate.

There are parts of California where the dewberry is reported to be very important and to yield surplus. The honey is of good flavor and a light amber color.

Everett George of Independence, Oregon reports good crops of honey from blackberry in that locality and describes it as almost as clear as water.

In the British Bee Journal, September 27, 1888, Edward J. Gibbons describes blackberry honey as "light in color (amber), rather thin, strong but pleasing odor and excellent in flavor."

From England, F. N. Howes, in his book "Plants and Beekeeping," reports that honey from blackberry is dense and slow to granulate but that the flavor is not of the best.

In an article, "Blackberry and Raspberry Improvement," (Yearbook USDA 1937, G. M. Darrow states:—

> "Normally the wild blackberries of the east are entirely or nearly self-sterile, and those of the Pacific Coast have male and female organs on

separate plants. All need cross-pollination. In the clearings and pastures bees and other insects have crossed the blackberry species for the last 100 to 300 years."

(See Raspberry).

BLACK BRUSH (Acacia amentacea).

This shrub is common to hillsides from the Guadalupe River to the Rio Grande and west to the Pecos and from San Antonio south to Mexico. It is a beautiful chaparral bush blooming in late winter from February to April, when its masses of cream colored blossoms are very attractive. While many beekeepers report it as a source of honey, it is probably much over-estimated. H. B. Parks writes that it is the most over estimated honey plant in Texas and that he is unable to find any evidence that the bees get any nectar from the blossoms, though he states that its extra floral nectaries are sometimes visited by the bees. He places but little value on it for pollen. (See acacia.)

BLACK GUM, see Tupelo.

BLACK HAW (Viburnum prunifolium).

The black haw belongs to a group containing several valuable honey plants. It occurs from Connecticut to Michigan and south to Georgia, Alabama, Mississippi, Louisiana and Texas. It is a shrub or small tree, blooming in the North in May and June and in the South in April and May. Its principal value lies in stimulating early brood-rearing, since in few localities it is of sufficient abundance to be important as a source of surplus.

School lists it as yielding well, early, in Texas.

BLACK LOCUST, see Locust.

BLACK MANGROVE (Avicennia nitida).

Black mangrove (*Avicennia nitida*) is also known as black-

Mangrove bloom.

wood, or blacktree. It is an evergreen tree, growing along the seashores of the coast of Florida. It is said also to occur to some extent along the gulf coast to Texas and throughout the coasts of Tropical America. It varies from a bushy shrub to a tall tree 60 or more feet in height in tropical regions. The wood is coarse-grained, hard and very durable in contact with the soil. The tree is to be found only in the vicinity of salt water.

The honey from the mangrove is light in color, mild in flavor, and is generally regarded as of first quality. According to E. G. Baldwin it was the heaviest yielder of nectar known in the south prior to the big freeze of 1895. In one year he reports Harry Mitchell, of Hawk's Park, as having secured an average of 380 pounds per colony from mangrove alone. Following the freeze it failed to yield nectar in surplus quantity for about fifteen years, and reports since that time have indicated that it is not up to its former importance.

The blooming period opens about the middle of June and usually includes the entire month of July. The flow usually lasts from six to eight weeks.

Florida beekeepers report that in dry weather the black mangrove accumulates a coating of salt from the fogs which discourages the bees and causes them to stop visiting the flowers. Following a rain which washes away the salt the bees begin working again vigorously. Reports indicate that yields are very sensitive to weather conditions. When everything is favorable the yields are very heavy, in fact among the heaviest known. When the weather is unsuitable the crop may be almost an entire failure.

BLACK MEDIC (Medicago lupulina).

Black medic is a widely distributed plant in California, but not very common. It blooms from April to June. It is the source of some nectar, but is probably not of much importance.

It is also common in waste places in many eastern localities, but apparently is not often of much value to the bees. In an old copy of the British Bee Journal the statement is made that in England it produces honey during the severest drouths.

BLACK TREE, **see Black Mangrove.**
BLACK TWIN-BERRY, **see Honeysuckle.**
BLACKWOOD, **see Black Mangrove.**

BLADDER-POD (Lesquerella Gordoni).

Under date of February 4, 1926, W. C. Collier sent to the author a plant from Tucson, Arizona, from which his bees were storing surplus honey at that time. He wrote as follows:—

"This is a peculiar country, freezing at night and a honeyflow in the daytime. Last night we had thin ice and at eleven o'clock (today) nectar would shake out of the combs. Some of my bees have a super of new honey which they are sealing."

This interesting plant from which Arizona bees were storing honey was identified as bladderpod, (*Lesquerella Gordoni*). It is a common spring annual in West Texas, southern New Mexico and Arizona. It belongs to the mustard family, a group of plants which furnishes

The black mangrove tree.

many good honey sources, though some of them are of rather poor quality. For related plants see mustard, Prince's plume, etc.

From H. B. Parks we learn that *Lesquerella recurvata* is abundant from SanAntonio to Del Rio, Texas and that an average of a super per colony of honey is often secured from this source in spring.

Lesquerella ovifolia found from Nebraska and Kansas to Arizona is reported by J. B. Douglas of Bonita, Arizona as of value in freezing weather.

BLAZING STAR (Liatris) Button Snakeroot.

Common to the prairies are the blazing stars, perennial herbs which bloom in late summer. The purple flowers on a long spike make

Blossoms of blazing star in author's garden.

a fine show during the season of bloom. There are areas in northern Nebraska, in the wet valleys and on the margins of the sand hill region, where the blazing stars are still abundant. Also in other western prairie sections.

Bessey lists four species as yielding both nectar and pollen: *L. punctata, L. pycnostachya, L. scariosa,* and *L. squarrosa.*

In American Bee Journal (page 620), September 1888, *L. scariosa* is called gay feather. A beekeeper writing from Hamilton, Illinois, says that the bees work on the blossoms all day.

BLOODROOT (Sanguinaria canadensis).

The bloodroot is a common wild flower in the moist woods of all our Northern States. It blooms early in April, and is eagerly sought by the bees for pollen.

BLUEBERRY (Vaccinium).

There are at least four species of blueberry which give surplus honey in localities where they are abundant. According to Sladen, the

The bloodroot is a source of early pollen.

dwarf or early sweet blueberry, (*Vaccinium pennsylvanicum*), and the sour-top or velvet-leaf blueberry (*Vaccinium canadense*), often give surplus in northern Ontario, northern Quebec and eastern Manitoba.

Blueberry honey is frequently reported from New England. W. J. Sheppard reports that *Vaccinium ovalifolium* is of importance in British Columbia.

There are several other species in the Northeastern States that probably yield some nectar. Sladen reports blueberry as important in Nova Scotia.

According to Lovell *Vaccinium corymbosum,* the high bush blueberry, is important in southeastern Massachusetts, Rhode Island and Connecticut, in some localities beekeepers being principally dependent upon it. The flow comes late in May or early June and lasts for about ten days. Strong colonies store as high as 50 to 90 pounds of surplus from it. He reports *V. pennsylvanicum,* the low bush blueberry already mentioned, as common throughout northern New England, in pastures and on rocky land. It is important in the blueberry barrens of Maine. The honey is amber and of good flavor. (See Farkleberry.)

An Oregon beekeeper writing in Western Honeybee states that there are four varieties of huckleberries native to his locality. He states that the blue huckleberry is very abundant and one of the earliest wild flowers, blooming in March. It furnishes an abundance of nectar when weather conditions will permit the bees to gather it. He states that the honey is amber, but of good body and flavor. About the middle of April the scarlet huckleberry appears. He reports this variety as even more abundant than the former, growing in the bottom lands as well as on the mountains and furnishing abundant bee pasture for about six weeks. He described the honey as very light amber, mild and exceedingly sweet with a pleasing flavor peculiar to itself. When fully ripened it is so thick as to be difficult to extract. The scarlet huckleberry, he states, is one of the best honey plants of which he has any knowledge.

Blossoms of blueberry.

Although no scientific names are given in connection with the above reference, from the meagre descriptions it is assumed that some species of *Vaccinium* are referred to. There are at least nine species of this group known to the Pacific Coast.

According to T. A. Merrill, "Pollination of the highbush blueberry," Michigan Agr. Experiment Station, Tech. Bul. 151, "Bumblebees and honeybees play a very active part in blueberry pollination."

BLUEBONNET, see Lupine.
BLUEBOTTLE, see Centaurea.

BLUE CURLS (Trichostema lanceolatum).

Blue curls is a plant known by a great variety of names in California. It is known as vinegar weed, camphor weed, turpentine weed,

The blue curls.

flea weed, bastard-pennyroyal, etc. It abounds over a large portion of California. According to Jepson, the range is throughout the Coast Ranges, Sierra Nevada foothills and southern California. The Western Honeybee lists it as the best fall honey plant in California. Both foliage and flowers have a pungent, penetrating odor. The plant is found mostly in stubble fields, where it appears after grain harvest.

> "Under favorable atmospheric conditions it yields abundantly of a very white honey, that granulates quickly and with a fine grain. Sometimes it granulates before the bees have time to seal the cells. One peculiarity of this plant is that it continues to yield honey for several hours after falling to the ground. The quality of the honey is good."—Western Honeybee. October, 1914.
>
> "It is often claimed that rain will end the honeyflow from blue curls. I find by several seasons' observation that this is an error. If a rain is followed by favorable atmospheric conditions—warmth and humidity—the nectar secretion is increased instead of diminished."—Western Honeybee, November, 1916.

The flow from blue curls begins in August and continues till frost. Richter reports that very large yields are sometimes secured from this source, tons of honey sometimes being stored in the vicinity of Fresno. A report of an average of 80 pounds per colony from this source reached the author when visiting at Visalia.

The blue curls of the east (*T. dichotomum*), also known as bastard pennyroyal, is found in sandy soils from Maine to Florida and southwest to Missouri and Texas. The small flowers are blue or pink, rarely white, and appear in summer or autumn. This species is reported as the source of some nectar in Texas, though nowhere important as a source of surplus, as far as the author can ascertain.

BLUE DAISY, see Chicory.
BLUE DANDELION, see Chicory.
BLUE DEVIL, see Aster.
BLUE GUM, see Eucalyptus.
BLUE LUPINE, see Lupine.
BLUE MALLOW, see Mallow.
BLUE SAILORS, see Chicory.
BLUET, see Centaurea.

BLUE THISTLE (Eryngium articulatum).

The blue thistle is listed by Richter as the source of a dark honey of good flavor in California. He reports it from the Suisun Marshes, along the Consumnes River and the Alvarado Marshes, blooming from August to October.

The purple thistle (*Eryngium Leavenworthii*) is reported as furnishing a good yield of honey of very poor quality in dry seasons in the mid-coast area of Texas.

There are several species of *Eryngium* of wide distribution, but they are not apparently recognized as of much value to the bees. Prof. S. W. Bilsing, of the Texas College of Agriculture, reports that *E. Leavenworthii* is an important source of honey in the vicinity of Bay

City, Texas. He describes the honey as of a dark color and not unpleasant taste.

Several species are common to the southern states, especially from Carolina to Florida and Texas. One or two species are found on low grounds as far north as Connecticut and Minnesota.

BLUEVINE (Gonolobus laevis).

This plant is also known as devil's shoestring, climbing milkweed, sand vine, wild sweet potato vine and anglepod. It is a vine of luxu-

Seed pods and leaves of the bluevine, or sandvine.

riant growth, common on low lands from Virginia and Tennessee westward to Missouri and south to Texas. It is common in Southern Ohio, Indiana and Illinois, where it is a persistent and troublesome weed. It is especially troublesome in the corn fields, where it may be found climbing the stalks. A single vine will run for many feet on fence wires or other support. It blooms freely through July and August, and is the source of large quantities of surplus honey of good quality. The honey is light in color, mild in flavor and does not granulate readily. S. H. Burton, of Washington, Indiana, reports a yield of 60 pounds per colony in three weeks and an average of as high as 80 pounds per colony has been reported from Southern Indiana. From Missouri come similar reports, W. L. Wiley, of Brunswick, reporting as much as 100 pounds surplus from strong colonies.

The plant may be readily recognized by the abundant clusters of small, white flowers, which are followed by seed pods similar to those of the milkweed. When the pods are dry they split open and the seeds are widely scattered by means of their cottony parachutes.

The honey is clear, heavy-bodied and of excellent flavor. The plant grows chiefly in cornfields in river bottom land and is perennial. It blooms before smartweed, but the smartweed honey is usually mixed with it, as it comes in later. It yields best in dry years.

It is reported as valuable near Buffalo, Texas.

BLUEWEED or VIPER'S BUGLOSS (Echium vulgare).

The blueweed, or viper's bugloss, is a weed naturalized from Europe. It is common in the meadows and roadsides of the Eastern States. It has showy purple or blue flowers and grows about two feet high. June is the flowering period. Sladen lists it as an important source of nectar in Southern Ontario. Probably not sufficiently abundant to be very important in many places.

This or related plant was introduced into Australia from Europe by a settler named Patterson, and has become widely spread there, where it is known as "Patterson's curse." Rayment states that it yields honey, but hardly sufficient to store much in the supers. (Money in Bees in Australasia).

According to Dr. E. F. Phillips, blueweed, commonly called blue thistle, is a reliable source of surplus in the Shenandoah Valley and in parts of the higher valleys of Maryland.

BONESET (Eupatorium).

There are 475 species of *Eupatorium* known, many of them found in Tropical America. Some are found in Europe, Asia and South America, so that the plants have a wide range. Forty-five or more species are common to North America. White snakeroot is common from New England south to Tennessee and Georgia. According to J. M. Buchanan, it is common over the State of Tennessee, but only yields honey in the northern part. He reports the honey to be a light amber, of strong flavor. The yield comes in August and September. The white snakeroot (*E. urticaefolium*) is also common to the woodlands of the Middle West. Although the bees visit this species freely, from September until killed by frost, usually in October, the honey yield is probably rather small.

The Joe-Pye weed or turnip weed (*E. purpureum*), is frequently reported as yielding honey, and it is one of the most widely distributed species. It is found from New Brunswick to Manitoba and south to Colorado, Texas and Florida.

The boneset of commerce is *E. perfoliatum,* which is one of the best for honey in the Northern States and Canada.

All the bonesets are autumn bloomers, and the honey is usually mixed with that of heartsease, asters, goldenrod, Spanish needles and other plants blooming at the same period. Several years ago Professor Beal made the statement, in the American Bee Journal, that there were twenty species or more valuable to the bees. The wide distribution of the group, together with the regularity of its yield, make it one of

White snakeroot.

special importance to the beekeeper, although the amount of surplus gathered from boneset is not often as large as that of many other well-known plants.

Some species are to be found in almost any kind of situation. While white snakeroot grows in shady woodlands, other kinds delight in open, sunny situations, along roadsides and in pastures or waste places. Some do best on high lands, while others thrive in low, wet land. In places in the Mississippi River bottoms in Illinois, boneset,

E. *serotinum* yields in late summer and fall.

together with heartsease and Spanish needle, cover acres of the richest land like a waving grain field.

An Illinois beekeeper reports *Eupatorium serotinum* to be one of the best honey plants he knows. Also valuable on alluvial soils in Louisiana.

BOOTJACK, see Spanish Needle.

BORAGE (Borago officinalis).

Borage is a European plant which is cultivated for honey and as an ornamental. Its blue flowers are very attractive to the bees. The following quotations indicate its value to the beekeeper:

> "The period of blooming is from June 20 to cold weather. Where there are no plants for bees to work upon, borage does very well; but when white clover and basswood are in bloom, bees will forsake the borage for them. As cold weather begins to come, they swarm to the borage. It is a good honey plant, when there are no plants of greater importance in bloom."—Fisk Bangs, American Bee Journal, page 84, 1878.
>
> "In Practicher Wegweiser, page 280, Herr Wilhelm says that in response to the general cry, 'Sow borage,' he has been sowing it for years and now has it in abundance. How the bees do hum upon it! But, alas! now that he has it in such abundance that it shows its character in the surplus honey, he finds it such as no customer wants, and says it is as black as a certain 'gentleman' with whom beekeepers do not generally care to have dealings. The task of getting it now rooted out is a difficult one."—American Bee Journal, page 103, 1908.

Staminate blossoms of the box elder.

BOSTON IVY (Ampelopsis tricuspidata).

The Boston ivy is a well-known climbing vine, clinging to the walls of brick and stone buildings in all our northern cities. The flowers are very attractive to the bees in midsummer and they store some honey from this source. The quality is rather inferior. (See Ivy).

BOX ELDER (Acer negundo).

The box elder, or ash-leaved maple, is a near relative of the maples, and is sometimes included with them. As in the willows, the stamens are borne on one plant and the pistils on another.

The box elder is found from New England and Southern Canada west to Dakota and southward. It is also common in California. Apparently its range does not extend as far southward as other maples. It is very commonly planted for windbreaks and shade in the prairie States of the Central West. Some honey is yielded by the blossoms, and honeydew is often secreted by aphis feeding on the leaves. While not generally regarded as especially valuable, its season is such that its addition to honey-producing flora is important. The blooms come very soon after soft maple, in April. In western Canada it is called Manitoba maple.

BOX THORN, see Lycium.

BRASSICA, see Mustard.

BRAZIL or LOGWOOD (Condalia obovata).

Brazil or logwood is a spiny shrub, or small tree, common to Western Texas. It occurs in the lower Rio Grande and is reported as a source of honey at Brownsville. There it is smaller than farther north and west. In places it forms very dense thickets. Beekeepers report it as important in the fall, yielding a dark honey. The honey is said to be of fine flavor, despite its dark color. The flow at Beeville is reported as being very rapid. At Goliad, W. C. Collier reports Brazil as the best all-round honey plant. He states that it blooms sometimes in spring and sometimes in fall. Again, it sometimes blooms several times, and yields at irregular periods. At Crystal City, the honey from Brazil is reported as rank flavored, exactly opposite from reports of quality in eastern parts of its range.

E. G. LeStourgeon says that Brazil honey is always dark in color—about the color of light sorghum molasses, and of somewhat similar flavor. He states that it blooms after every rain, no matter how many times or how light the rainfall, which accounts for the apparent confusion in the reports about its season. It is often the source of considerable surplus.

BREASTWEED, see Lizard's Tail.

BRITISH COLUMBIA—Honey Sources of.

British Columbia is a region of high mountain ranges and narrow valleys. But little of its area is suited to agriculture and wild plants are accordingly the source of the greater part of the surplus honey. Since lumbering is the principal industry, forest fires are of frequent occurence and these are followed by a luxuriant growth of fireweed or willow-herb which covers immense areas. It is by far the most important source of nectar, taking the province as a whole.

In the wet belt on the lower mainland, a considerable area of the Fraser Valley is under cultivation and here we find several of the forage crops which are important to the beekeeper. Alsike and white Dutch clover occur in this section, as in all the valleys where there is sufficient moisture. Raspberries. Labrador tea, salal and cascara yield surplus.

In the Irrigated valleys of the interior, fruit growing is important and the bees get nectar and pollen from the fruit trees as well as from alfalfa and hairy vetch grown as cover crops. Sweet clover is found along ditch banks and margins of fields. Dogbane, locally known as "milkweed," is well known as a source of honey. In the eastern sections snowberry, wolfberry and wolf-willow or silverberry, are important.

On Vancouver Island there is a wide variety of honey plants with most localities offering poor average yields, due to the lack of sufficient acreage of nectar plants. In some of the more northerly sections, fireweed insures a good yield of surplus. The madrona tree is found in many neighborhoods on the Island. Maples are the source of early honey both on the Island and the lower mainland. Dandelions are coming to be important in most well settled sections and are spreading with the settlement of the country. Bulletin 92 of the British Columbia Department of Agriculture by Sheppard, Finley and Roberts lists the following additional plants common to the province: Oregon grape, bearberry, bird-cherry, blueberry, Canada thistle, Choke-cherry, crab-apple, goldenrod, huckleberry, willow and salal.

BROCCOLI; (Brassica oleracea italica).

Broccoli is one of the cauliflower tribe which does not form a solid white head but rather a small greenish head on the main stalk and numerous smaller heads on thick branches. In the Rio Grande Valley of Texas broccoli is extensively grown as a winter crop and blooms freely from the branches after the heads are cut. During the months of January to March the bees work broccoli freely and apparently harvest considerable honey from this source.

BROOM (Cytisus).

There are about fifty species of *cytisus,* which are ornamental shrubs native to Southern Europe, North Africa and Western Asia. Most of them produce yellow flowers in profusion and several species are grown in the warmer parts of America.

The broom is well fitted by nature to grow in unfavorable situations where water is scarce and the soil is poor. The stems develop green tissue which apparently serves in part the same purposes as the leaves. These plants belong to the legumes and the large bright yellow flowers resemble the blossoms of the pea.

Scotch Broom, (*Cytisus scoparius*) has become naturalized on Vancouver Island where it seems very much at home. It is reported by W. J. Sheppard, (Gleanings April 1926, page 233) as an important source of early pollen in the vicinity of Victoria. Bailey also reports it as naturalized along the Atlantic Coast from Virginia to Nova Scotia. (Cyc. of Hort.)

G. Clark Nuttall in his book "Beautiful Flowering Shrubs" states that while there is no honey in the flowers they are visited by many kinds of bees for pollen. Ferd. Von Mueller in "Extra Tropical Plants," lists it as among honey plants.

The White Broom (*C. proliferus*) is common in some parts of California. (See Tree Clover.)

Several other species are grown as ornamentals, especially on the Pacific Coast.

Baccharis sarothroides, known in Arizona as Desert Broom, belongs to an entirely different group of plants. (See Baccharis.)

BROOMWEED (Gutierrezia texana).

Broomweed is a common fall plant over a large portion of Texas, on the prairies. According to Scholl it yields well in September and October. The honey is dark and strong and valued mostly for winter stores. In the November 1, 1906, issue of American Bee Journal he writes as follows concerning this plant:

> "Broomweed is still in bloom, the pastures being one sheet of golden yellow. Cold nights and cool, windy days have interfered with the bees somewhat, but there are yet many warm days when the bees are very busy. Some of my bees have stored a good deal of surplus from this plant, for this time of the year—about 20 pounds per colony. The honey is a golden yellow and has a somewhat strong taste, a little bitter, and hence is not a suitable honey for market."

A related plant, *Amphiachyris dracunculoides* is also known as broomweed. It is found on the plains from Nebraska and Colorado southward. Specimens received from N. H. Barnett of Howard, Kansas, are accompanied with the statement that the plant yields nectar there but that surplus in quantity has been harvested but once in ten years.

BRUNNICHIA.

The tendril-bearing smartweed, or ladies' ear drops (*Brunnichia cirrhosa*) is a perennial climbing vine, common to the southeastern States. It ranges from southern Illinois and Arkansas to South Carolina and Florida. An Arkansas beekeeper reports that the plant is abundant in his locality and covered with bees.

H. B. Parks reports it grows on the low lands of Southeast Texas, where it blooms from March till August and yields some surplus.

Brunnichia is common on the Brazos River bottoms in east Texas and is the source of surplus. Prof. Bilsing states that the honey is dark and strong and not in demand in the market.

Doctor Oertel reports that Brunnichia yields heavily in Louisiana and is the source of surplus in the swamps where it is abundant. He regards the honey as rather light in color and not strong.

Roy Weaver of Navasota, Texas says: "We know of no plant that is a better yielder or more dependable. The honey although greenish is one of the best flavored honies to be found."

BUCKBRUSH or BUCKBUSH.

The term "Buckbrush" is applied to a number of different shrubs. Some of these are good sources of honey. In Central Arkansas the name is given to smooth dogwood (*Cornus paniculata*), which is of little value to the bees. (See dogwood).

In Iowa and surrounding states Indian currant (*Symphoricarpos orbiculatus*). The Indian currant is an important source of honey in the woodland borders throughout this region. The snowberry (*S. racemosus*) is also known as buckbrush in many places. (See Indian currant, also snowberry).

In California the name buckbrush is commonly applied to *Ceanothus cuneatus*. This shrub, with whitish bark, grows on the dry rocky slopes of the mountains and foothills from Southern Washington south to Central California. It is abundant in the thickets on the mountains near the western coast in many localities. The blooming period is March and April, yielding a light amber honey and a plentiful supply of pollen.

On the North Pacific Coast *C. sanguineus* is known as buckbrush. This species is reported as an important honey and pollen source in the uplands of Washington. (See Mountain Lilac).

In northeastern Oregon the antelope brush (*Purshia tridentata*) is called buckbrush. It is the source of dark, strong honey. (See Antelope Brush). Wolfberry is known as buckbrush in some localities. (See Wolfberry).

BUCKEYE (Aesculus).

The buckeye or horse chestnut is widely distributed and well known because of the poisonous properties of the peculiar nut-like fruit, everywhere called buckeye. There are several species, with minor differences. The photograph is of the blossoms of the Ohio buckeye (*Aesculus glabra*). This species occurs from New England west to Iowa, Kansas and Oklahoma and south to Georgia, Alabama and east Texas. There is a species common on the Pacific Coast known as the California buckeye (*Aesculus californica*). This species is reported as yielding considerable honey in some localities in California, and some beekeepers think it is poisonous to the bees.

Blossoms and leaves of buckeye horse-chestnut.

The buckeye is widely mentioned as a honey plant, though there are few localities where it is sufficiently abundant to be important as a source of surplus.

A report from Ohio in the American Bee Journal (Aug., 1869) states that strong colonies had built combs and stored as high as thirty pounds of surplus honey from buckeye that season. The quality of the honey is described as good, very thick and about the color of basswood honey, though not equal to it in quality.

In the American Bee Journal for December, 1925, George H. Vansell published an extended account of the poisoning of bees by the California buckeye. The following extracts are of interest in this connection:

> The effect upon the whole colony is most demoralizing. Many of the young worker larvae are killed outright and are devoured by the adults or thrown out of the hive. Others develop into adults of very abnormal appearance, being wholly unable to fly on account of the crumpled conditions of the wings. The legs are often abnormally small and the whole body may be dwarfed. Queen and drone brood are affected similarly.
>
> Many of the adult field bees feeding upon buckeye become very much swollen, as in a dysenteric condition. Their intestines are gorged with a very sour, melodorous yellow substance to such an extent that a slight pressure will cause them to burst with a snap. Many of the affected bees become shaky and are picked almost free of hairs by other workers so that they appear shiny, as in the case of paralysis. Apparently none of the poisoned

mature bees or the emerging young are able to fly and they crawl out of the hives by the hundreds and die. I have seen dead bees in front of single colonies to the depth of three or four inches.

BUCKTHORN (Rhamnus).

The common buckthorn (*Rhamnus cathartica*) is a hedge plant introduced from Europe and commonly cultivated. It has become naturalized in some localities.

In California there are three species reported as important sources of honey. The coffee berry (*Rhamnus californica*) is an evergreen shrub 4 to 6 feet high, with olive-like leaves, common to the Coast Ranges and the Sierra Nevada Mountains and southward. Called also pigeon berry.

Richter reports this species as yielding an amber honey of very heavy body in the foothills of the Sierra Nevada Mountains and also in San Diego County. Honey of good flavor, slightly cathartic.

Concerning honey from coffee berry E. O. E. Klipphann wrote in

Blossoms of redberry.

Western Honeybee for May 1917. "It yields about half as much as Manzanita. The honey is rather dark, has a very nice, mild flavor and granulates in one or two weeks after extracting."

In the July issue of the same magazine editor Bixby said: "Our native coffee berry is one of the heaviest nectar producers."

The redberry (*Rhamnus crocea*) occurs in the Napa Range and southward near the coast to southern California, according to Jepson. It is tree-like, with a distinct trunk, sometimes several stems clustered, 5 to 12 feet high. Richter reports it of special value for early breeding.

The Cascara Sagrada or Chittam (*Rhamnus purshiana*) occurs in northern California, where it is reported as an important source of amber honey. In the timbered portions of western Oregon, Washington and British Columbia, it is reported as one of the chief sources of honey. The honey is amber, with a delightful aroma. When fully ripened it is too thick to extract readily, and there is much breakage of combs. The flow begins in May and the honey is usually mixed with that from other plants. The blooming period lasts about a month.

> "We get more honey from cascara than from any other one plant in this vicinity. It is so dark as a comb honey that it is a poor seller to those who go on looks alone. We prefer it on our table to any other honey. I have customers who will take no other. It is not purgative, but one of the best remedies for chronic constipation known. I have never known any of the pure article to granulate under any conditions."—A. D. Herold, Gleanings, Jan. 1, 1910.

Rhamnus Caroliniana, the yellow buckthorn, commonly called Indian cherry or polecat-tree, is a shrub or small tree found on river banks and moist hillsides from New Jersey to Kansas, Texas and Florida. Indiana beekeepers report that it yields nectar for about three weeks in May and June. From E. Martin of Goodland, Missouri, comes a report of a surplus of about forty pounds per colony of fine flavored honey thought to come from this species.

R. lanceolata is common in similar situations from Pennsylvania to Iowa and eastern Nebraska and southward.

There are many localities in the southeastern states where one or another of the buckthorns is sufficiently common to be of importance to the beekeeper, though probably seldom the source of much surplus.

Rhamnus alaternus is listed by Ferd. Von Mueller as a splendid honey plant in the countries surrounding the Mediterranean Sea. It is probable that all species of Rhamnus are valuable as sources of nectar, and since some are found in Europe, North Africa, Northern and Western Asia as well as North America they are of interest to beekeepers generally. (See also Lotibush).

BUCKWHEAT (Fagopyrum esculentum).

Buckwheat is a native to Asia, which was early introduced into America from Europe by the colonists. It has become an important field crop, and buckwheat flour is a staple in American markets. It is often sowed as a catch crop on lands that have not been ready for early sown crops, or where the first sowed crop failed to secure a stand. It requires a short season in which to reach maturity and is usually sown in June or July. It needs a cool, moist climate for best results and often fails to yield a satisfactory crop of grain or to secrete nectar in the hot and dry atmosphere of Iowa and Nebraska.

It is well suited to sandy or other light soils and is grown extensively in the sandy lands of northern Michigan.

Buckwheat is an important source of surplus honey in the region of the Great Lakes. Ontario, New York, Pennsylvania, Ohio and Michigan are the States from which come the great crops of buckwheat honey. Although the crop is often grown further west, the amount of honey secured is very disappointing in most cases. The author corresponded with a number of prominent Iowa beekeepers, and found only one who had secured surplus honey from buckwheat to any extent, and his was mixed with other sources to such an extent as to have a very different color and flavor from that which is secured unmixed.

It was given a trial at the Texas Agricultural College for a period of three years. It failed to meet expectations as a honey plant on account of the hot and dry weather which prevails there during most summers. It was reported as blooming profusely, but not yielding nectar.

In New York it is regarded as one of the best honey plants. The late E. W. Alexander, writing in Gleanings, stated that he had kept 200 colonies in a location with scarcely 100 acres of buckwheat within four miles, yet had harvested 15 to 20 pounds of section honey from buckwheat per colony. He stated that it yielded best with cool nights followed by a clear sky and a hot sun, with little or no wind. Under such conditions it secreted nectar freely from about 9 a. m. to 2 p. m. No bees would be seen on it earlier or later in the day. On one occasion, when there were 1,500 acres of buckwheat within reach of his bees, they were gathering fast at the beginning of the August harvest. A thunder storm caused the temperature to drop 21 degrees in less than half an hour. The weather remained cloudy and windy, with a temperature of about 65 degrees, for eleven days. During that time the bees gathered no honey, and destroyed much of their brood.—Gleanings, March 15, 1907.

The Ontario Association's Crop Report Committee reported an average per colony production of 23 pounds from the more than ten thousand colonies belonging to its members for two years in succession. A writer in the American Bee Journal asserts that in a favorable season an acre of buckwheat will yield 25 pounds of honey daily, and that a strong colony within half a mile of a field, will store six to eight pounds per day. Alexander kept as high as 700 colonies in one yard, at his home near Delanson, New York, and secured satisfactory crops, though, of course, there was a large acreage within reach.

J. E. Crane, of Vermont, writes that buckwheat yields best on light or sandy soils. He states that living in a section which is largely clay soil he has known but two seasons in which his bees have secured surplus from buckwheat in nearly sixty years.

Quality of the Honey

Honey from buckwheat is very dark and has a strong flavor. People who are accustomed to light and mild honey of the clover type seldom like it. On the other hand, residents of the west, who were raised in the buckwheat country of eastern New York, regard the clover honey as insipid and not to be compared to the dark honey with which they were familiar in childhood. Buckwheat honey has a peculiar flavor, slightly nauseating to one unaccustomed to it.

During a heavy flow from this source there is a strong odor present in the apiary which can be detected for some distance. J. L. Byer, in the American Bee Journal, tells of a case where a farmer and his wife spent some time looking for dead chickens in the vicinity of the hives, mistaking the odor of the new nectar, which the bees were bringing in, for that of a dead fowl. (Page 306, October, 1908.)

Buckwheat varies somewhat in density, according to weather conditions at the time it is gathered. When fully ripened on the hives it is sometimes so thick as to be hard to extract. Some beekeepers report honey from this source which weighs as much as fourteen pounds to the gallon. On the other hand, there are numerous reports of very thin honey, probably because of being extracted before fully ripened.

Buckwheat honey is used largely in France to make a gingerbread "pain d'epices," in which the peculiar odor and flavor of the buckwheat is very noticeable and is much liked by those who are accustomed to it.

In a private letter to the author under date of Feb. 6, 1935, F. W. Lesser of Fayetteville, New York writes:

> "I have kept bees and observed the crops in the sandy sections of Saratoga, Onondaga and Oswego Counties of this State and the yield of buckwheat there is always better than in nearby sections where the soil is not sandy. I have had bees on the heavy soils of the St. Lawrence Valley and buckwheat did not yield there and old timers there told me that it did not. Buckwheat will yield very little on heavy limestone soils in this State. I have been fooled time and again by expecting a flow from good sized fields in this locality but never have had it."

> "The number of acres required for thirty colonies of bees will be about four, if it is a good season for securing honey. I have known one acre of buckwheat to furnish enough food for bees, so that 800 pounds of honey and 85 bushels of grain were made from it. This was, however, an unusually favorable season."—J.D., American Bee Journal, July, 1875.

BUCKWHEAT TREE, see Ti-Ti.

BUFFALO BEAN, see Loco Weed.

BUFFALO-BERRY (Shepherdia argentea).

The buffalo-berry is one of the hardiest of native shrubs and its currant-like fruit was an important article of food with the Indians of the plains region. In the wild state it is commonly found from Manitoba southward to Kansas. Since it withstands both cold and drouth it offers special opportunity to the plant breeder who will devote the time necessary to its improvement.

It is one of the first honey plants to bloom in spring and is valuable for early stimulation in the Dakotas and Minnesota.

BUGLE-WEED see Indigo-weed, also Water Horehound.
BULL BAY, see Magnolia.
BUMELIA, see Coma, also Gum-Elastic.
BUM-WOOD, see Poisonwood.

BUR CLOVER (Medicago hispida).

Bur clover is a relative of alfalfa which is very common over much of California. The burs are produced abundantly and the plant is prized as a stock forage. The plant, although spreading like a weed, is valuable both to live stock and for bees. The principal blooming period is from March till June, though it blooms to some extent at all seasons.

Richter lists it as especially valuable to stimulate early breeding, but states that surplus is occasionally harvested from this source and that it is fully equal to filaree or pin clover as a honey plant.

Burdock.

BURDOCK (Arctium lappa).

The burdock is a course, disagreeable weed, introduced from Europe and Asia. It is now common over the United States. The burrs fasten themselves to the clothing or to passing animals, and in this manner the seeds are spread. It is a biennial, common in barn lots and waste places.

The burdock is one of the many plants on which the bees work to some extent that never count for very much in the total production of the hive. The sources of surplus are comparatively few in number, but there are hundreds of plants from which the bees get a taste of honey or pollen. The presence or absence of these minor plants makes a great difference in the value of a locality for honey production. If there are enough of them to keep the bees busy, and sustain the colony between flows when the good yield comes, the bees are in the best possible condition to take advantage of the opportunity.

BUR-MARIGOLD, see Spanish Needle.

BURNET (Sanguisorbia minor).

Burnet is a native of Europe that has been cultivated in England since about 1760. It is a deep rooted perennial that will stand heavy pasturing and which in mild climates continues to grow during winter. It is used in pasture mixtures, especially on rather poor soil. It has been tried to some extent in California but has not become popular in America.

In the American Bee Journal test gardens it has been found difficult to maintain a stand under mid-west conditions. So many plants winter kill as to make it of little use even in a mixture.

The bees visit the plant freely during its period of bloom but get only pollen which is provided in abundance.

Since burnet has become naturalized in some New England and nearby areas it may succeed better as a field crop in that region.

BUSH HONEYSUCKLE (Diervilla Lonicera).

The bush honeysuckle is a common bush shrub in the northeastern States. It is to be found from Newfoundland south to North Carolina and west to Minnesota. It is reported as common in New England and Ontario. In Northern Minnesota, where it is abundant, beekeepers report that the bees work it eagerly when the weather will permit. The flowers are not showy, but the profuse bloom is rich in nectar. The blooming period begins in June and, in some places, continues as late as August.

BUTTERCUP (Ranunculus).

There are numerous species of buttercups, some of which may be of considerable value for pollen. Coleman reports that the common buttercup of California (*R. californicus*), which is very abundant over a wide area of that state is the source of considerable early honey as well as pollen. (See Western Honeybee, April, 1921).

"The pollen of buttercups has been proved to be actually injurious to bees in Switzerland and elsewhere in Europe and responsible for a kind of 'May sickness.'"—F. N. Howes in Plants and Beekeeping.

BUTTER WEED or GROUNDSEL (Senecio).

The groundsels are herbs with alternate leaves and mostly yellow flowers, common to the northeastern States. The bees are reported as working upon them freely, but they are probably of minor importance.

S. lobatus is common on wet soils from North Carolina west to Missouri and south to Florida and Mexico. It is reported as very common on the prairie soils of Mississippi, where it is the source of considerable surplus honey of an amber color and fair quality. The bees get much pollen from it also.

Bessey lists the ragwort (*Senecio aureus*), as the source of honey in Nebraska.

S. Mollis is common in South Texas in late winter where it provides some bee pasture. *S. globellus* is also common in Texas and Louisiana. Doctor Oertel reports this species as rather important in Louisiana as a source of both nectar and pollen in spring. It comes up abundantly in rice fields which are left uncultivated.

F. N. Howes, (Plants and Beekeeping) describes ragwort honey as deep yellow in color possessing a strong almost bitter flavor.

BUTTON-BUSH (Cephalanthus occidentalis).

The button-bush, also called button willow, is a bushy shrub growing in marshy places, stagnant shallow water, and along streams, from New England to Texas and west to California. This shrub, or in places a small tree, has a very wide range and is found in most of the States where honey production is important. Bulletin No. 102 of the Texas Agricultural College, reports it as common throughout Texas, and the bulletin relating to honey plants of California (217 Experiment Station), records it as a good honey plant in California. It is listed in the catalog of plants of nearly every State and of Canada, which the author has consulted. It is also said to occur in Asia, and possibly Africa.

Our readers who live in the vicinity of wet lands are likely to find specimens near at hand. In a few sections it is sufficiently abundant to be an important addition to the midsummer flora. It is reported as more particularly valuable in the overflowed lands along the Mississippi River. The bees seek it eagerly when in bloom, and in places where it is plentiful it is regarded as of considerable value as a honey plant.

The honey is light in color and mild in flavor, according to published reports. Flowers are crowded together in dense heads, giving them the appearance of cotton balls.

The shrub is very bushy, with an abundant foliage. It is reported as reaching a height of 40 feet in California. In Alabama it is recorded as a shrub of from 6 to 15 feet in height, which is more like its appearance in Iowa. Here it is rather a small bush, not much higher than a man's head, and as far across, with many branches from the ground.

The blooming period is July and August, according to locality, a season when additions to the honey-producing flora are most welcome.

Blossoms of the button-bush.

Robert E. Foster, Florida Apiary Inspector, reports that in some sections of Florida button-bush grows in abundance and that he has seen the bees working it heavily and gathering a surplus of light colored and heavy bodied honey of good flavor.

Everett Oertel reports that in his opinion button-bush is overrated and that the corolla tubes are too deep for the bees to reach the nectar consistently in Louisiana.

L. A. Schott of Benton, Missouri reports a nice crop of honey from button-bush which grows along the drainage ditches in southeast Missouri, and that a pleasant aroma pervades the apiary at nightfall when bees are working this plant.

BUTTONBALL-TREE, see Sycamore.
BUTTON-SNAKEROOT, see Blazing Star.

BUTTON-WEED (Diodia teres).

The button-weed occurs on sandy lands from New England to Florida and west to Kansas and Texas. Scholl lists it as yielding honey well during drought in Texas, but not as a source of surplus.

BUTTON WILLOW, see Button Bush.
BUTTONWOOD,. see White Mangrove, also Sycamore.

C

CABBAGE (Brassica oleracea).

Cabbage belongs to a group of valuable honey plants, including mustard, turnip, etc. In the seed belt of California, where grown for seed on a large scale, cabbage is valuable. The late J. S. Harbison said of it:

> "Cabbage blossoms afford a considerable amount of honey of a fine quality and flavor."—Beekeeper's Directory.

CABOMBA, see Water Shield.
CACHINILLIA, see arrow-wood.
CACTUS, see Prickly Pear, also Giant Cactus.

CAJEPUT-TREE (Melaleuca Leucadendra).

The cajeput-tree is of comparatively recent introduction into this country. It will stand slight frost so can be grown only in the warmer portions of Florida and California. It was brought from Queensland, Australia, althought it is common also in India.

The tree attains to large size, a height of 80 feet and a diameter of as much as four feet. It is well adapted to swampy areas and will grow on land where eucalyptus will not live. The wood is said to be hard and close grained and to last indefinitely underground, hence is often used for piling, ship building and other uses where it remains in contact with water.

The leaves are the source of the cajeput-oil which is similar to eucalyptus-oil. In southern Florida the tree seems to be very much at home and is rapidly spreading its range.

Frank Stirling of Davy reports that it is the heaviest yielding source of nectar with which he is familiar. It blooms in early winter, usually in November and gives a six weeks flow. In neighborhoods where the tree is common he reports an average yield of a shallow ten frame super, per colony, per week, during the flow. He describes the honey as amber in color, mild but distinct flavor, with considerable aroma. It candies quickly. The sample shown the author met the above description fully and was of a quality which would grow in favor when available in quantity in the market.

There is every reason to believe that the cajeput-tree will in time become an important source of nectar in south Florida.

In "Honey and Pollen Flora of New South Wales," W. A. Goodacre refers to this species as "Broad-leaved Tea-tree" and makes the following comment concerning its honey:—

> Like other species of tea-tree it is a good honey producer, and flowers fairly consistently every season. The honey extracted is rather dark in color, strong in flavor, and candies rapidly. It is not generally favored as a table honey, and care has to be exercised when using it in a blend, the flavor and aroma being so strong.
>
> Bees build up wonderfully when this species is flowering, and the colonies are put into the best condition to secure supplies from the more valued honey flora to flower later.

CALICO BUSH, see Laurel.

CALIFORNIA—Honey Sources of.

Because of the great diversity of its soil and climate, California has a greater variety of honey plants than any of our states. Mt. Whitney, with an altitude of 14,502 feet, is the highest point in America and in Death Valley, with its depressions below sea level, we find the lowest. Between these two extremes we find all possible gradations in altitude, with consequent variation in climate. In the northern part of the state the thermometer has been known to fall to 40 degrees below zero in winter, while summer temperatures above 100 are not uncommon in Death Valley. The highest temperature registered there is 134 degrees on a standard thermometer in the shade of a ventilated shelter.

California soils are equally varied, with a corresponding variety of products which include nearly every vegetable product which can be grown in any part of America. The state has been divided into five climate states, as shown in the accompanying map.

In August (1922) American Bee Journal, M. C. Richter, a well-known California beekeeper, gives an outline of the nectar resources in these different areas. From this article much of the following information is gleaned.

The Northern Coast Region

In the north coast region there is little good beekeeping territory, owing to the absence of suitable flora in the large forests. The summers are cool and damp. Away from the coast and in parts of Siskiyou and Shasta Counties, some fine beekeeping opportunities are found in the fertile valleys. There the seasons are short and flows rapid, conditions which are favorable to big crops. The crops are secured from the clovers, tarweed, alfalfa, etc., with spring stimulation from ceanothus, manzanita, filaree and bur clover.

Central Coast Region

The northern portion of this region, excepting Lake County is not good beekeeping territory, although at times good crops of eucalyptus honey are harvested.

In the central portion, however, there are some very good locations where good crops are secured from alfalfa, sweet clover, asparagus, onion, mustard, turnip and fruit bloom. Large crops of honeydew are also harvested at times.

The southern portion of this region is very good for honey production. Santa Clara County is occupied by well-known commercial queen breeders who find the long season and slow flows well adapted to their purpose. In addition to the early fruit bloom they find manzanita and black sage a good dependence.

The four southern counties are mountainous and include some of the best locations in the state. Sage, manzanita, wild alfalfa and wild buckwheat are abundant. The heavy rainfall is favorable to good flows in many locations at high elevations which are at present difficult of access.

Southern Coast Region

In this region we find more beekeepers extensively engaged in honey production than in any similar area in America, if not in the world. It is a famous region which first came to public attention when Harbison shipped his trainload of comb honey to the eastern market. Large crops of sage and orange honey are gathered here. Surplus is gathered also from manzanita, eucalyptus, lima beans, mustard, wild alfalfa, wild buckwheat, sumac, tarweed and blue curls. Migratory beekeeping is practiced to a large extent, the beekeeper moving to a new field at the close of a honeyflow.

The Interior Valley Region

The agriculture of the interior valleys is largely dependent on irrigation. Alfalfa is the principal source of honey, although large crops are also gathered from other sources. Tulare County is famous for its orange honey. Star thistle yields heavily in some of the northern counties of this region. The seasons are long and it often happens that the beekeeper will get some honey in ten or eleven of the twelve months of the year. Fruit bloom, sweet clover, melons, smartweed, eucalyptus, button-bush, carpet-grass, jackass clover, alkali-weed, tarweed and blue curls are all important. In the Imperial Valley cotton and cantaloupes are valuable sources of nectar. In the high altitude of the northern portion there is heavier rainfall and correspondingly more dependable flows.

Mountain Plateau Region

Although there are known to be many good locations in this region, much remains to be learned concerning its beekeeping possibilities. Alfalfa furnishes the main crop, followed by sweet clover and wild buckwheat. Rabbit-brush yields in the fall. In the spring fruit bloom, willows and locust furnish important sources of nectar. The heaviest

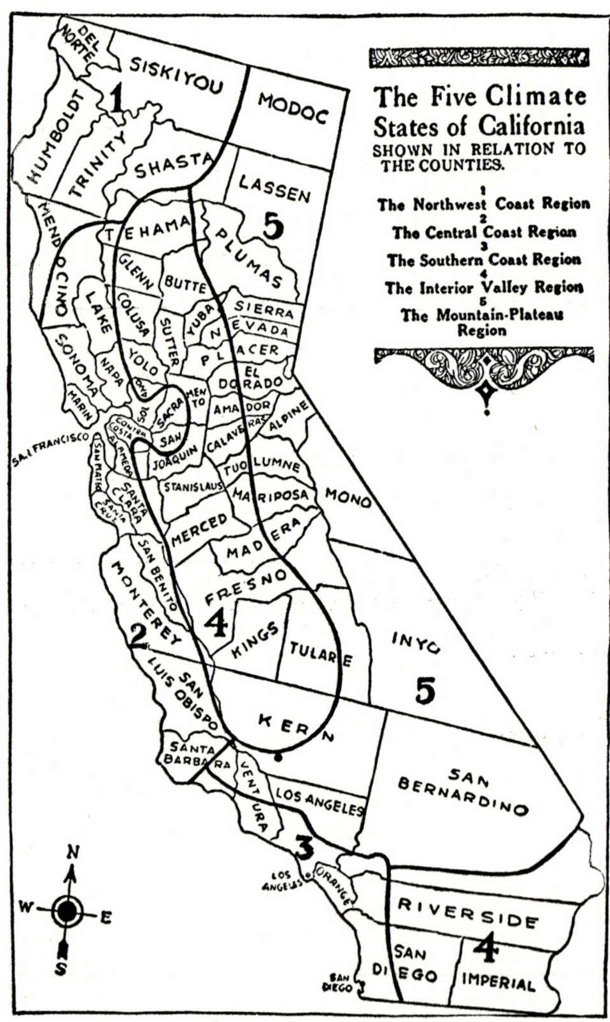

Map showing the principal natural climatic divisions of California.

rainfall is found at altitudes between 5,000 and 6,000 feet. Richter states that there are many good locations along the entire western foothill slope of the Sierra Nevada Mountains and cites northern Lassen County as a promising district.

Many of the national forests are located in this region. In the national forests a beekeeper can lease a location at a nominal rental and be protected against others encroaching on his territory.

Sources of Nectar for the State as a Whole

The principal sources of surplus during an average season are, willow, wild buckwheat, California barberry, mustard, Cleome, Jackass clover, Christmas berry, fruit bloom, wild alfalfa, bur clover, alfalfa, sweet clover, beans, pin clover, buckeye, pepper-tree, coffeeberry, cascara sagrada, eucalyptus, blue thistle, manzanita, phacelia, carpet grass, horehound, sage, blue curls, button willow, Napa thistle, tarweed, goldenrod and rabbit brush.

To these may be added a number of others which occasionally or locally yield surplus such as white clover, basswood, tamarix, asparagus, mistletoe, wild radish, black locust, mountain lilac, cotton, madrona, yerba buena, figwort, melons, Spanish needles and numerous garden plants grown for seed in limited areas.

The list of plants which yield pollen is a very long one, for California with its wide climatic range has a varied flora including most of the plants common to the United States as a whole.

CALIFORNIA DANDELION, see cat's ear.
CALIFORNIA HOLLY, see Christmas Berry.

CALIFORNIA LAUREL (Umbellularia californica).

The California laurel is also known in Oregon as myrtle. It has a variety of local names, including pepperwood, bay-tree and bay laurel. It is usually found as a tree of from 25 to 50 feet in height, but sometimes is much smaller and shrub-like in appearance. It is common to the canyons among the mountains near the coast from southern Oregon to southern California. It reaches its greatest size and abundance in the northern portion of its range. It blooms in late winter and early spring. Coleman (Western Honeybee, May, 1921) reports it as the source of considerable dark amber honey of a syrupy consistency, which is mostly used for brood rearing, although considerable surplus is sometimes stored. Richter (California Honey Plants) lists it among the sources of nectar not known to produce surplus. It is probably of far more importance in northern California and Oregon than farther south.

CALIFORNIA POPPY, see Poppy.
CAMASS, see Wild Hyacinth.

CAMPANILLA (Ipomoea).

The campanillas, or Christmas bells, are plants similar to our American morning-glory, which are common to Cuba and adjacent countries. The vines are perennial and grow to considerable size, covering wayside fences, trees, etc. The name Christmas bells, applied to white bellflower, comes from the fact that the height of the blooming period is reached near the holiday season. There are two species, known as white campanilla (*Ipomoea sidaefolia*), and the pink bellflower (*Ipomoea triloba*). The latter blooms a month or two before the white variety.

The honey, according to Root's A B C, is equal to alfalfa or sage in color and flavor, the comb built during a campanilla flow being pearly white and the wax, when melted, as white as tallow. See Morning glory.

Ordetx in his "Plantas Melliferas de Cuba" lists the white campanilla as the most valuable honey plant in that country. He states that it is the source of a large portion of the surplus crop estimated as high as one third of the total. It is commonly called "aguinaldo" and is said to yield a white wax and a pearly white honey that can be compared in quality with the best honeys of the world.

The rose colored campanula flowers at a time when the colonies are not sufficiently strong to make the most of the flow and for this reason is not so favorably known. It comes into bloom a month before the white variety.

CAMPHOR WEED, see Blue Curls.

CANADA THISTLE (Cirsium arvense).

The Canada thistle is a most troublesome weed, naturalized from Europe and widely spread through Eastern America. It is a perennial plant, growing from one to two feet high. The roots creep extensively, thus gradually spreading the plant to surrounding areas. It is very persistent and difficult to eradicate. It is common in fields, pastures and along roadsides.

It is the source of a light honey of good quality. It is most frequently reported as a source of surplus in Ontario and the Eastern States.

In connection with an account of the Toronto Convention of the North American Beekeepers' Association, in 1883, A. I. Root wrote in Gleanings (page 594) concerning Canada thistle honey as follows:

> "One of the funny surprises was to find tons upon tons of the most beautiful white honey, both comb and extracted, that it has ever been my good fortune to taste, all made from Canada thistle. The flavor is so much like basswood that I insisted that some late blooming basswood trees had yielded enough to give it flavor, though my Canadian friends think I am mistaken. Friend Jones thinks that a great part of the wonderful yields that they have

had in Canada are from this same Canada thistle. * * * * * For whiteness, transparency and beauty of flavor, I have never met anything anywhere like it."

CANADIAN HEMP, see Dogbane.
CANDLEBERRY, see Wax Myrtle.

CARAGANA or PEA TREE (Caragana arborescens).

The Siberian pea tree is a hardy ornamental shrub which is widely grown, especially on the northern prairies. In the vicinity of Winnipeg, Manitoba, it is very common and some beekeepers report it as the source of surplus honey in favorable seasons. There are miles of hedge of this species at the Government farm at Mordan, Manitoba. W. E. Lake, the apiarist there, reported that it bloomed about the middle of May and gave about two weeks' flow. It gives a good yield of light honey of fine quality, according to his report.

CARELESS WEED (Amaranthus).

When visiting beekeepers in the Salt River Valley in Arizona I was told that careless weed was of some local importance to the bees. Never having heard the amaranth mentioned elsewhere as a bee plant I was somewhat puzzled by these reports. Since that time it has been impossible to verify the plant as a source of nectar, although the bees gather pollen from it at times.

The following letter from E. V. Walter from San Antonio under date of September 27, 1927 is of interest here:—

> "A few days ago I happened to be in a patch of careless weed soon after daybreak and found the bees swarming in great numbers around the fresh bloom on the plant. I watched them carefully and found that they gather large quantities of pollen quite readily, but in no case was I able to find them even attempting to gather nectar. The pollen seems somewhat powdery and easily shaken from the plant. It is of a distinct greenish yellow color."

CARPET GRASS (Lippia).

The carpet grass of California (*Lippia repens*) is a native of Chili which has been introduced on the Pacific Coast and widely cultivated as a lawn plant. It is known as lawn plant, carpet grass and lippia. It is much sought by the bees and is the source of some honey, a surplus of two cases per colony being not unusual, according to C. D. Stuart. The honey, he states, is light amber, heavy body, and similar in quality to alfalfa. The plant is of trailing habit, spreading by rooting of runners like strawberries.

Lippia nodiflora, mat grass or fog fruit, is native to California, and, according to Richter, as the principal source of surplus honey in the vicinity of Sacramento. Three-fourths of the surplus honey from Sutter County he reports as from this source. There it begins to bloom in May and lasts till frost. According to Jepson, the plant is esteemed as a covering for levees, to resist erosion. It is especially valuable on overflowed lands after the water recedes. The honey is said to be light in color, mild in flavor, and to granulate readily.

CARROT (Daucus carota).

The carrot is a well-known garden vegetable which came from Europe and is everywhere cultivated in gardens. It has escaped and become naturalized over a wide scope of country. It yields some honey, and where grown on a large scale for seed, it is valuable for the bees.

Carpet grass or lippia.

A sample of carrot honey received by the author from Cary W. Hartman, Oakland, Calif., was produced in a large field of carrots by strong colonies. It was very dark in color with a strong flavor.

Tom Davis of Sacramento says: "Honey is dark amber and strong flavor. It is somewhat turgid and granulates within two weeks after extraction. It foams quite a bit when heated denoting the presence of yeasts. The average production of carrot honey is between 100,000 and 150,000 pounds. Carrots produce best on heavy soils and at temperatures exceeding 90 degrees and will not yield any surplus when maximum is below 80."

The wild carrot, also called Queen Anne's lace, bird's nest plant, devil's plague, and sometimes lace flower, is a common weed in fields and waste places in the northeastern states. It yields nectar apparently at very irregular intervals and is valuable only in an occasional season when conditions are especially favorable. A Pennsylvania beekeeper writes that he had never seen a bee on it prior to 1921, when it yielded freely for a time in his locality.

CASCARA SAGRADA, see Buckthorn.
CASSIA, see Partridge Pea.

CASSIOPE.

In the high altitudes of the Cascade Mountains and other ranges of the northwest coast *Cassiope mertensiana,* a low branched evergreen shrub is very common. Near the tree line it is said to occur in very large areas. A somewhat similar species, *C. tetragona,* is found in similar situations northward to Alaska.

Since these plants are found mostly in unsettled regions they are not always available as bee pasture. They are reported by western beemen as the source of nectar and the honey along with that from *Phyllodoce,* is known as "heather." (See heather).

Cassiope is also known as false white heather or moss heather.

CASTOR BEAN (Ricinus communis).

The castor bean, or castor oil plant, is often cultivated for the oil contents of the beans. Large areas were planted during the late war, especially in Texas. The plant has escaped from cultivation and become naturalized in the Southeastern States. It is generally planted for ornament over a large scope of country. It is reported as very attractive to the bees and of some value for honey, where sufficiently abundant. Scholl lists it in the Texas bulletin as yielding well in favorable seasons. Richter lists it also for California.

> "I observed that my bees are getting honey from a saccharine substance that exudes from the immature capsule of the castor-oil plant (*Ricinus communis.*) It appears that the sweet watery juice exudes nocturnally; but on exposure to the sun it becomes hard and has the granular aspect of brown sugar. It is very sweet and my bees consume it voraciously. They also get pollen from the same plant." Wm. A. Smith, Texas. Gleanings 1882, page 407.

The nectar is secreted by extra-floral nectaries on the stem, the leaf-stalks and margins of the leaves.

Lovell in "Honey Plants of North America" credits the reports of honey from castor bean as erroneous. He mentions extra-floral nectaries as present which are without function.

Bees are not attracted to it in author's Iowa garden.

CATALPA (Catalpa speciosa).

The catalpa tree produces a great profusion of bloom. The blossoms are so large that a bee can readily crawl right into the heart of the flower.

The testimony of competent observers gives an unqualified indorsement of the catalpa as a nectar producer, though there is slight mention of it in our literature. The fact that large areas of these trees are being planted for timber, in many places, makes them of special interest to the beekeeper. The catalpa, or Indian bean (*Catalpa speciosa*) is a native of the woodlands of southern Indiana and Tennessee, west to Arkansas. This form, known as the hardy catalpa, is also widely planted in Iowa, Illinois, Kansas, Nebraska and other States. There is another similar species closely resembling it which occurs further south, and is common in the Gulf States.

The leaves are heart-shaped and the blossoms are large, nearly white, and grow in large clusters. The tree grows very rapidly, furnishing desirable timber for fence posts, telephone poles, railroad ties, etc. In Kansas, large areas have been planted by the railroad companies for the purpose of growing ties. Beekeepers situated near such plantings should find the trees of material value.

In American Bee Journal, July 1938, page 319, appeared an article by Dr. J. Schiller concerning *Catalpa bignonioides* in Austria. This species native to the southeast from Georgia and Florida to Mississippi is commonly cultivated in Europe.

We quote as follows:—

> "Nectar formation in the weak scented flowers is abundant and they are worked by both bumblebees and honeybees. It is unique, however, in that for a week before the blossoms appear a peculiar fragrance is common from the leaves and there is an abundance of nectar from the numerous nectaries on the under sides of the leaves during the entire period of blooming and for eight to ten days afterward.
>
> The vigor of the nectar secretion of the leaf nectaries is dependant upon the warmth and humidity of both the air and soil. Early in the morning large drops occur on the leaves which have not yet been visited by the bees. The nectar also issues abundantly through the day, when two or three bees may be seen on each large nectary quietly sucking.
>
> These leaf nectaries are distinct from the blossoms. The leaf nectaries are long lived and the blossom nectaries short lived. The nectar secretion ceases in the blossoms as soon as the substance causing osmosis is washed away with water. In contrast, rain may wash the leaf nectaries without affecting the nectar secretion so it is often long continued."

Commenting upon the above article, Dr. Oertel reported, page 403, September Journal, same year, that he had observed the bees working extra floral nectaries of catalpa in Louisiana about a month after blossoms had fallen.

CATNIP (Nepeta Cataria).

Catnip, or catmint, was introduced from Europe, and cultivated in herb gardens. It is thus an escaped introduction and has become very widely naturalized in the United States. It is usually found only in the vicinity of buildings and gardens, and seldom spreads into the fields to any extent. Catnip tea was popular among the grandmothers of an earlier generation. The plant is a perennial, growing from 2 to 4 feet high, with flowers in clusters, the more conspicuous ones being in a terminal spike. The blooming season is rather long, and the bees visit it very freely. Apparently, the plant yields much nectar, although it is seldom present in sufficient quantity to test its real value as a honey producer. If it had sufficient value for other purposes to justify its cultivation, it would probably be an important source of nectar.

Concerning catnip, J. C. Wallenmeyer in American Bee Journal, page 144, August 1894, stated that he preferred catnip to any other honey plant because it yielded more than any other. He stated further that it yielded excellent honey. From his article the following extract is taken:

> "I had visited my plants in the garden at every hour of the day, from early morn until night, and always found the blossoms covered with bees. My diary last year shows that the bloom lasted from July 1st until frost on October 10th, making 100 days of continuous bloom. No other plant will do this. * * it yields honey incessantly."

Catnip.

In 1902 Dr. J. L. Gandy of Humbolt, Nebraska attracted considerable attention by an article in Gleanings telling of his experience with catnip. He had a clause in his leases requiring his tenants to sow a certain amount of catnip and claimed substantial honey crops as a result.

The Persian catnip, (*Nepeta mussini*) is a popular favorite in rock gardens. The vigorous plants spread over a width of two to four feet in old specimens. In our test gardens there is a long period of bloom from early April until late June and another in autumn with a scattering of flowers all summer. The plants are alive with bees during the time of flowering.

Another species from Siberia, *Nepeta macrantha,* blooms freely in midsummer and is likewise attractive to the bees.

CATSCLAW (Acacia).

Because of their curved thorns, several species of acacia are known as "catsclaw." There are two species well known to Texas beekeepers under this name. The first *Acacia greggii,* also called Devil's claws or Paradise flower, is a small tree or shrub common on dry soil in Texas, New Mexico, southern Arizona and Mexico. It is commonly reported as one of the most important sources of honey in that region. At Brownsville the author found reports of some early honey from this source. At Mercedes it was reported that it often fails in extremely hot weather. The first flow was reported as coming as early as March. Beekeepers at that point report as high as 10 pounds per colony average from catsclaw, mesquite and huajillo, in a favorable season. There were few beekeepers visited between San Antonio and Brownsville who did not mention the three sources together. Apparently the flows interlap so that the honey is so mixed that it is difficult to determine the proportion to be credited to each separated source.

The round-flowered catsclaw (*Acacia roemeriana*), is also common in south Texas and is regarded as an important source of honey, but is not as common as the first named species.

The tree catsclaw (*A. Wrightii*), also called Texas catsclaw, is re-

The round-flowered catsclaw.

ported to be the source of the greater bulk of catsclaw honey in southwest Texas, from the Medina River to Mexico.

Catsclaw honey is famous for its fine quality and at times the yield is very heavy. (See also Acacia).

CAT'S EAR (Hypochoeris radicata).

Cat's ear, sometimes called California dandelion, is abundant west of the Cascade Mountains in Washington and south to California. It is a naturalized European weed in pastures and fields, blooming in midsummer. According to H. A. Scullen, it supplies considerable nectar in Washington. It is amber color and in some localities darkens the firewood honey.

The following notes concerning the behavior of this plant in New Zealand are taken from the Smallholder of Nov. 16, 1934:—

> "The honey is amber colored, thin in consistency and the pollen deep orange yellow."
>
> "I have known seasons when the bees gathered little from any other source and still have given a fair surplus and gone through the winter in fine condition. I have found the honey very slow to granulate—some was still liquid eighteen months after extraction, but this was almost entirely from a late blossoming of flatweed at the end of a decidedly poor season."

An annual variety, (*Hypochoeris glabra*) is also naturalized in Western North America. This variety like the former is called "flatweed" in New Zealand and according to the Smallholder above quoted blooms in late spring and early summer and produces a little pollen and honey.

CELASTRUS Scandens, see Bittersweet.

CELERY (Apium graveolens).

The blossoms of the cultivated celery yield nectar abundantly, and where grown for seed, it is a valuable source of honey. In the seed belt of California it yields well, as the following will indicate:

> "I saw the hives stacked five or six high and on opening them we found them jammed full of honey; in fact, the bees should have had room long before, but Mr. Gear had had difficulty in getting help, and the bees had got ahead of him. The colonies were so crowded that the space between the frames and the tops of the hives were built full of burr combs. All this honey was from celery and parsnip. * * * On our arrival at the field it was easy to see that there was honey in the blossoms. In the sunlight the little drops of nectar gleamed like myriads of little diamonds. * * * I tasted some of the raw nectar from the celery. Sure enough, there was quite a strong suggestion of celery flavor."—E. R. Root, Gleanings, Nov., 1919, page 712.

CENTAUREA.

There are several species of centaurea which yield nectar. The common cornflower of the gardens (*Centaurea Cyanus*), also known as blue-bottle, bluet, ragged sailor or batchelor's button, is a good honey plant. The bees work upon it from morning till night, though it is seldom sufficiently abundant to be important as a source of surplus. Richter lists this species from California as commonly cultivated. He also lists the Napa thistle, or tacalote (*Centaurea melitensis*), as yielding some honey of light amber color, good flavor and fair body. This species is known in the southeastern States as Lombardy star thistle. (See Star Thistle.) *Centaurea repens,* Russian knapweed yields surplus honey.

CENTURY PLANT (Agave).

The agaves are an important group of long-lived perennial plants native to Tropical America, Mexico and, to some extent to the southwestern United States. Each plant has a cluster of numerous fleshy leaves and, when blooming, a tall flower stalk. There are at least five species native to the United States. The range is southern New Mexico, Arizona and California. The best known is the common century plant (*Agave americana*), an introduced species, cultivated for ornament. This is a conspicuous figure of ornamental planting in southern California.

The century plant.

In Mexico some species of agaves furnish fibre, while others are the source of pulque and mescal, intoxicating drinks much used in the country below the Rio Grande.

The plants do not bloom until they are several years old. The flower stalks grow very rapidly and reach a height of 25 or 30 feet within a few weeks' time. When in bloom they secrete nectar in abundance, and the bees swarm over them in great numbers. The flower stalks support innumerable blossoms, so that a single plant will yield a considerable amount of honey.

In Arizona the author met several beekeepers who reported good honeyflows from century plant, locally called "mescal." The honey is said to be very dark and of inferior quality and unsalable except at a low price. One beekeeper secured 75 cases of it one year. Some reports are to the effect that bees are inclined to be unusually cross when working this plant. It is the practice to extract very closely ahead of the flow from mescal and allow the bees to fill their hives with this poor honey for winter, thus saving the honey of better quality for sale. The honey requires a long ripening period on the hives before it is fit for use in any case, and the quality is always poor.

A. J. Crawford who spent some time in Mexico reported an average yield of 90 pounds per colony in that country from mescal at a distance of four miles from the plants on which the bees were working.

CEPHALANTHUS, see Button Bush.

CERINTHE.

C. W. Wood reports three species of cerinthe as favorite bee plants. Because of garden value, ease of culture and attraction for bees he suggests that they are worthy of more attention than they have thus far received. (American Bee Journal, Nov. 1935).

CHAMISE (Adenostoma fasciculatum).

Chamise is an evergreen bush or shrub of spreading habit about two to ten feet in height. Jepson describes it as "the most abundant and characteristic bush of the higher coast ranges and the Sierra, Nevada, commonly gregarious and occupying, to the exclusion of other shrubs, extensive and especially abrupt slopes and mountain ridges. It often forms a distinct zone between the foothills and the yellow pine belt."

Richter lists it as eagerly sought by the bees in several California counties, though no mention is made of it as a source of surplus.

Coleman states, (Western Honeybee, May, 1921) that the bees often store considerable surplus from it and that honey is light amber and of good quality.

The author has found that beekeepers in southern California regard it as of little value in that region and state that bees seldom visit it there. G. H. Vansell, reports that in northern California beekeepers sometimes move their apiaries within reach of it at blooming time and secure considerable surplus honey.

In many California localities it is known as "greasewood" and chamisal.

Chamise in bloom.

CHAPMAN HONEY PLANT (Echinops sphaerocephalus).

The Chapman honey plant was introduced from France about 1885. The bee journals of 1886 and 1887 devote a large amount of space to a discussion of this plant. It was brought prominently to the attention of American beekeepers by Hiram Chapman, of Versailles, New York, who planted about three acres of it at that place. He made such glowing reports of the plant at the National Beekeepers' Convention that a committee of prominent men was appointed to visit the Chapman home and report on the new plant at the convention of the following year. They made a lengthy and very favorable report, which is published in full on page 28 of the American Bee Journal for January 5, 1887.

Numerous beekeepers secured seed, and so attractive did the plant prove to the bees that favorable reports appeared frequently in the columns of the journals for the next few years. However, the great expectations were not realized, for it soon disappeared, and is seldom mentioned in current literature. The following quotation from Dr. C. C. Miller, appeared in Gleanings, in December, 1918.

> "After reading the British Bee Journal of September 26, I should have made a vigorous effort to secure a supply of seed of *Echinops Sphoerocephalus,* if I had no previous experience with the plant. No bee plant that I have ever grown was so attractive to the bees. Whenever the weather was favorable the heads were crowded. I have counted fourteen or fifteen bees on one at the same time."
>
> This is the Chapman honey plant that had a big boom in this country a number of years ago; but it is not heard of now, and is not included among the honey plants in the bee books. Upon its introduction I planted quite a patch of it, and like Mr. Harwood, I never saw the bees so thick on any other plant. But close observation showed that the bees were not in eager haste in their usual way when getting a big yield, but were in large part idle. It looked a little as if the plant had some kind of stupefying effect on them. At any rate, I should not take the trouble to plant it now, if land and seed were furnished free.

Chapman honey plant is known to the public as globe thistle.

To other than beekeepers the plant is known as globe-thistle. The name Chapman Honey Plant is no longer well known even among the beemen since the boom of planting it for honey has subsided.

An extended account of the value of globe thistle or blue thistle by N. Pankiw appeared in American Bee Journal, June 1942. He reports a blooming period from July 15 to August 25 at Dufrost, Manitoba, and stated that "From the time when flowers opened till the end of the blooming period, the bees have been sucking nectar all the time." The bees started gathering before five in the morning and as late as nine at night with no interruption during cloudy weather.

In the above mentioned article Pankiw quotes from several Ukrainian publications indicating the use of the plant for forage as well as bee pasture. The fat content of cream from cows fed on it was much higher than those on other feed. Apparently this plant is held in high esteem in that region.

A related species, *Echinops ritro* is equally attractive to the bees and a more desirable garden plant. It grows to a height of about three

Echinops is said to be grown for forage in some Balkan countries.

feet and is perennial in habit. The flower heads are smaller and of a sky blue color.

CHAPARRAL BROOM, see Baccharis.

CHAYOTE (Sechium edule).

Chayote is a vegetable of the squash family, commonly grown in the American Tropics. It is found in Mexico, Cuba, Porto Rico and other countries in that latitude. There has been some discussion of the plant in the bee magazines and occasionally the plants have been grown in American gardens. That it is a valuable honey plant there can be but little doubt. J. J. Siebert, writing from Porto Rico to the American Bee Journal, says that it is all that it is claimed to be, and that as a source of nectar it has only one rival, the banana, blossoming all the year round. The following is from the U. S. Department of Agriculture, Bulletin No. 28:

"As in other vegetables of the squash family, the stamens and pistils are in separate flowers, pollination taking place through the agency of insects. To attract these, the flowers of both kinds, but especially the pistillate, yield abundant nectar, which is secreted in ten glands, two at the base of each of the lobes of the corolla. In most of the countries into which it has been introduced, beekeeping has not been a regular industry, and the value of the chayote as a source of honey has not been noticed, but the reports of experimenters in New South Wales contain very emphatic statements on the subject:

'When the plant is in flower I have noticed that the vines were swarming with bees, and as flowers are scarce in the autumn, the plant will no doubt be valuable as a honey-producer.

'The plant, which spreads over a large area, commenced flowering at the close of the year, and has been well laden with mellifluous blossoms ever since. The bees are extremely fond of the chocho, and with the apiarist the newly-introduced plant must become a strong favorite.'

"The chayote differs from many cucurbitaceae in producing numerous flowers on each fertile branch. It has long been known that the flowers of this family are rich in honey, but from the standpoint of the beekeeper they have been considered of little importance, because seldom accessible in sufficient amount, though in the United States fields are recognized as good bee pastures. The chayote seems to make up by numbers what the flowers lack in size, so that the yield of honey may be larger than in related plants. In addition to this there is the fact that Sechium is a perennial bloomer in the Tropics, and in the subtropical regions has a very long season. It is thus possible that in the regions like parts of Florida, where beekeeping is already an established industry, the honey-producing qualities of the chayote may be found of practical account in connection with its other utilities."

CHEESES, see Mallow.

CHERRY (Prunus cerasus).

The cultivated cherry is closely related to the plum and is equally attractive to the bees. Both bloom at about the same time. In Cali-

Cherry blossoms.

fornia the cherry is reported as one of the best of the fruit trees for honey production. In the East it is valued principally for stimulating early brood-rearing.

When the weather is warm and bright during its period of bloom a great variety of insects may be found on the blossoms. Bees are seldom strong enough to store surplus from cherry. (See also Wild Cherry.)

Concerning the pollination of cherry blossoms, F. S. Lagasse (Delaware University Agr. Extension bul. 15) has written:—

> "Interplanting of several varieties, better cultural practices, and the placement of bees in the sour cherry orchard is recommended for increasing the set of fruit."

Regarding sweet cherry, Tufts and Philp, (Cali. Ex. Sta. Bul. 385) say, "Pollinizing agencies, such as honeybees are necessary to set a good fruit crop."

CHESTNUT (Castanea dentata).

The chestnut was an important timber tree from northern New England and Ontario to Michigan and along the mountains to Georgia. It is a tall and slender tree in the forests, but a magnificent spreading shade tree when grown with sufficient room. It sometimes reaches a height of 90 to 100 feet. It has been largely destroyed by blight.

As a source of nectar it is probably nowhere important, though it yields pollen in June or July. E. E. Hasty, writing in the American Bee Journal, from Toledo, Ohio, (Sept., 1906), reports that there the bees roar on chestnut bloom, even when basswood blooms at the same time.

It is frequently listed as a source of honey, but the author can find no authoritative records of surplus stored from it.

In Gleanings, 1887, page 25, two beekeepers, one from Pennsylvania and one from Massachusetts, write to say that honey from chestnut is bitter in taste, dark in color and has a rank smell. One writer states that chestnut yields at irregular intervals and often fails for several years at a time.

A. Alfonsus, of Austria, in the Bee World for August, 1921, writes as follows: (He may refer to Horse Chestnut which is common in Vienna.)

> "Chestnut honey is of very light yellow color, but as the chestnut trees produce nectar only after warm nights, which are rare at the time when the trees are in flower, it is seldom met with."
> Others list it only as a source of pollen.

In "Flora Apicola de Espana," M. Pons Fabregues states concerning *Castanea vulgaris*, "very melliferous but honey of poor quality."

CHICKWEED (Stellaria Media).

Chickweed is a widely distributed weed which is native to Europe. It is included in the lists of honey plants from many widely separated localities, although but little information concerning its value is avail-

Chicory in bloom.

able. Apparently the bees are attracted to it under almost any climatic conditions, since it is reported from such extremes as Oregon and Maryland.

Doctor Oertel reports that chickweed is the source of pollen in late winter in Louisiana where it grows in cultivated fields until hot weather or until land is plowed.

CHICORY (Cichorium intybus).

Chicory is a well-known plant native to Europe, northern Africa and southwestern Asia. The roots are much used as a substitute for coffee, as high as 5,000 tons being imported into Great Britain in a single year. It is also mentioned as a fodder plant in the old world, where it is valued especially for sheep.

It has been introduced into America and has become established as a weed in fields and waste places from Nova Scotia to Minnesota and south to North Carolina and Kansas. It is also known to occur in Florida, Colorado and on the Pacific Coast from Washington to California. In some localities in Michigan it is grown to a considerable extent for commercial purposes. The plant was first brought to Massachusetts from Holland in 1785. Since it readily runs wild, it has since become established over a wide area. In many places it has become so troublesome that the U. S. Department of Agriculture issued a circular dealing with its control and eradication.

There are numerous references to it as a honey plant in the early numbers of American Bee Journal. It flowers over a long period in late summer and yields both pollen and honey. It is said to be one of the most attractive plants to the bees.

It is locally known by a variety of names as blue dandelion, blue daisy, coffee-weed, wild succory and blue sailors. It is also sometimes called bachelor's-buttons.

The following comment on honey from chicory is from the Bee World, March, 1935:—

> The honey is of a peculiar markedly yellow color, even when granulated, slightly greenish, and has a flavor which might well be that of chicory as met in coffee.

CHINABERRY, see Wild China.

CHINA TREE (Melia azedarach).

The China tree, also known as pride of India, is a native of the Far East, probably coming from China. It has become naturalized in the Southeastern States from the Atlantic Coast to Texas. It is found as far north as Arkansas and Virginia.

H. B. Parks reports that in Texas it is a fairly good honey plant, blooming very early in the season. On account of its early blooming, it is principally valuable to stimulate early brood rearing.

This tree should not be confused with the Wild China, which see.

CHINESE SUMAC, see Varnish Tree.

Bloom of china-tree.

CHINQUAPIN (Castanea pumila).

The chinquapin is a shrub or small tree common from New Jersey and southern Pennsylvania southward to Missouri and Texas. It is well known to the beekeepers in parts of Alabama, Georgia and north Florida, where it flowers in May. In Arkansas, it is a large tree, reaching a height of fifty feet in some cases.

The chinquapin.

In some localities beekeepers report good crops of honey from chinquapin, but the quality is inferior. It is dark and strong, with a bitter taste. Some use it for feeding to replace the better grades of honey which may be taken from the bees. In color it looks like New Orleans molasses, and a sample, in the author's collection for several years, shows no tendency to granulation.

In the west the giant chinquapin (*C. chrysophylla*) is a large tree reaching a height of 100 feet or more. It is found in the mountains of northern California and Oregon. Coleman reports in Western Honeybee (March, 1921) that it is an important source of honey for food in late summer and fall and that the honey is amber in color and of good quality. He also reports the golden chinquapin which is regarded as a variety of the above as an important fall source. The bush chinquapin (*C. sempervirens*), a small spreading shrub found on arid slopes at high altitudes, is likewise mentioned as important where sufficiently common.

The Christmas berry, or California holly.

CHITTAM, see Buckthorn.
CHOCTAW ROOT, see Dogbane.
CHOKE CHERRY, see Wild Cherry.

CHRISTMAS BERRY, or CALIFORNIA HOLLY (Heteromeles arbutifolia).

The Christmas berry is known also as toy-on or tollon berry, as well as California holly. It is common along the streams and on the mountainsides of California, where it flowers in June and July. It has white flowers and the bright red berries ripen in late autumn. The berries, according to C. D. Stuart, are acid and slightly astringent, though not unpleasant. He states that the berries were eaten by the Indians as a kind of salad, and that a wine-red drink is sometimes made from them after an old Spanish-Californian recipe. The plant is much used for Christmas decoration on the Pacific Coast.

According to Richter, the plant is the source of a thick amber honey of decided flavor, which candies with a coarse grain, within two or three months after extraction. He reports surplus from this source in Monterey, Colusa and Nevada Counties, California.

Jepson gives the range as "Throughout the coast ranges and Sierra Nevada, and southward to southern and Lower California. Frequent along streams and gulches in the lower hills, and also abundant on stony slopes at middle elevations, especially from Napa to Humboldt Counties."

At Visalia the author heard reports of an average of a case per colony of surplus honey from this source in the mountains.

CLEMATIS.

The white clematis (*clematis virginiana*) is commonly known by the name of virgin's bower, but also has several other local names, such as love vine, traveler's joy and devil's hair.

The range of the plant is from Nova Scotia and Ontario west to Lake Winnipeg and Nebraska and south to Louisiana and Florida. It may be expected almost anywhere east of the Mississippi River. It is a slender climbing vine on the borders of woods, roadsides and hedgerows. The blossoms are white and fragrant, blooming in midsummer. It is much sought by the bees, and apparently produces considerable nectar. It is doubtful whether the plant is anywhere sufficiently abundant to make an appreciable difference in the production of the hive.

Blossom and leaf of white clematis.

Richter, in his "Honey Plants of California," reports a related species, the hill clematis, (*Clematis ligusticifolia*), as common in the hilly districts almost throughout California. It is said to produce "a great deal of pollen and some honey." Common also in the Rocky Mountains.

"Four tons of straight wild clematis honey were obtained by Mr. W. H. Turnbull of British Columbia, from one of his apiaries. The honey is light amber in color, somewhat cloudy in appearance, and of a most pleasant, butterscotch flavor. It is of exceptionally heavy body, the combs requiring more than half an hour in the extractor and then not being entirely clear of honey. Even at room temperature the honey is so heavy that no bubble will rise in an inverted jar."—N. N. Dodge, page 37 American Bee Journal, January, 1930.

CLEOME, see Rocky Mountain Bee Plant.

CLEOMELLA (Cleomella angustifolia).

The Cleomella is very similar to the Rocky Mountain bee plant or cleome. The flowers are small and yellow and the plant does not grow so tall as the cleome.

An Oklahoma beekeeper reports as follows:

"It seems to be a remarkable honey plant. It was in bloom for more than ten weeks during the dry season, and bees worked upon it freely every morning. The blossom is very fragrant, sweet, yellow, and is at the branches. It keeps crowding out a new growth and blooming, forming small purse-shaped seed pods as the blossoms drop. The growth is much like sweet clover or yellow mustard, but forming a larger spreading top. Some plants grow four feet tall and three feet across and an inch through at the butt. It is an annual, and no stock will eat it."—M. S. Hubbell, Helena, Okla. American Bee Journal.

Cleomella is found from Nebraska to Utah and south to Texas.

CLETHRA, see Pepperbush.

CLIFF ROSE, (Cowania mexicana, or C. stansburiana).

The cliff rose, sometimes known as quinine-bush, is a spreading shrub three to six feet in height common to dry mountainsides from Colorado to Nevada and south to California and Mexico. Before the coming of the white race the Indians of Utah and Nevada obtained material for clothing, mats, sandals and ropes from this shrub. The bark was removed in strips and woven or braided together.

It is reported as common at the foot of the mountains in the vicinity of Flagstaff, Arizona, where it is said to be a valuable source of good quality honey. S. M. Campbell of that place states that it is valuable for both pollen and nectar and that the honey is white in color with the finest kind of flavor. He reports a tendency to bloom after every rain from June until frost.

CLIMBING BONESET or DUCKBLIND (Mikania scandens).

The climbing boneset occurs in low, damp thickets and swampy places from New England and Ontario to Florida and Texas. It is very common in the Kankakee swamps, also in the Mississippi River bottoms in the region of Memphis, Tennessee, and in low grounds of the Gulf States. Mohr states that it occurs from the mountain region to the coast plain of Alabama, but that it is most abundant in the lower pine region and along the coast. The author found it common in many localities in Mississippi.

It is a luxuriant herbaceous climber with blossoms similar to the other bonesets and with a similar blooming time. The flowers appear from August until November. The bees work it eagerly and apparently it is of considerable value as a source of nectar. There is little reference to it in beekeeping literature, probably due to the fact that the bees work on so many different plants in swamps and lowlands that few beekeepers recognize its importance.

In the Kankakee swamps it is locally called "Duckblind," due to the heavy festoons which it forms over the bushes, furnishing a good hiding place for hunters. It is known also as climbing hempweed.

There are many related species in South America, especially in Brazil and also in the West Indies.

A letter from Jes Dalton, of Bordelonville, stated that it was one of the best honey plants in swampy sections of Louisiana. There it is called wild potato vine and pome-de-terre.

CLIMBING HEMPWEED, see Climbing Boneset.

CLIMBING MILKWEED, see Bluevine.

CLOVER (Trifolium).

The clovers are by far the most important American honey plants. If we include the closely related alfalfa and sweet clover, they probably are the source of more surplus honey than all the other plants together. They are to be found in nearly every part of America and yield nectar more freely than most plants. The quality of clover honey is of the best and in quantity of yield it ranks high, under favorable conditions. If the whole group was to be removed, honey production as a commercial proposition would decline to a very large degree.

Clover seems to yield most heavily in the northern part of its range and gradually declines southward. White clover is the most important of the group. Alsike is quite as valuable where equally abundant. The corolla tubes of red clover are usually too deep for the honeybee to reach the nectar, but occasionally some honey is secured from this source. (See Red Clover). Each of the clovers is considered separately. (See White Clover, Alsike, etc.)

Trifolium ambiguum, named "Pellett Clover" by the Iowa Beekeepers Association is a new clover which has attracted wide attention in the American Bee Journal test garden. It spreads rapidly by underground rhizomic roots. It yields nectar freely and is eagerly visited by the bees. It seeds abundantly and in case it proves suited to American environment bids fair to become widely used. (Am. Bee J'nl, Nov. 1946).

CLOVER BROOM, see Indigo-weed.
COCOANUT, see Palm.
COCOA PLUM, see Gopher Apple.

COCKLEBUR (Xanthium canadense).

The cocklebur is a coarse weed, common in fields and waste places from Minnesota to Texas and eastward. The burs are annoying in clinging to clothing and to the hairs of horses and cattle.

There is an occasional report of honey, (probably honey-dew), from this source, especially from Louisiana. In Texas, Scholl lists the

Clover field at blooming time.

plant as a source of pollen in late fall. It is reported as valuable for pollen in California.

In Australia, Rayment lists *X. Spinosum*, there known as Bathurst burr, as the source of large quantities of ill-flavored honeydew.

COFFEE BERRY, see Buckthorn.
COFFEE-WEED, see Chicory.

COGSWELLIA (peucedanum).

Rydberg, in his "Flora of the Rocky Mountains and Adjacent Plains," lists 29 species of *Cogswellia* common to that region. The plants belong to the parsley family and are locally known as whisk-broom parsley, biscuit root, or cous.

R. A. Bray sent the author a specimen of *C. villosa,* from Big Timber, Montana, in May, 1920, with the statement that it seemed to be of great importance to the bees there. Hilly pasture land was yellow with it over thousands of acres. He mentions it as the earliest source of pollen, blooming about two weeks ahead of the dandelion in his locality. This is a rather low growing parsley with yellow flowers.

COLIMA (Xanthoxylum pterota).

Colima is a species of prickly ash common to the valley of the lower Rio Grande River. It is a small shrub with zigzag branches, armed with short curved thorns. The flowers are in axillary clusters.

Colima is reported as yielding but a light flow, of principal value for stimulative purposes. At Mathis, Texas, Wm. Atchley reports that he had a good surplus flow from colima in 1900. The honey was golden and very thick, weighing 12½ pounds to the gallon. The flavor was good. Although he kept bees in that vicinity for a number of years he secured surplus from Colima but the one time. (See also Prickly Ash).

COLORADO—Honey Plants of.

Practically all surplus is secured from alfalfa and sweet clover, with an occasional crop from cleome. The rosin weed (*Grindelia squar-*

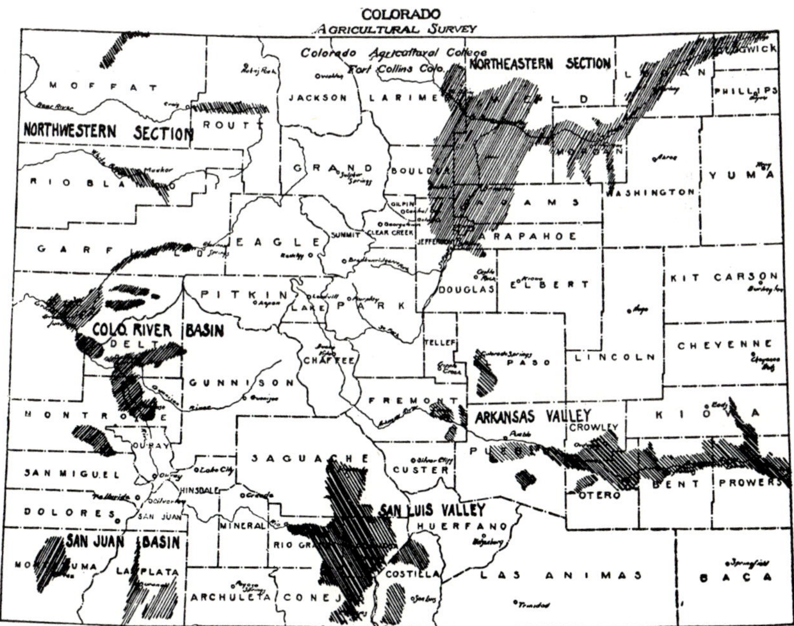

Most of Colorado's surplus honey is produced in the irrigated regions, indicated by the shaded areas.

rosa) yields considerable honey of low grade which often spoils the grade of the white honey by being mixed in the supers. Narcissus, parsley and prairie clover are prairie plants which attract the bees freely. Wild onion yields some honey. Dandelion and fruit bloom are important for spring brood rearing and canteloupes yield some surplus in the Rock Ford region. Sunflowers, mentzelia, lupines and loco weeds add something to the sum total brought to the hives. The white clematis is very common along the streams, as are willows. Wild currant is common in the mountain canyons. *Gaura coccinea,* the red gaura or ragged lady, is much sought by the bees for both honey and pollen. Oreocarya is a desert plant which yields surplus, but which is rapidly disappearing through cultivation of the land on which it grows.

Colorado is one of the highest and dryest of the states. The greater portion of its area is more than 6,000 feet above sea level, with numerous mountain peaks as much as 14,000 feet high. Less than one-fourth of the total area is below 5,000 feet. The Rocky Mountain range runs north and south across the center of Colorado and about two-fifths of the state is included in the great plains region which lies to the east of the mountains. This region is drained by the Platte and Arkansas Rivers and their tributaries. (See Colorado, by Newton Boggs. Am. Bee Journal, June, 1923).

Since the annual rainfall of Colorado is low, averaging only about 17 inches, beekeeping is largely confined to the irrigated valleys where alfalfa is the principal crop. The high altitude and abundant sunshine make an agreeable climate, favorable to life in the open. The bees have frequent opportunity for flight during the winter months and outdoor wintering is the rule, even though the temperature at times drops to a low point.

In the valleys on the western slope fruit growing is an extensive industry and here the beekeepers sometimes complain of losses through the spraying of the trees with arsenates while in bloom. About three cuttings of alfalfa are harvested each year in these valleys and where the crop is grown for seed good honey crops are secured. In the San Luis Valley, in the south central region, both alfalfa and sweet clover are extensively grown. The valley has a short season, due to its elevation of about 7,500 feet, but it offers favorable conditions for honey production.

COLUMBINE (Aquilegia).

The columbines are rich in nectar, but because of their deep corolla tubes the bees cannot reach it under ordinary conditions. Bessey lists both the wild columbine, which is native to the state, and the cultivated form introduced from Europe in his list of honey plants of Nebraska.

There are instances recorded where the honeybees have gathered nectar from columbine through punctures in the tubes which were cut by other insects. These instances are hardly sufficient to justify including it in a list of honey plants, since it could never be important to the beekeeper.

COLUMBO, **see Monument Plant.**

Coma in bloom. This species (*Bumelia augustifolia*) is abundant from Pearsall, Texas, to the Rio Grande River, blooming from October to February, and is the best of the group.

COMA (Bumelia lycioides).

Coma is the Mexican name for southern buckthorn (*Bumelia lycioides*), which is found from Florida north to Virginia and west to Texas. It is also known as ironwood. It is a spiny shrub with flowers in dense clusters. In South Texas it is also known as coma and is frequently reported as an important source of honey. It blooms there from October to February and produces nectar freely. At Rio Hondo, in the lower Rio Grande Valley, beekeepers reported to the author that a flow of six weeks from coma, from September to November, was common, and that swarms issuing as late as December had gathered sufficient honey from this source to carry them through. The honey is said to be light amber in color and of good quality. The flow varies greatly, depending upon the rains. (See also Gum-elastic.)

Bumelia angustifolia, is common in the valley of the lower Rio Grande and is reported as far north as Pearsall, Texas. It is said to be the best of the group as a source of honey.

COMFREY, see Prickly Comfrey.

CONE FLOWER (Rudbeckia).

The cone flowers are not often mentioned as honey plants, yet apparently they are the source of some nectar. The plants are widely distributed east of the Missouri River and are probably of limited local importance.

Bessey (Honey Plants of Nebraska) lists three species as yielding both pollen and honey. The purple coneflower or "Nigger head," (*Rudbeckia angustifolia*), coneflower, (*R. columnaris*) and *R. pinnati,* are common to that state. Several others are common to the plains region.

In the author's garden the coneflowers have shown very little attraction to the bees. Only one, *Rudbeckia laciniata,* a tall growing perennial species widely distributed on low ground appears to be of any importance. Improved varieties of this species with deeply cut leaves are cultivated in gardens under name of golden-glow.

CONNECTICUT—Honey Sources of.

Connecticut is a small state with an area of less than 5000 square miles, less than some counties in western states. It is a region of long and cold winters, usually with heavy snowfall. It has rather short summers with an occasional short period of extreme heat. Weather changes are frequent and often extreme. The annual precipitation is about 45 inches.

Beekeeping is a minor enterprise and but few commercial honey producers are located in this state. While crops are not as large as in more favored regions, nearby markets offer very favorable opportunity to dispose of the surplus at good prices.

While the Connecticut River Valley is a prosperous agricultural region, the greater part of the state has rather poor soil and offers little inducement to the beekeeper. Sumac is probably the most dependable source of nectar.

The clovers, alsike and white clover, occur all over Connecticut and yield nectar freely. Goldenrod and asters are also important. Buckwheat, wild raspberry and milkweeds all yield surplus under favorable conditions. Basswood, locust, maple and clethra are among the valuable trees and shrubs. Fruit bloom, willow, etc., are important in early spring. Some honey from tobacco is reported in Connecticut. See Maine and Massachusetts for additional plants.

CORAL BEAN or FRIJOLILLO (Sophora secundiflora).

The coral bean is a small tree common along the streams of Texas and New Mexico, and southward to the interior of Mexico. The Mexicans call it frijolillo. The beans are said by Coulter to be used by the Indians as an intoxicant. The beans are round and red and as large as small marbles. They contain a powerful poisonous alkaloid. Some beekeepers have expressed a fear that the honey might be poisonous to the bees, though apparently there is little grounds for such fear. The tree is abundant along the Nueces River in the shade of larger timber and was yielding nectar very freely at the time of the author's visit to that section on March 10, 1918. There had been a long dearth of nectar and the bees were extremely short of both honey and pollen. The new nectar was coming in in considerable quantity and the honey stored had a peculiar flavor. The tree is particularly abundant in the vicinity of Matagorda Bay.

It is locally called laurel.

CORAL BERRY, see Indian Currant.
CORALLITO, see Coral Vine.
CORAL SUMAC, see Poisonwood.

CORAL VINE (Antigonon leptopus).

The coral vine or Rosa de Montana, also known as pink-vine, corallita and San Miguelito, is a very attractive climbing vine common to Mexico and Central America. It is common also in Florida, where it flowers from early spring till late autumn. On favorable soils it grows 25 to 50 feet in height. The flowers are deep rose red or rose pink. It is commonly grown as an ornamental on fences, porches and trees throughout the American tropics.

Frank Stirling writes that in his opinion this vine is more valuable for planting for honey alone than any other plant in Florida. Since it blooms for such a long period and yields nectar freely, it is invaluable where abundant.

Alfred H. Pering states that the honey is almost water white and has a very peculiar flavor that cannot be adequately described. It is very thin which he calls its most distinguishing feature. He mentions the long blooming season, from mid-season until January, and states that the bees work upon it frantically from early to late.

The coral vine is extensively grown on the island of Cuba where it flowers throughout the year. Ordetx describes the honey as almost white, of good taste and strong aroma. He suggests that it might be profitable for cultivation as a source of honey.

COREOPSIS.

Coreopsis is a group of plants closely resembling the *Bidens* or Spanish needles. *B. trichosperma* and *B. laevis* are quite commonly known by the name of *coreopsis*. The two groups are quite commonly confused and common names are nowhere sufficient to separate them. The tickseeds are usually classified as *Coreopsis* while the Spanish needles are classified as *Bidens*. Many good honey plants are included under the common name of coreopsis in popular usage. They are late flowering and as a rule have yellow flowers. Heavy flows are often secured in the fall in river bottoms and marshy places. (See Spanish needle).

CORIANDER (Coriandrum sativum).

Coriander is an annual plant of the parsley family which comes from Southern Europe. It has long been grown to some extent in American gardens and has escaped and run wild in some places.

The plant is the source of oil and the aromatic seeds are much used in flavoring confectionary and as a condiment in bakery goods.

The fact that more than a million pounds of the seed are imported annually indicates the extent of commercial use in this country. Recently there have been attempts to grow the plant more extensively in order to meet the domestic demands and this may lead to further increase in acreage.

As a bee plant we have little information aside from the fact that the bees visit the flowers freely. As one beeman says, the bees go a bit frantic over the blossoms but there is not enough of it to make a difference in the honey crop.

CORN, see Indian Corn.
CORNEL, see Dogwood.
CORN-ITCH, see Cow-itch.

COSMOS.

The cosmos are common annual garden flowers native to Mexico but widely grown as ornamentals. They are tall growers and free blooming and being of easy culture are very popular. Flowering in late summer they attract bees in large numbers. Since the number of flowers within reach are rarely sufficient to become important they are of incidental interest to the beekeeper.

COTONEASTER.

The cotoneasters are ornamental shrubs grown for their decorative fruits, which hang on during the winter months or for the foliage, which with some species turns to bright colors in autumn. There are numerous species, some of which are quite hardy. They are native to Europe and Asia.

Horticultural writers mention the attraction which insects find to the blossoms in search of abundant nectar, and beekeepers have called the author's attention to the attraction which it offers to the honeybee. W. G. Merritt, a nurseryman at Augusta, Michigan, wrote that he has found the bees working on *C. integerrima* late in fall when all other blossoms had perished and when nights had become quite frosty. Likewise L. T. Floyd stated that he had never seen anything which seemed to attract bees in such large numbers for the amount of bloom available as the cotoneasters on the college grounds at Winnipeg, Manitoba.

COTTON (Gossypium).

Although the cotton plant is found growing wild in many warm countries, in the United States it is known only as a staple field crop. It was brought to this country as early as 1621, and has been the most important plant grown on southern plantations since the early development of the country.

The plant thrives in a warm and humid climate, and needs five to six months of warm weather. However, it is grown successfully under semi-arid conditions in parts of Texas and other Southern States. The so-called cotton belt extends from the northwest corner of Texas south to the Rio Grande, and east to the Atlantic seaboard.

In the Southern States, cotton is the one field crop grown on a sufficient scale to offer ideal conditions for the beekeeper. However, cotton is fickle in its behavior, and cannot always be depended upon to produce nectar, no matter how abundant the crop. In some cotton-growing districts the beekeepers swear by cotton, while in other localities they declare that it is of little value. The character of the soil seems to be a very important factor in the secretion of nectar by this plant. The vigor of the growth and the amount of available plant food in the soil are also important. Reports from different sections indicate that the quality of the honey varies in different sections.

W. D. Null, of Demopolis, Ala., wrote to the author as follows:

> "This, you know, was for sixty years the heaviest cotton-growing section in the nation. Bees will not work cotton if they can work anything else, even bitterweed. It yields honey of very poor quality, and never very much, some years none at all. Weather conditions must be just right, and that don't come often. The honey is the same grade as the most honeydews."

In contrast, we find the following report of good honey and abundant yield in American Bee Journal for 1907, page 267:

> "Cotton blossoms furnish a great deal of excellent honey, and the theory that it explodes or ferments is all bosh. It makes an excellent rich honey, oily, and it is not liked so well by some until they get used to it."—Jules Belknap, M. D., Sulphur Springs, Ark.

When the writer made his first trip through Georgia he was much puzzled by the different reports of apparently good observers in different parts of the State. The matter was finally explained by a beekeeper who had lived in different localities, by the variation in behavior of the plant under different conditions. There is perhaps no important honey plant which varies so much, in the quality of its nec-

Blossoms of the cotton plant. (U. S. Department of Agriculture.)

tar, as does cotton. The poor quality in some places can doubtless be explained by the fact that the flow is not abundant, and is mixed with other low-grade stores. However, honeydew is also sometimes reported from the plant itself.

"Sometimes, during a damp spell, the cotton gets covered with vast numbers of aphis, and the upper side of the leaves will first get gummy and then will even drip a kind of dirty-looking sweet fluid. If there is anything else on hand the bees will not touch it."—W. H. Alder, Callahan County, Texas, page 334, American Bee Journal, 1899.

It is needless to say that this would make a poor product, and it is not improbable that honeydew is sometimes secured from cotton in localities where it seldom yields nectar. The secretion is apparently dependent far more upon soil than upon any other condition. Upon the black, waxy lands of Texas and upon rich soils, it reaches its highest development. The boundary of the belt, where cotton yields freely and where it does not, is very marked in Texas. North of the escarpment which runs across Bexar County, Texas, near San Antonio, it is an important source. South of that line few beekeepers report it as dependable. North of this line the soil is black and heavy; south it is sandy. Wherever the writer has found beekeepers on sandy soil, they have reported the yield from cotton as uncertain; while on the heavy soils they report it as fairly constant, with suitable weather conditions. In east Texas cotton is reported as yielding well on river bottom lands and but little on the hills. In the southern sections, and also in other States, an occasional crop is reported where it does not yield regularly:

"We had a very dry, sultry spell here the latter part of last August, and up to that time the bees were living from hand to mouth. All at once they began storing from the cotton bloom, though it looked as though cotton was going to die in the fields from drought and heat, yet it yielded until the bees had stored from 30 to 60 pounds per colony."—J. J. Wilder, Cordele, Ga., American Bee Journal, page 141, 1906.

On suitable soils it is one of the most dependable sources of nectar:

"The apiarist who has his bees located within range of extensive cotton areas can count on at least an average crop year after year, with more certainty than many of the other numerous honey yielders which we have."—Louis Scholl, page 652, Gleanings, 1912.

"My main sources for surplus are mesquite trees, the cotton fields being the second of importance in the central and northern parts of the State, or throughout the black land region. On sandy or light soil cotton yields very little honey. * * *

"The yield is good, averaging about 73 pounds of bulk comb honey per year. One year it was over 100 pounds. Honey from cotton is very light in color, the comb very white, and of excellent flavor when well ripened. As soon as cool weather sets in this honey fairly draws out in long strings, when handled with a spoon."—Gleanings, page 1313, 1907.

From the above it will be seen that cotton honey is of good quality, at least in some localities. Samples, said to be from cotton from Georgia, are strong and of rather poor quality, while cotton honey received from Texas is light in color, of mild and rather pleasing flavor. The honey from Cotton granulates very quickly. That produced in the Southeastern States also has the effect of bursting the containers, possibly from the effects of fermentation. The humidity of the atmosphere evidently has a marked effect on the quality of the honey from this plant. The following reports indicate the quality:

> "As to the quality of cotton honey, I can say from my own experience, that it varies in color from light amber to almost water white. While I do not consider it equal to white clover in flavor, it is superior to basswood. * * * The flow increases toward the last of the season, and if we can get two weeks of nice weather after frost it amounts to a considerable increase in the crop."—J. D. Yancey, Hunt County, Texas. Gleanings, page 162, 1910.
>
> "It did well on our rich bottom land and yielded a fair crop of the finest honey it was ever my pleasure to see. It was so thick that it was almost impossible to extract it, and entirely out of the question to strain it through a single thickness of cheese-cloth. It was light in color, mild in flavor, and very heavy, and in my opinion superior to any honey ever shipped to this locality, not excepting huajillo. The long drought and consequent absence of all other bloom, enabled us to get a purer cotton honey than we had ever been able to get before. Again, in the late fall, when the weather began to get cool, our cotton took a second growth, soon blooming profusely, and by accident we got also a fair fall crop."—O. Saunders, Trenton, Texas. Page 734, Gleanings, 1910.

One great advantage of the cotton flow is its long continuation. In Texas it begins to bloom in May or June, and the bees work it steadily until late fall, often November. Extra cultivation or fertilization of the soil increases the vigor of the plant and the nectar flow is increased accordingly:

> "I can remember when the bees gathered only enough nectar from it to stimulate brood rearing, and now we get from one to three supers of surplus from this source alone. * * * On land where we used to make a bale of cotton to 4 or 5 acres, now we make 1 to 2 bales per acre, using high grades of commercial fertilizer and more prolific varieties of the plant. It yields more where it grows best, and of a much longer duration."—J. J. Wilder, Cordele, Ga. Page 237, American Bee Journal, 1911.

Bees get nectar not only from the cotton blossoms, but from extra floral nectaries as well. At times almost entirely, and to gather freely they seem to neglect the blossoms for the extra-floral nectaries. Some of these are located under the flower and begin to secrete nectar before the blossoms open. Others are located on the under sides of the leaves, and vary from one to three on each leaf. When atmospheric conditions are favorable, these glands secrete abundantly and the nectar gathers in drops. At times it is so abundant that the men cultivating get their clothes saturated with the nectar, from the brushing of the leaves against them. Later in the day the heat of the sun evaporates most of the moisture, leaving the clothing sticky. In hot and dry weather the flow is on in the morning and again in the evening, while in cloudy or damp weather it lasts all day.

When first gathered, the honey is said to be very thin and clear, with a strong and nauseating taste, resembling the taste of the plant itself. As the moisture is evaporated and the nectar ripens in the hive this disagreeable taste is lost to a large extent. During a heavy flow a strong odor is frequently present in the apiary, which can be noticed at some distance from the hives. Scholl compares this odor to that of crushed cotton leaves. He reports that at times it becomes so strong as to have a sickening effect on the apiarist, even interfering with his work, on calm days.

The heaviest flows come from rank-growing plants on rich soils, during warm and wet weather. At such times the honey is lighter in color and superior in quality, while the honey stored from plants growing on light soils during dry weather is darker and strong in taste.

Pollen from the cotton plant is white in color, and is produced in abundance from the large bell-shaped flowers. When the bloom first opens it is white, later turning pink.

COTTOM-GUM, see Tupelo.
COTTONWOOD, see Aspen.
COUNTRY CHEESE, see Mallow.
COUS, see Cogswellia.

COW-CLOVER, (Trifolium involucratum).

Cow-clover is reported as common to the valleys of middle-western and northern California where it appears to be native. Beekeepers report it as of first importance as a source of nectar in some localities. Apparently it is of limited distribution but of considerable value where found.

The trumpet creeper is often called cow-itch.

COW-ITCH.

Beekeepers from various southern localities report honey from cow-itch, but like other common names for plants, there is some uncertainty as to just what the beekeeper means. In the July, 1922, number of the American Bee Journal Mr. H. B. Parks pointed out the confusion in this case, together with a number of other similar ones.

The well-known trumpet creeper or trumpet flower (*Tecoma radicans*), is common in moist soil from New Jersey to southern Iowa and south to Florida and Texas. It is also widely cultivated as an ornamental, not only in the region already mentioned, but northward and westward. The large trumpet-shaped flowers are showy and much visited by hawk moths and humming birds. It is commonly called "cow-itch," in some localities, also, cross-vine. The blooming period comes in late summer from late July until September. The blooms are rich in nectar, but it is not always accessible to the bees. At times bumblebees or other insects cut holes in the base of the corolla and thus expose the nectar. In such cases the honeybees are quick to take advantage of the opportunity. However, it is doubtful whether the trumpet creeper is ever an important source of nectar for the commercial beekeeper.

Another plant known as "cow-itch" is (*Cissus incisa*), which, as far as the writer can ascertain, has no other well-known common name outside of Texas, where it is known as "yerba del buey." It is a climbing vine common to open sandy woods from Florida and Texas north to Missouri and Kansas. This is a good honey plant, but it is difficult to determine the extent of its value, owing to the confusion with *Ampelopsis arborea,* the seven-leaved ivy. This species is not only called cow-itch, but also peppervine. It is found in swampy and moist places from Virginia to Missouri and south to Texas and Florida. It is a stout climber with few tendrils and yields nectar freely. At Buffalo and Palestine, Texas, the author found reports that cow-itch, which evidently referred to this species, was a heavy yielder of nectar, the bees beginning on it as soon as the basswood bloom closed. It blooms from June to October, with fruit in all stages. The honey was reported to be light amber in color and of good flavor. R. A. Nestor reported yields of from 35 to 50 pounds surplus per colony from this source, mixed with partridge pea.

In the Texas Bulletin Sanborn and Scholl list the *Cissus incisa* as the source of surplus where abundant.

COWPEA (Vigna sinensis).

The cowpea is widely cultivated in the warmer regions of the old world and in our own Southern States. It is grown for forage and for green manure. The plant is more closely related to the beans than to the peas.

R. A. Nestor reports that it yields freely in east Texas, and where planted in sufficient acreage yields surplus. The honey is very dark in color, but of mild flavor, according to his report.

The nectar from cowpeas is secreted by extra floral nectaries and beekeepers are often mystified because the bees are working at the "joints" instead of on the flowers. Some report that bees gather nectar from the flowers, also.

The cow-itch of the South is *Cissus incisa,* an important source of surplus honey.

The following reports indicate the value in different localities:

"There is no finer honey plant than the cowpea, while it lasts, but it blooms only about a week. During this time, if the weather is fair, the bees swarm over the fields from early morn till dewey eve."—J. D. Rowan, Tupelo, Miss. Gleanings, Sept. 15, 1909.

"The cowpea is one of our most abundant sources of honey for late summer. The crop is planted here from May 1 to August 1, and furnishes nectar through a considerable period of otherwise scarcity. Unlike other plants, the stems, and not the blossoms, secrete the nectar as the young pods are forming. These the bees work upon excessively. The honey is of good body, thick, deep, approaching dark yellow in color, and of strong taste like that of tulip-poplar, only stronger, with a somewhat slight, wild-green-bean-like flavor."—C. C. Gettys, Hollis, N. C. Gleanings, Sept. 14, 1909.

"A small patch of peas was covered with bees from morning till night. Nearly all of them were working on the stalks, as usual; but here and there

I saw a few Italians pushing their tongues down into the blossoms. I have never noticed any pollen from the field peas."—Mrs. Ameda Ellis, Fremont, Mo. Gleanings, June 1, 1910.

"The peas bloom when there is a honey dearth and the bees gather honey from them. However, I notice they do not work on them much if there is a better honey plant blooming at the same time. My bees get a good deal of nice honey from them."—G. H. Latham, Jr., Rapidan, Va. Gleanings, May 15, 1910.

COYOTE MINT, (Monardella villosa).

The coyote mint is found in the dry rocky hills of the Coast Ranges and Sierra Nevada mountains of California. It has declining stems and densely flowered heads which are the source of dark honey in summer. The plant is abundant in some areas and is reported as the source of considerable honey and some pollen.

COYOTE WEED, see Turkey Mullein.

CRAB APPLE (Malus).

There are several species of wild crab apples native to America. The fruit is small and sour and of little value. The blossoms are rose colored and very fragrant, making the tree worthy of cultivation as an ornamental. As a source of honey the wild crab apple ranks with the cultivated apple and other fruit trees. The blossoms appear from March to May and serve to stimulate spring brood rearing, though where abundant, strong colonies may gather some surplus.

The southern crab apple (*Malus angustifolia*) occurs in open woods from southern Pennsylvania south to Florida and west to Louisiana and Missouri.

The American crab apple (*Malus coronaria*) is common from eastern Canada to Michigan and south to Alabama and South Carolina. It is a small tree and often grows in dense thickets, where it furnishes ideal bee pasturage.

The Soulard crab apple (*Malus soulardi*) is much like the American crab apple in tree and flower, but has a larger fruit. It is sometimes cultivated for its fruit, which is used as a substitute for quince. It occurs in the wild state from Minnesota to Texas, but is not very common.

The Oregon crab apple (*Malus diversifolia*), is a native of the Pacific Coast from Alaska through Western Canada south to California. The fruit was formerly dried by the Indians for winter use.

All the wild crap apples are valuable sources of nectar.

CRANBERRY, (Vaccinium macrocarpon).

The cranberry grows in wet bogs and marshes of Eastern Canada and the Northeastern United States. The berries are grown in large commercial plantations known as cranberry bogs, especially in New England and Wisconsin. The honeybee is recognized as important in securing proper pollination. No reports are available which indicate that the cranberry is an important source of nectar, although the bees work the blossoms freely.

A variety of the small cranberry (*V. Oxycoccos*) is found in the sphagnum bogs of Washington and nearby regions near the coast. It is found also in the far north of Canada, from Newfoundland westward, and south to Michigan and New Jersey. See also Blueberry, Huckleberry, etc.

CRANBERRYBUSH (Viburnum americanum) Highbush Cranberry

The cranberrybush or highbush cranberry is a common shrub in northern woodlands. It is also cultivated extensively as an ornamental. It is sought by the bees and is frequently mentioned as a honey plant. In some sections of Manitoba it is regarded as important. It has a wide range, being found from New Foundland to British Columbia and from New Jersey to eastern Iowa.

There are several other species of *Viburnum,* including sheepberry or nannyberry, arrow-wood, etc. Since these shrubs grow sparingly in woodlands there is little information available as to extent of the nectar yield. (See black haw.)

Creosote bush in Arizona desert.

CRANESBILL, see Geranium.

CREAM CUPS (Platystemon californicus).

Cream cups is a low, spreading annual 3 to 6 inches high, with creamy-yellow petals. It is common in the lower altitudes throughout the state of California. The blooming period is early in spring, usually April. Coleman states that the honey is light amber in color and that it is important in the Sierra foothills. (Western Honeybee, May, 1921). Richter, on the other hand, lists it as a honey plant not known to yield a surplus. (Bul. 217 Cali. Ex. Station).

CREOSOTE BUSH (Covillea glutinosa).

Creosote Bush is the most common shrub of the dry mesas from western Texas, through southern New Mexico and Arizona to southern

California. To the beekeepers of this region it seems very generally to be known as "greasewood." The Spanish name is "Hediondillo."

Creosote bush is an evergreen shrub 3 to 5 feet in height; under especially favorable conditions it grows six or more feet. It is an erect, spreading bush much branched near the ground. The resinous leaves have a decided odor of creosote, especially evident when the plant is wet. The Spanish name "Hediondillo" means stinking and is likewise derived from this characteristic.

Blossoms and leaves of creosote bush.

The bright yellow flowers about half an inch in diameter appear in early spring. The author has found it blooming in southern Arizona in February. New Mexico beekeepers report that it blooms at rather uncertain periods, sometimes more than once.

J. W. Powell of Messilla Park, New Mexico, states that it sometimes yields a small amount of surplus, but that its principal value is for early spring stimulation. Most reports throughout its range are that the bees use it mostly for spring brood rearing. It also yields much early pollen.

CRIMSON CLOVER (Trifolium incarnatum).

Crimson clover is grown in the Southern States. It is an earlier bloomer than the other clovers. The blossoms are more showy than either alsike or red clover. The plant is an annual, and must be resown to perpetuate a field.

The honey yield is reported to be good and the quality similar to that of the other clovers. It is nowhere grown on the scale of the others, so is not so well known as a source of honey.

Bonnier gives it third rank as a honey yielder, while the British Bee Journal states that it is about on a par with buckwheat, and that neither is satisfactory when honey of later yield is worked for.

Niswonger lists it as a very important plant in Kentucky, and states that the honey is of a very light yellow color of good quality.

CROCIDIUM.

Crocidium multicaule is a small annual herb with alternate leaves and yellow flowers which blooms in early spring, common to the plains and hills on the open ranges of the northwest. It is found from British Columbia to Washington, Oregon, Idaho and California.

J. Skovbo, of Hermiston, writes that in northwestern Oregon it blooms about the middle of March along with the willows and continues for about three weeks. He regards it as an important source of pollen and occasionally some nectar. In 1920 he estimated that his bees stored an average of seven pounds of honey per colony from this source.

This plant may be regarded as of limited local importance in the Pacific Northwest.

CROCUS.

A group of early spring flowers native to the Mediterranean region of Europe, widely cultivated in gardens. Among the first flowers to bloom, they are very attractive to the bees.

CROSSVINE, see Ampelopsis, also Cow-itch.

CROTALARIA

There are several species of crotalaria which are native to our southern states. They are commonly known as rattlebox or rattlepod. If they are important to the bees the fact is not generally known.

Of late, however, some introduced species have come into common use as cover crops on sandy soils, especially in Florida. They are especially useful in orange groves where they furnish an important source of nitrogen.

One species is reported as of some importance as a source of honey in the citrus districts. Dr. Waldo Horton of Winter Haven, Florida, reports that bees get surplus honey from *Crotalaria striati,* usually in early winter. He has had a strong colony to fill a fifty pound super from this source. He reports the honey to be dark amber, with a maroon tinge and strong, not too pleasant flavor. He could classify it as a baking honey rather than a table honey. He states that bees also work on *Crotalaria spectabilis* to some extent.

The fact that crotalaria is being grown in constantly increasing acreage and that it is a winter bloomer, makes it of much interest to beekeepers in the region where it is extensively cultivated.

CROTON

Scholl lists four species of croton as yielding pollen and nectar in Texas, though none of them are of much importance. Richter states

Crocus blossoms are among the first to furnish pollen in spring.

that the bees visit the small blossoms of *Croton californicus* in large numbers. The author can find no records which indicate that the plants are of special value anywhere.

The woolly croton, commonly called goatweed, is found from southern Illinois, Missouri and eastern Kansas, south through Tennessee and Arkansas to Mississippi, Louisiana and Texas. It is very common along waysides, borders of fields and in pastures in northern Mississippi. In the vicinity of Greenville the beekeepers regard it as of some importance. It is also found in Alabama and Georgia.

The plant is silvery green with small and inconspicuous flowers and is sometimes called sage, also hogwort.

CROWNBEARD (Verbesina). Also called Wingstem.

There are several species of this plant, some of which are very attractive to the bees. They may be found in the borders of open woodlands and partially shaded situations of the region east of the Missouri River. Where sufficiently abundant the crownbeard is the source of a considerable quantity of nectar. A coarse weed, growing four to eight feet tall, with winged stems and yellow, or sometimes white, blossoms.

Victor Vinson of Irvine, Kentucky wrote to the author that in July, 1926 his bees had not made enough honey to extract and it looked liked he would have to feed to carry them through the winter. During the latter part of that month and through August they worked freely on the crownbeard with the result that he secured 2026 pounds of honey from 22 colonies, leaving ample honey for winter. The honey was of a beautiful golden color and of very good quality. Most of it was sold at thirty cents per pound. See Yellow Top, also Golden Honey Plant.

CUCUMBER (Cucumis sativus).

The cucumber is dependent upon bees for pollination of the blossoms. The flowers are imperfect, the male organs being contained in one flower while the female organs are in another. For this reason it is necessary that insects carry the pollen from the staminate blossoms to the pistillate ones. Where pickles have been grown under glass, they have proved unfruitful until bees are given access to the bloom. Formerly the pickle growers fertilized the blossoms by hand to some extent. This was very laborious. According to B. N. Gates (in 3rd Report Iowa Bee Inspector), one grower had forty acres under glass in Massachusetts and the industry requires about three thousand colonies of bees annually to serve in the cucumber greenhouses.

In some sections, cucumbers are grown extensively for pickles. At Marengo, Ill., Doctor Miller reported that about 600 acres were planted to pickles. He reported that, whereas he formerly had no fall flow, his bees did gather some fall crop, part of it eveidently from cucumbers. There are numerous localities where cucumbers are of some importance to the beekeeper. Lovell states that the honey is pale yellow or amber and has at first a rather strong flavor, which largely disappears in time.

CUCUMBER-TREE, see Magnolia.

CULVER'S ROOT (Veronica virginica).

The Culver's root is found from New England and Ontario to Manitoba and southward to Arkansas and Georgia. There are reports to the effect that the bees fairly swarm on this plant, but apparently it is not of much importance to the beekeeper.

CUP PLANT (Silphium perfoliatum).

Cup plant, also called rosin weed, is a common square-stemmed plant with leaves grown together at the base forming a cup. It grows from four to eight feet high and is abundant on rich lands, along streams and in woodside borders, in the Mississippi Valley. It produces numerous large yellow flowers and, where plentiful, furnishes

Blossoms of the Culver's root.

considerable forage for the bees. It is probably seldom important as a source of surplus.

CURRANT (Ribes).

There are several varieties of the cultivated garden currant and many species of the wild currants which are valuable sources of nectar. There are at least seven species of wild currant native to New Mexico. It is probable that some species of wild currants are to be found in every State where bees are kept. In the vicinity of large plantings of cultivated currants they are an important source of nectar, but generally speaking, they are a minor source and of chief value for pollen and for stimulating early brood rearing along with most other fruits.

The yellow flowers of the buffalo currants (*Ribes aureum*) are very fragrant and apparently contain much nectar. The writer has often noticed the bees working on the blossoms, but since the corolla tubes are half an inch or more in length he supposed they were getting only pollen. A close examination showed that the bees were unmistakably getting nectar from this source and that the tubes had been slit entirely down one side by some unknown agency.

CURRANT-TREE, see Serviceberry.

CYMOPTERUS.

A specimen of *C. acaulis* was sent to the author by H. A. Mark of Oshkosh, Nebraska with the following note:

> "This plant I have come to consider one of our most important pollen plants. During my experience, with the exception of last year, it has been the first plant in the spring to furnish pollen, and throughout the spring bees will be seen with their pollen baskets filled with its purple pollen."

It is a member of the parsley family found on the prairies from Minnesota to Arkansas, and westward across the plains. This family of plants appears to be of special importance for pollen on the dry western plains. (See also Cogswellia.)

CYNOGLOSSUM, see Hound's Tongue.

D

DAHLIA.

The cultivated dahlias of the garden are attractive to the bees The author has observed as many as five bees at a time on a single dahlia blossom.

DAHOON, see Holly.
DAISY, see Ox-eye Daisy.

DALEA (Parosela).

There are twenty to thirty different species of Dalea, by some authorities classified as *Parosela,* in the desert region of Texas, New Mexico, Arizona, California and adjacent Mexico. These shrubs are reported as the source of honey of splendid quality. Owing to the fact that they mostly grow in localities where beekeeping is little practiced, they are not often mentioned. Rolla Kellogg, a California beekeeper, expressed the opinion that were there bees within reach of those plants in sufficient numbers, this honey would become famous. (See Indigo Bush).

In the desert region of southern New Mexico *D. scoparia,* is a rather common shrub, known to local beekeepers as purple sage. It is reported as an exceptionally attractive plant to the bees.

DANDELION (Taraxacum officinale).

The dandelion is one of the most widely distributed plants in America. Originally introduced from Europe, it has been naturalized over practically the entire continent. As each plant will produce hundreds of seeds, which are borne for long distances on the wind, its wide distribution is not surprising. The plant is sometimes used for medicinal purposes, serving as a mild laxative and tonic. The tender shoots are very popular as a table delicacy in early spring, with those who are fond of greens. The bright yellow flowers are very showy, and if the

The much despised dandelion is a valuable source of nectar.

plant was not so abundant, would be considered attractive. The warfare against the dandelion is as relentless and as continuous as the campaign against the house fly. Little is to be accomplished by digging the plants from one's own lawn, when a whole pastureful are going to seed a mile or two away.

The beekeeper has little to complain of from these weeds, as there is nothing of greater value during the short period of bloom. While the honey gathered from dandelions is dark and strong, most of it will be consumed for brood rearing. Occasionally a small surplus will be secured from this source, but it blooms so early that surplus is unusual. Large quantities of pollen as well as nectar are produced, so that a large acreage of dandelions within reach of the apiary is much to be desired.

DATE-PLUM, see Persimmon.

DATURA.

There are several species of *datura* which are coarse weeds of wide distribution. *D. tatula* is commonly known as purple thorn-apple and is naturalized from tropical America. It occurs from Ontario and New York to Iowa and Missouri, south to Florida and Louisiana. It is probably native to the South Atlantic and Gulf Region. It is found also on the Pacific Coast from Washington to California. *D. stramonium,* commonly called Jimson weed or Jamestown weed, is even more widely distributed. The blossoms are very large and showy and since they open principally at night they are regarded as hawk-moth flowers. This group is probably not often of importance to the beekeeper, although bees sometimes visit the blossoms. Mrs. Florence B. Richardson, of Hughson, California, writes that she found bees in the corolla of the wild datura each morning sucking up the drops of nectar in the lower part. She states that the blossoms of this species, commonly called angel's trumpet flower, stay open until nearly noon and are visited by many insects.

In "Annales de la Societe Entomologique de France," 1905, appears Monographie des vespides du genre Nectarina, by Robert du Buysson. In describing the honey-gathering habits of these honey-storing wasps he writes as follows:

> "The honey is sold in the villages by Indians around the lakes of Zocoalco and Chapala. It is sometimes toxic when daturas are in bloom. Cases of intoxication, especially among the "vanqueros," (cowboys), are seen. This is in dry seasons when flowers are lacking, the Daturas being the only plants furnishing food to the nectarinas. There are many species of Datura in Mexico, most abundant being *D. ceratocaule* Orteg., one of the most beautiful and largest, of a penetrating fragrance."

In the American Bee Journal for July, 1866, Jamestown weed is mentioned as a source of poisonous honey. The author has seen the bees at work on this species, but has not regarded it as important to the beekeeper.

DEAD NETTLE (Lamium).

There are several species of Lamium which have been naturalized from Europe and been widely disseminated in this country. The white dead nettle or white Lamium has escaped from gardens. In Sweden this plant is said to be commonly boiled and eaten as greens by the common people. The following information regarding the Lamiums is from Deutsche Illustrierte Bienenzeitung, September, 1921:

> "It is often asserted that from both *Lamium maculatum* and *Lamium album,* the spotted and white Lamiums, the bees often obtain pollen but no honey, because the nectar is so deep in the tube of the corolla that the bees, with their short tongues, cannot reach it.

> "Thousands of times have I lain on the ground beside Lamium to watch as the bees searched for nectar from these blossoms. The bee alighted on the lip of the blossom and inserted her tongue with much force as deep as possible, and as the corolla at the upper end opened, the bee could force in her head and thus advance her tongue to the bottom of the corolla, where the nectar awaited her."—Okonomiera Wust.

The red dead nettle (*Lamium purpureum*) is found in waste places and cultivated soil from Newfoundland to North Carolina, where it has established itself as an escape from Europe. It is found as far west as Missouri and is abundant in some sections of the southeastern part of that state. L. A. Schott writes that in the south end of Scott County there are areas where it is found in large quantities and the bees leave everything for it when it blooms in April. He states that they work in the bloom during the entire day, but get only nectar, gathering no appreciable amount of pollen from this source.

Lamium amplexicaule is reported as one of the earliest sources of nectar and pollen in Louisiana coming into bloom in March in the Shreveport vicinity. Further south it comes up abundantly in the fields of sugar cane and blooms through January and February. This species, known as "henbit" came from Europe and has been widely naturalized from Ontario to the Gulf of Mexico and in California. It is one of the commonest winter annuals in the southern states. It appears everywhere in cultivated fields.

DEATH CAMAS (Zygadenus venenosus) Poison Camas.

The death camas is a plant of the lily family common to wet meadows from Montana and British Columbia to California. It is reported as the cause of death of large numbers of honeybees during its period of bloom, in the Minidoka project in Utah. The bees die in the hives after their return from the fields for honey, sometimes in alarming numbers.

The plant is often reported as poisonous to cattle and sheep, occasionally causing death.

DEER CLOVER, see Wild Alfalfa.
DEER'S EARS, see Monument Plant.
DEER PLUM, see Gopher Apple.
DEER-TONGUE, see Vanilla-Plant.
DEERWEED, see Wild Alfalfa.

DELAWARE—Honey Sources of.

Clover, tulip-tree, willows, maples, fruit bloom, dandelion, heartsease, buckwheat, blueberry, huckleberry, boneset and asters are among the well-known sources of honey in Delaware. Beekeeping is not highly developed in this State, and few pay serious attention to honey production. See New Jersey for further information.

DESERT BROOM, see Baccharis.
DESERT BUSH, see Jerusalem Thorn.
DEVIL'S CLAW, see Acacia, also Catsclaw.
DEVIL'S CLUB, see Hercules Cub.
DEVIL'S PAINT BRUSH, see Hawkweed.
DEVIL'S PLAGUE, see Carrot.
DEVIL'S SHOESTRING, see Bluevine.
DEWBERRY, see Blackberry.
DIERVILLA, see Bush Honeysuckle.
DOCTOR-GUM, see Poisonwood.

DODDER OR LOVE-VINE (Cuscuta).

There are a dozen or more species of dodder which are leafless annual herbs with thread-like stems of a reddish or yellowish color. They are parasitic on the plants on which they are found, adhering by means of suckers. The flowers are small and appear in late summer and fall. They are found twining about numerous plants including the clovers, alfalfa, heartsease, goldenrod, viburnum, ragweed, etc. A few are found on such shrubs as hazel or buttonbush.

The author has received numerous reports to the effect that bees get nectar from dodder, but apparently it is of minor importance.

> "Beekeepers of Zavala County, (Texas) report many losses of bees from dodder vine, usually known as 'love-vine.' It is very poisonous, usually resulting in the immediate death of the bees feeding on it.
>
> "In some apiaries the loss is reported to run as high as 50 percent, due to the fact that many of the flowers on which bees ordinarily feed have dried up from the drouth which has prevailed over so much of the south."—American Bee Journal, page 483, October, 1930.

DOGBANE (Apocynum).

When not in bloom the dogbane resembles the milkweed. There are several species found in Europe, temperate Asia and North America. In the United States there are two common species, *Apocynum cannabinum,* known as Indian hemp, Canadian hemp or Choctaw root, and *Apocynum androsaemifolium,* the spreading dogbane.

Dogbane can be distinguished from milkweed by the finer stem and smaller leaves. The stems are usually reddish in color. The flowers are very different. At times the bees work on this plant very freely. It is especially abundant in some localities along the Missouri River in Kansas and Missouri.

W. J. Sheppard, of British Columbia, writes as follows concerning spreading dogbane in the northwest:

> "In many places in the interior of British Columbia there are hundreds of acres of a little appreciated wild flower commonly called 'milkweed,' which rivals the well-known fireweed in the quality of honey it is capable of producing. It is not a true milkweed, however, but the spreading dogbane. Fireweed can only be depended on to yield nectar freely when it is

growing on sufficiently moist land. Spreading dogbane, on the other hand, usually produces quantities of honey in dry seasons, and in districts where there is not a large amount of precipitation. The plant seems to prefer poor soil and dry climate. During 1922 there has been a very large yield of honey in some districts from this source. It remains in bloom for a long period and the flowers have a perceptibly strong and pleasant aroma.

"Honey from dogbane is water white, like that from fireweed, and cannot be distinguished from it in color. It has, however, a superior flavor and usually more density."

DOGFENNEL, see Mayweed.

DOGWOOD (Cornus).

The dogwoods or flowering cornels are a large group of shrubs with showy flowers. Some species are common over all of Eastern America, from Nova Scotia and New Brunswick south and west to Texas. The group is not important to the beekeeper, although occasionally some honey is reported from them.

Dogbane.

DOLL CHEESE, see Mallow.

DOMBEYA.

The dombeyas are not well known in the United States as yet, although several species have been introduced into Florida and California, where they are used for park and ornamental planting. The species commonly used are natives of Madagascar and Africa. They are shrubs and small trees. *D. Wallichii* is found growing in middle and south Florida and, according to Frank Stirling, is a very valuable mid-winter honey plant in that region. Coming as it does after the late fall flows and before the early citrus flows it fills a gap and serves to stimulate early brood rearing.

DOORWEED, see Heartsease.

DOUGLAS FIR (Pseudotsuga mucronata).

In the dry belt of British Columbia, and probably in some other limited sections of the northwest the Douglas fir secretes a purely vegetable sugar in large quantity. When this phenomena was first brought to public attention it was thought to be the result of insect exudation similar to that occurring on the Norway spruce. (See Spruce). However, later investigation disclosed the fact that it was in fact of vegetable origin. (Douglas Fir Sugar, by J. Davidson, Canadian Field Naturalist, Vol. XXXIII, No. 1). We quote as follows:

"The sugar appears as white masses in size from ¼ to 1½ to 2 inches in diameter. The smaller, formed like white drops at the tips of the leaves, occasionally two or three leaf tips are embedded in larger drops, while the largest masses are usually scattered irregularly over the leaves and branches.

"The sugar tastes decidedly sweet. * * * * It is completely soluble. When collected it is hard and dry, with no tendency to be sticky. A slight rain is sufficient to dissolve the sugar off the trees and patches of re-crystallized sugar may be found at the base of the trees or on the ground. Frequently however, in this situation, it does not re-crystallize, but is found in a fluid or semi-fluid state, which is attractive to flies and other insects."

Distribution

"The region in which the sugar-bearing Douglas firs are most abundant lies between the 50th and 51st parallels and between 121-122 longitude. This includes the driest and hottest part of the dry belt of British Columbia. Within this area they are rather common and in the Thompson Valley west of the mouth of the Nicola River, also near the junction of the Fraser Rivers at Lytton. They have been found a little above Lilloet in the Fraser Valley, but according to the present information are not known to occur north of Clinton in this region.

"About ten miles north of the apex of the angle formed by the junction of the Thompson and Fraser Rivers, lies Botani Valley, at an altitude of between 3,500 and 4,000 feet. Some years sugar is comparatively abundant on trees in this region; the geology and flora are very different from that of the adjacent Thompson or Fraser Valleys. Here one may find sugar-bearing Douglas firs growing on southern and southwestern slopes having the greatest sun exposure.

"Suitable habitats are found at intervals over a considerable area of the dry belt region. Douglas fir sugar has been reported from around Kamloops and Savona also from the Nicola and Similkameen Valleys, and is said to be found in the southern part of Okanagan Valley. In so far as chief of the Kootenay Indians is aware, it is not known in the Kootenay country, although it is reported by an Indian as being found in the eastern part of Washington State, United States.

Habitats

"The habitats in which sugar-bearing firs are found are usually on gentle slopes facing east or north in that region of the dry belt where the Douglas

Masses of fir sugar exuded by the Douglas fir tree in the dry belt of British Columbia.

fir is encroaching on the dry belt flora. The trees are in comparatively open areas, with abundant exposure to the sun.

"As a rule sugar is not found on trees situated on fully exposed southern or western slopes, nor on areas where Douglas fir forms a dense forest. Southern and western slopes exposed to the full heat of the sun dry out sooner than ground gently sloping to the east or north. The greater abundance of soil moisture in the latter is a point to be kept in mind."

Where bees are kept within reach of the sugar-bearing firs it is gathered in large quantity and stored in the hives. It crystallizes readily, and for this reason is unsafe for winter stores. Lovell mentions a case where a beekeeper in the Olympic National Forest secured 50 pounds of fir sugar honey in a dry season and the following winter lost heavily from dysentery. (American Bee Journal, March, 1921, p. 93).

This sugar does not appear every year, as it is secreted only during extreme hot and dry weather.

Analysis of samples shows that it contains about 50 per cent of pure crystalline melezitose, which is a comparatively rare product.

Honeydew is secured from several of the conifers, but it is usually of insect origin, while the Douglas fir sugar is purely vegetable. (See also pine, spruce, fir, etc.)

DROUTH-WEED, see Turkey Mullein.
DUCK ACORN, see Water Chinquapin.
DUCKBLIND, see Climbing Boneset.
DUST FLOWER, see Chicory.
DUTCH CHEESE, see Mallow.

E

EBONY, see Texan Ebony.
ECHINOPS, see Chapman Honey Plant.
ECHIUM, see Blueweed.

ELDERBERRY (Sambucus canadensis).

The American elder or elderberry is a common shrub from New Brunswick west to Saskatchewan and south to Arizona and Texas. Since the plant blooms late in May and June, there is usually an abundance of pollen in most localities. The bees, however, gather the pollen freely at times, and it is of value where pollen is not plentiful at this season.

The berries are used for pies and wine. The flowers and bark are used to some extent for medicinal purposes.

Richter lists the blue elderberry (*Sambucus glauca*) as important for pollen in California, but as yielding no nectar.

ELM (Ulmus).

The elms are very attractive to the bees for pollen. The American or white elm is more especially valuable, and a large tree will attract so many bees that the humming sounds like a swarm.

There are numerous reports of nectar from elm in Texas. Scholl lists the winged elm (*Ulmus alata*) as giving a good yield of honey, surplus sometimes being secured from it. He described the honey as amber in color, with strong and characteristic aroma. He also lists the American elm as a source of nectar.

It is probable that occasional crops of honeydew honey are secured from elm.

August P. Beilmann reports that at the Arboretum of the Missouri Botanical Garden, *Ulmus Parvifolia,* blooms in September and yields both pollen and much nectar. It blooms for about ten days and during the entire time the bees are in an uproar.—(Am. Bee J'n'l, Jan. 1947).

EPHEDRA.

There are several species of *Ephedra*, common to the dry regions from Texas north to Utah. They are known as joint fir or Brigham tea and one species, *E. antisyphilitica,* is known as Mexican ground pine, in Texas. The latter species blooms in late winter and is important for stimulating early brood rearing in the region south and west of San Antonio.

Ephedra is also reported as important to the bees in the desert regions of Utah. Since these plants are common to Nevada, Arizona and New Mexico, it is probable that they add something to the forage of these states, also. Portions of California, Colorado and Wyoming are also within the range of one or more species of *Ephedra.*

J. W. Powell, of Messilla Park, New Mexico reports that ephedra sometimes yields surplus in that region. The honey, he states is of strong flavor and indifferent quality, being of a greenish color somewhat resembling cylinder oil but of heavy body.

The Spanish names of Popotillo and Canutillo are in common use in New Mexico.

In many desert regions some species of ephedra is the most common plant, but in such places there is seldom a sufficient variety of nectar bearing flora to support beekeeping profitably.

EPILOBIUM, see Fireweed.
ERODIUM, see Pin Clover.
ERYNGIUM, see Blue Thistle.

EUCALYPTUS.

There are about one hundred and fifty different species of eucalyptus trees, most of which are native to Australia and Tasmania, where they are the most characteristic and important timber trees. Many of them secrete resinous gums, hence are called "gum trees." A number of commercial products are derived from them. They have been widely introduced into California, and, to some extent, also into Florida, Texas and other Southern States. The various species are known as sugar gum, blue gum, mahogany gum, red gum, stringy bark, white iron bark, red box tree and various similar names. Richter lists 21 species as yielding honey in California. According to this author there is a great variation in the quality of honey from the different species. While some species seem to yield water-white honey of good quality, others produce an amber product of low value. The blue gum (*Eucalyptus globulus*) is said to produce "honey, amber, of acid flavor, heavy body and granulating within a few months. The blue gum is very constant in nectar secretion, even in spite of unfavorable weather, and since it is of wide distribution, considerable quantities of honey come from this source. On account of the pronounced flavor of eucalyptus honey there is little or no demand for it in the retail trade."

On the other hand, he describes white ironbark (*Eucalyptus leucoxylon*) "a great honey producer, with a beautiful flavor much like vanilla extract."

Blossoms of the blue gum of California (*Eucalyptus globulus*).

The blooming period of the different species varies so that there are some in bloom at all times during the year. The blue gum, already mentioned, blooms from December until June, while the sugar gum blooms from August to November. Several species bloom during the winter months, when they are especially valuable in sustaining the bees until the time of the main honeyflows.

At the California short course, at Davis, in 1918, M. H. Mendleson, of Ventura, spoke as follows concerning this source:

"Every winter the gum, or eucalyptus, comes into bloom. Last winter they filled two or three stories high on eucalyptus. I got 17½ cents per pound for it; previously I never could get more than 4 cents, but it was seldom that I got a surplus from this source. The most eucalyptus honey I have ever known was gathered last spring.

"The scarlet bloom (*Eucalyptus ficifolia*) is one of the greatest honey producers I have ever known. It grows from 10 to 25 feet and has a brilliant bloom, in clusters, a beautiful sight. * * * The honey is of fine flavor, water white. You can see the nectar wave in the flower cup when you shake it. The flow sometimes lasts a month. It will stand a light frost, but not a heavy one. Our hot east winds are very hard on it, as they scorch the trees."

EUPATORIUM, see Boneset.

EUPHORBIA, see Snow-on-the-Mountain, also Poinsettia.

EVERLASTING (Anaphalis margaritacea).

The Pearly Everlasting is a woolly perennial common on dry soils from New Foundland to Alaska and south to northern California and Pennsylvania. It is known by a great variety of names. In Washington it is commonly called straw-flower by the beekeepers. In the east it is known as life-everlasting, moonshine, ladies' tobacco, silver-button, etc.

It is important as a source of honey in some localities on the Pacific Coast. At the Washington beekeepers' convention, attended by the writer in 1919, it was much discussed and some beekeepers present reported an average of 20 pounds or more of honey from this source. It was described as a bitter honey coming at the close of fireweed. The color was reported as amber and the flavor strong as well as bitter. The body was said to be thin. Some beekeepers reported that they had never known the bees to work it.

F

FAIRY CHEESE, see Mallow.

FALSE HELLEBORE (Veratrum californicum) Corn Lily.

The western false hellebore has come to public attention through discovery that the nectar is apparently poisonous to bees. It is a characteristic plant of wet meadows in the Sierra Nevada mountains at from 4500 to 8500 feet. It has also been brought into cultivation to some extent in gardens. The stems from three to six feet in height are somewhat suggestive of a cornstalk, hence the name, corn lily.

False hellebore is not commonly regarded as important as a honey plant but G. H. Vansell and W. G. Watkins have found numerous bees dead or dying on or below the blossoms. (Journal of Economic Entomology Feb. 1933.)

FALSE INDIGO (Amorpha).

There are several species of amorpha common to America, but the one which is probably of most importance to the beekeeper is the shrub commonly called false indigo or river locust (*Amorpha fruticosa*). It is also known as bastard indigo in some localities. It grows most commonly in damp, shady bottom lands and on the banks of streams. It is occasionally found in upland woodland borders where the soil is deep and rich.

Blossoms and leaves of false indigo.

It is from 5 to 8 feet in height under most conditions, but occasionally reaches a height of 15 to 18 feet. It is widely distributed, being found from New England, where it is rare, west to Minnesota and Saskatchewan and south to Florida and Mexico. In Colorado it is reported at altitudes of 4,000 feet in Logan County and in the river flats east of Ft. Collins. In Texas it is found on the river banks apparently throughout the state. In Alabama and Georgia it is common as well as in the middle west. In the southern portion of the range, the flowers appear in April and May, while in the northern regions it blooms as late as July. In Nebraska and Kansas, where it is of greatest importance to the bees, the blooming season is early June or late May. The flowers are deep blue or purple and are borne in long spike-like racemes. In the Arkansas Valley beekeepers report that it yields both nectar and pollen in abundance. Nebraska values it since it fills in the gap between fruit bloom and white clover.

The lead plant or shoestring (*Amorpha canescens*), also known as wild tea is a bushy shrub 1 to 3 feet high which is very common on the plains from Manitoba to Texas and New Mexico. The flowers are very similar to the false indigo and the blooming period is in mid-summer. The name, lead plant, comes from its color. There is a common saying among farmers in Nebraska that where the shoestring is found alfalfa will succeed. Although beekeepers report that the bees work on the shoestring or lead plant, it apparently is not of great value, even on the prairies where it is common.

In addition to the above species there is the dwarf false indigo (*Amorpha nana*), which is found from Manitoba to Iowa, Nebraska, Colorado and New Mexico. This little shrub, found on the open prairie, is seldom more than one foot in height. The smooth amorpha (*Amorpha glabra*) is found along the coast from North Carolina to Florida.

There is one representative of the group on the Pacific Coast, the California false indigo (*Amorpha californica*). This species is found in southern California, Arizona, New Mexico and also in Mexico.

Although, as will be seen from the above descriptions the group is widely distributed, the writer has not been able to find any localities outside the states of Nebraska and Kansas where they are of special importance to the beekeeper.

Recent research at the Oklahoma Experiment Station indicates the possibility that Amorpha fruticosa which they know as "Cat Willow" may come into use as a source of an insecticide to substitute for rotonone. (Chemurgic Digest Nov. 15, 1944).

FALSE LOOSESTRIFE, see Ludwigia.

FARKLE-BERRY or SPARKLE-BERRY (Vaccinium arboreum).

The farkle-berry or sparkle-berry is also known as winter huckleberry and tree huckleberry. It is a tall shrub 6 to 25 feet in height and common from southern Illinois and Missouri east to North Carolina and south to Texas and Florida. In parts of Arkansas it is one of the main sources of honey. W. C. Brass writes that in Lonoke County, near the center of the State, it is very abundant and takes the place of clover further north. It blooms in May and the bees roar on the sparkle-berry bushes like swarming time. It belongs to the group of plants to which the blueberries belong and most of these are good honey plants.

H. B. Parks states that tree huckleberry occurs in large thickets in east Texas and is the source of good surplus in April and May. He describes the honey as light amber and of good quality.

In Gleanings for August, 1934, he comments on tree huckleberry as follows:—

> "The tree huckleberry made a wonderful bloom this spring and in many places where this honey is almost unknown local beekeepers have small stores of almost pure huckleberry honey. Because of the few beekeepers in the region where this tree grows and the long periods between the heavy blooms, this honey is prominent where honey yields are reported. Some of the old timers in East Texas say that such yields come about once in fifteen years."

FEVER-BUSH, see Holly.
FIDDLE NECK, see Phacelia.

FIGWORT (Scrophularia marilandica).

Simpson honey plant, or figwort, is a widely distributed plant. It is common in the woods from Maine to the Rocky Mountains and south to the Gulf. It is also said to occur on the Pacific Coast. The same, or a similar plant occurs in Europe and Asia.

Figwort, or Simpson honey plant.

It is a tall growing plant from 3 to 6 feet high, with numerous small branches. The stem is four angled, with rather long-pointed leaves. The flowers are very numerous and quite small. It blooms in the late summer and is freely visited by the bees.

In the eighties when there was much interest in the subject of planting crops for the bees, figwort was much discussed in the bee magazines and seed was widely distributed under name of Simpson's Honey Plant. In Gleanings for November, 1882, Dr. C. C. Miller gave an extended account of his experience in trying to establish a plot of this plant.

From Gleanings for Feb. 1882, the following note by H. A. March is taken:

> "There were thousands of blossoms and there appeared to be two or three bees to each blossom, pushing and crowding for the nectar, and such a humming and roaring one would think they were on a first class robbing expedition. I call the Simpson's a decided success; nine weeks of steady work for the bees, when nearly all other bloom is dried up. I am satisfied

that an acre of plants as thrifty as those in my garden would produce from 400 to 500 pounds of honey."

James A. Simpson for whom the plant was named by the beekeepers described the honey in Gleanings, January 1878 as follows:—

"The honey is of excellent quality, as clear as clover, making a beautiful comb, but lacking any distinct aroma as linden or buckwheat."

In the same issue the editor mentions having received a sample of the honey from Simpson and states that he would call it clover from looks and taste, unless it be that it lacks the mild flavor of well ripened clover honey. This raises the question as to whether Simpson may not have been mistaken in the source and whether it did not in fact come from white clover. Honey samples from the minor sources are always somewhat in doubt.

Scullen and Vansell list *S oregana* and *S lanceolata* from Oregon. They report nectar concentration as only about half that of fireweed under similar conditions, ranging from slightly more than 18 to 32 per cent. (Bul. 412 Oregon Ex. Sta.)

FILAREE, see Pin Clover.

FILBERT (Corylus avellana).

Filberts are grown commercially in Oregon. They are reported as an abundant source of pollen and also as occasionally yielding honeydew in that state.

FIR (Abies).

Writing in the American Bee Journal J. A. Herbele (Nov. 1916, p. 381) states that in Switzerland about 40 per cent of the honey crop is from honeydew, mostly from *Pinus abies,* a fir tree. From this fir tree the beekeepers of the Vosges Mountains, the Black Forest and in parts of Switzerland, harvest large crops of honeydew, called "waldhonig." Notwithstanding its greenish-black color it is much esteemed by the population. Herbele expresses the opinion that this waldhonig is of plant origin. Since the Douglas fir secretes sugar freely in the Pacific Northwest, this seems very probable. He mentions a record of the production of 385 pounds of honey by one colony in 1900 and an average production of 90 pounds per colony in the apiary of a German beekeeper in 1915. See also Douglas Fir.

Scullen and Vansell report white fir, (*Abies concolor*) as the source of honeydew in California and probably in Oregon. (Bul. 412 Oregon Ex. Sta.)

FIREWEED or WILLOW HERB (Epilobium angustifolium).

Fireweed is a common plant in the woodlands of the Northern States and of Canada. It is a tall herb with attractive pink blossoms on a long terminal spike. It springs up following forest fires and covers the burned district with a dense growth. The blooming period is long, lasting from July till frost, as new blossoms appear as the older ones fade. It is important as a source of honey in much of eastern Canada, Minnesota, parts of Michigan, Wisconsin and on the Pacific Coast where it is also valuable in British Columbia, Oregon, Washington and parts of Montana. It is gradually crowded out by other growth. A locality may yield great crops of fireweed honey for two or three years and then little surplus be gathered from it for many years. The author has visited beekeepers in Northern Michigan who count on an average of fifty pounds or more per colony, with as high as 125 pounds, in locations where fireweed was yielding. As fireweed disappears in that

Fireweed in north woods.

locality, raspberry and milkweed follow, and these are also good sources, so that the location does not suffer from the change.

Honey from fireweed is very light in color and of high quality. The late W. Z. Hutchinson, who wrote much concerning beekeeping in the forest region of northern Michigan, styled it as the whitest and finest-flavored honey with which he was acquainted.

As the timber is removed, settlement gradually clears the land, and the wild growth gives place to cultivated fields and pastures.

In an extended article on production of fireweed honey in Oregon, (American Bee Journal Oct. 1935), W. L. Arant writes:—

"Reports of phenomenal yields, however true, are often misleading because one gets the idea that they are dependable or even predictable. We have seen great seasonable variation in the honeyflows of the past ten seasons—total failures, mediocre flows, good ones, and only one real "flood." Altogether the yearly average for the ten-year period is by no means phenomenal, nor perhaps even equal to that of many other nectar-bearing plants."

"The greatest gains are made during the hottest weather when the air is clear, humid and quiet."

Stephen Harmeling, in first annual report of the Division of Apiculture of Washington, states that fireweed is the most valuable wild honey plant. There it blooms from June till August and yields throughout the blooming season, unless the season is too dry. He states that there are millions of acres in the Olympic and Cascade foothills, where the plants are so abundant that they literally color the landscape with a curtain of purple. Thousands of tons of honey annually go to waste in this region for lack of bees to gather it. See also amsinckia.

FIREWEED (Erechtites).

This genus is composed of coarse annuals with rank odor. Two species, New Zealand fireweed, (*Erechtites arguta*) and Australian fireweed, (*E. prenanthoides*) have been introduced from abroad to California.

One native species *E. hieracifolia,* is found in moist woods and recently burned clearings in the northeastern states. The many flowered heads are whitish in color and appear in summer from July until September.

C. W. Wood writing in American Bee Journal, May 1939, mentions this plant as under discussion at the Michigan College of Agriculture short course, as a splendid source of nectar. He states also that although not recognized as a honey plant in the literature it has a good reputation in some quarters. Plants which he had under observation were worked steadily by the bees from July until September leading him to believe that it might be a source of much surplus where present in sufficient quantity.

FLAT TOP. See Wild Buckwheat.

FLAX (Linum usitatissimum).

Flax rarely attracts bees as far as this author has been able to observe but Scullen and Vansell report that it is freely visited by bees for both nectar and pollen in the Williamette Valley in Oregon. (Bul. 412 Oregon Ex. Sta.)

Ed Braun of the experimental farm at Brandon, Manitoba, reported finding the bees evenly distributed over a field of 100 acres of flax at a time when sweet clover was in full bloom near at hand. No pollen was visible.

A. F. Gubin in Bee World, (p. 30-31, 1945), reports flax in respect to honey yield as ranking lower than any other nectar yielding plant. Cross pollination by bees increased the yield in both quantity and weight of seed. Due to its feeble attraction for the bees it is necessary to have many hives near the plot to secure satisfactory pollination.

FLAX-LILY, see New Zealand Flax.
FLAX, NEW ZEALAND, see New Zealand Flax.
FLEA WEED, see Blue Curls.

FLEECE VINE (Polygonun Auberti) Silver Lace Vine.

The fleece vine is a twining perennial to twenty feet which comes from China and Tibet. It flowers profusely and blooms for several weeks in late summer.

Bevan L. Hugh of White Rock, British Columbia, reports that with him it blooms from spring until frost in October and that the bees work the flowers so eagerly that at times it sounds like there might be a swarm in the air.

FLORIDA—Honey Sources of.

Florida has a very different flora and quite different climatic conditions than any other state. Semi-tropical in its nature, it has long been

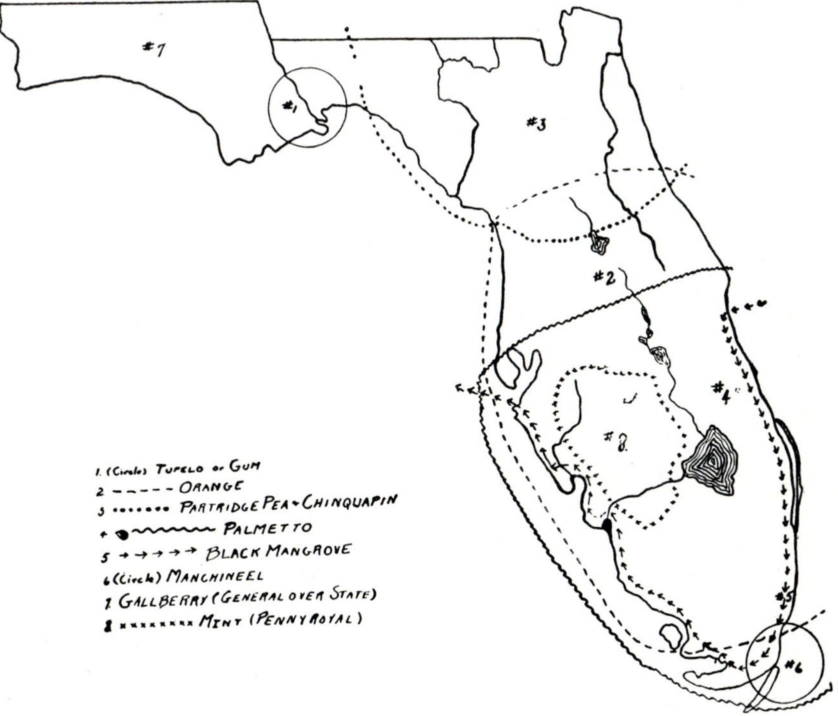

Map showing range of the principal sources of surplus honey in Florida.

famous as a winter playground. There is a great diversity of nectar sources in different sections of the state and some species are important only in limited areas. In the American Bee Journal of April, 1922, E. G. Baldwin outlined the principal regions of Florida as he has observed them, using the map which is reproduced herewith. The numbered areas show the approximate range of the principal sources mentioned.

In the north part of the state gallberry, tupelo, partridge pea and chinquapin are the important sources. Conditions there are very similar to southern Georgia and Alabama. Gallberry he mentions as the only source of nectar which is important over the entire state.

In Southern Florida the saw and cabbage palmetto yield important surplus. Orange, black mangrove, manchineel and mint yield large quantities of honey. The map shows at a glance where each is to be expected.

Wild pennyroyal which is found abundantly in the south central section is a winter bloomer and yields some surplus in favorable seasons. It is also valuable to insure early breeding to prepare the bees for the orange flow which comes somewhat later. The purple flowered mint is found in the region cocupied by the wild pennyroyal and also occurs farther north. It is a summer bloomer while the pennyroyal as already stated, blooms in winter.

Taking the state as a whole, the important sources of nectar may be listed as follows:

Tupelo (*Nyssa*), four species.

Orange, grape-fruit, partridge-pea, chinquapin, cabbage and saw palmetto, black mangrove, gallberry manchineel, wild pennyroyal, purple-flowered mint and ti-ti.

FLORIDA MAHOGANY, see Red Bay.
FLOWERING RASPBERRY, see Salmon Berry.
FOG FRUIT, see Carpet grass.

FOXGLOVE, (Digitalis purpurea).

The foxglove is a tall herb with alternate leaves and showy flowers well known in old fashioned gardens. It has escaped and become naturalized in meadows and pastures in a few eastern localities and also in the Pacific Coast from California to British Columbia. Introduced from Europe.

J. W. Winson writes that four British Columbia beekeepers insist that foxglove should be included among the honey plants of that province. He states that in a few isolated districts there is enough of the plant to make it a factor in the honey harvest. It requires something of a struggle on the part of the bee to force her way into the narrow tube far enough to secure the nectar.

Orvel R. Bassett of Raymond, Washington, reported a heavy

honeyflow of about five weeks, beginning about June 20 from this source. The honey was of very good flavor and of golden yellow color. Drops of nectar could be seen in the blossoms and a bee could secure a full load from one flower. An average of four added stories were filled from this flow.

It is from the dried leaves of Digitalis purpurea, collected from plants of the second year's growth that the digitalis used in medicine is secured.

FRAGRANT LAUREL, see Snow Brush.

FRASERA, see Monument Plant.

FRENCH MULBERRY (Callicarpa americana).

French mulberry is a common shrub in the Southern States. It is found on dry, sandy soil from Virginia to Arkansas and south to Florida and Texas. The leaves are opposite and the flowers in axillary cymes. The pink flowers appear in June and July.

It is reported as very attractive to the bees, though probably not common enough in many places to be the source of surplus honey.

FRENCHWEED, see Galinsoga.

FRUIT BLOOM.

All the orchard fruits are of more or less value to the beekeeper, and few differentiate between them when speaking of the sources of early nectar. In localities where a great variety of fruit is grown, the blooming period is longer than where there are large orchards of one kind. In most localities the beekeepers depend upon fruit blossoms for building up their colonies in spring, and when the weather is unfavorable at this time, feeding is often necessary to carry the bees until the next honeyflow. Apples, peaches, plums, apricots and cherries all secrete nectar abundantly and, given strong colonies of bees and favorable weather, large quantities of surplus honey should be gathered from these trees. If it were possible to bring the bees through the winter as strong as in the fall, there is no estimating

Leaves and fruit of French mulberry.

the amount of honey that would be gathered from the early blooming orchard fruits.

FURZE, see Gorse.

FUCHSIA.

Fuchsia is known in our northern states as a tender house plant but is grown outside to some extent in the milder sections. There are several species of shrubs or small trees native to the tropical and subtropical America, and a few to New Zealand. The following is quoted from British Bee Journal, Jan. 23, 1941:—

> "I received a completed section from a beekeeper friend of mine living in the wilds of Connemara, which she said was pure fuchsia honey. Fuchsia is in flower from August until Christmas and is a heavy yielder. In that part fuchsia hedges take the place of the hawthorn hedges of this country.
>
> "Beekeepers who tasted my section which was water white suggested that my friend had been feeding sugar. It was flavorless. The apiary being within 100 yards of the open Atlantic it is rare to get more than one degree of frost in any winter."

Writing to the author from New Zealand, P. J. St. John states that the bush fuchsia, known as "Konini" is a tree growing about 12 feet high, which can be relied upon to yield some nectar every spring. Occasionally it yields heavily. The honey, he states, is light amber in color, with mild and distinctive flavor, granulating quickly, at times before the combs are capped. In 1940, after a series of light crops, strong colonies stored as high as eight supers each from this source. When gathering pollen from fuchsia the bees have the appearance of carrying two small blue bags.

G

GAILLARDIA, see Marigold.

GALINSOGA or FRENCHWEED (Galinsoga parviflora).

Galinsoga is native to Peru and the warmer portions of South America and has become widely distributed as a weed over the continent from Maine to Ontario and westward to Oregon and southward to Alabama and Mexico. It is common to dooryards, roadsides and waste places.

The blooming period is from June to November and it is probably unimportant to the beekeeper, although there are a few reports of bees working it freely. Mr. George Gilbert, of Port Dickinson New York, in a letter to the author, writes as follows:

> "All summer long the bees do not notice this weed. It blooms from June until the last thing in the fall. In late September or early October the weed suddenly gives off an elusive odor, rather spicy, like alspice or the smell of sweet preserves being cooked. This in very dry weather. Then the bees

swarm on it for a few days. In the first week of October, 1922, our bees were on it as on clover in June, a regular cloud of workers humming cheerfully and getting a very little yellow pollen and nectar to each flower. They must get little, as they are on each flower so short a time. This weed could not rank as an important nectar source, but coming after almost everything else is gone it must give the bees a good boost toward winter stores."

GALLBERRY (Ilex glabra). See also Holly.

The gallberry, in some localities better known as inkberry, is usually heard of as a honey plant only in the South. However, it occurs as far north as Nova Scotia, on the seashore, and along the coast from Massachusetts to Virginia and Florida, and west to Louisiana. It is common in the low pine barrens of the Gulf States. It is a small evergreen shrub with small, dark leaves. It is an important honey plant in southern Georgia, where it is widely distributed over the sandy lands, especially of the coastal plains. It is important, also, in the Carolinas. It grows in dense thicket and rapidly extends over newly cleared lands.

"As a honey plant perhaps it has no equal in the southeast. We have never failed to get a surplus from it, even during the most unfavorable weather conditions. It begins to bloom the first of May and continues for 24 to 28 days. During this time bees disregard other bloom, working it up to about 8 o'clock for the pollen; then the flow comes on for the remainder of the day. * * * It is a great bloomer; even the stems are rolls of bloom. * * * We have never taken off a large crop of this honey, as 147 pounds of surplus is the best crop we have ever had from one colony. The honey is a light amber color, has a heavy body, a very mild taste, and is highly flavored. The demand for this honey is so great that we cannot furnish our local markets, consequently very little is shipped from the southeast to other markets.

"We have raised tons of this honey and have never seen a pound of the pure article, well ripened, that granulated.

"It has been said that it is impossible to overstock a good gallberry location. We do not know that this statement is true, but we have never heard of one being overstocked. We have had bees in a location where there were 362 colonies, with the same result as with 100 colonies. Good gallberry locations are nearly numberless and large quantities of this fine honey are wasted every year in localities where there is not a bee to gather it. The gallberry should be included in the list of the best honey plants."—J. J. Wilder, Cordele, Ga. Gleanings, page 1200, September, 1907.

GAURA.

There are several species of gaura of wide distribution. Although frequently mentioned as honey plants, they are seldom of sufficient abundance to be important.

Scholl reports that *Gaura filiformis* occasionally yields surplus in Texas, when conditions are favorable. The red gaura is reported as of value in Colorado.

The red gaura, also called ragged lady, is common from Montana to Arizona and Texas. It is much sought by the bees for both nectar and pollen. Its nectar secretion is abundant, but it is not sufficiently common to be important. A good stimulant.

According to H. B. Parks *Gaura rosea* is common over Texas, blooming in early spring. In some localities it is important as a source of early stimulation.

Concerning *Gaura coccinea* George H. Eversole of LaPlata, New Mexico wrote, (American Bee Journal Aug. 6, 1891):

> "It has proved to be by far our most productive honey plant—more so than the cleome, of which we have an abundance, or, considering the amount of it, better even than our alfalfa. The bees work on it about the first thing in early summer, and keep working on it every day until the last thing in the fall."

GAY FEATHER, see Blazing star.
GELSEMIUM, see Yellow Jessamine.

GEORGIA—Honey flora of.

Georgia is the largest state east of the Mississippi River and contains an area of more than 59,000 square miles. Because of its southern situation the summer is long and hot. The growing season averages from 190 days in the extreme north portion to 250 days in the southern section. The first killing frost in autumn is usually about October 27 in the north and November 15 in the south portion.

The rainfall is heavy. In the northeastern counties it averages about 70 inches annually, diminishing somewhat toward the interior where it averages from 47 to 53 inches. The periods of heaviest precipitation are July and August and February and March. The driest months are April and May and October and November, during which time there are the greatest number of clear and sunshiny days.

The altitude varies from near sea level in the Gulf Coast region and along the Atlantic seaboard to 1000 to 5000 feet above sea level in the mountainous regions of the northern section.

Soil and climate are adapted to the growing of a large variety of crops. Cotton, corn, peanuts and cereals are largely grown. Peaches are grown on an extensive scale in the Fort Valley region. Figs and grapes grow well over most of the state, and oranges and bananas in the extreme southeast section.

The important honey plants somewhat in order of blooming were given by Dr. J. H. P. Brown of Augusta, in the American Bee Journal, (Page 500, 1880) as follows:

The earliest blooming of our spring forage plants is the alder (*alnus*), which commences about the middle of January and lasts, some seasons, till the middle of February. It yields little or no honey, but during its time of bloom, its pollen-laden catkins are covered with bees. The amount of pollen that this plant affords is immense; and it comes in a time when breeding should be most encouraged.

In some sections of the South, particularly on light, sandy soils, there may be found some yellow jasmine (*Gelsemium sempervirens*). As its flowers possess very decided toxical properties, it is not a very desirable plant to have within range of one's bees. It blooms after the alder. While our native black bees are very seldom seen working upon it, the Italians, in some seasons, will work upon it quite briskly. I am inclined to think, from close observation, that it is mostly pollen they gather from it, though in some seasons it does yield some honey.

The wild plum (in some sections known as the hog plum) usually commences to bloom the last of February and lasts for two or three weeks. This is peculiarly a southern tree, and grows to great perfection nearly everywhere. Whole acres are often covered with it, forming a dense thicket, thus affording the bees rich pasture.

In March we have the peach, the apple (which continues into April), the mock orange, or evergreen wild cherry (*Cerasus carolinaensis*), the huckleberry, strawberry, and a few other plants of minor consideration. Further south they have the titi, the saw palmetto and the orange, all good forage plants.

The willow, wild cherry, hawthorn, blackberries, raspberries, locust, holly and tulip tree (*Liriodendron tulipifera*) bloom in April. The two latter are most valuable for honey. The holly blooms for about two weeks—the height of its flowering is about the first week in May. The tulip tree blooms for three weeks. This is the poplar tree of the south.

In May we have the black gum (*Nyssa multiflora*) and the persimmon, both excellent for forage. The blooms of these trees are dioecious, that is, the male flower is found on one plant and the female flower on another. Bees are very seldom seen working on the female tree, while on the male bloom they work in a continuous swarm.

In May, also blooms the bay (*Magnolia glauca*). This tree flowers for at least one month, and extends into June. It affords some of our best and most abundant forage. The *Magnolia grandiflora,* linden and China berry (*Melia azedarach*) bloom also in May. The magnolia blooms for six weeks, the linden from six to ten days, and the China tree for two weeks.

Sourwood, the varnish tree, Japan privet (*Ligustrum*) and a few other plants of less note embrace the principal forage in June.

I have now enumerated the chief honey-producing plants that go to make up our spring honey harvest. Take one season with another, our bees commence to lay up surplus about the last of April and continue until the first or middle of June. After this date but little honey is gathered from the holly, persimmon, black gum, bay and sourwood. Of course, some seasons there is considerable honey gathered from other sources. The color of the honey is usually a little dark, but of excellent flavor.

There is comparatively little forage during the summer months of July and August. The button bush (*Cephalanthus occidentalis*), sumach, *Asclepias tuberosa* (known as pleurisy root and butterfly weed), and *Yucca alnifolia* (Spanish bayonet), are the most important. The cotton plant, which generally commences to bloom about the first of

July, yields largely of pollen, but very little honey. Sumach is a rich mellifluous plant, but the warm, dry atmosphere evaporates the secretion very rapidly, so that the bees can only work on it very early in the morning, while the dew is on. The Spanish bayonet plant no doubt furnishes some nectar. It generally swarms with flies, various sorts of wild bees, and now and then a few honeybees will visit it.

Bees are generally able to gather sufficient stores during July and August to keep up brood rearing and the strength of the colony until the blooming of the autumn forage.

The first to bloom of the fall pasturage is the *Chrysopsis graminifolia* of Nuttall, a perennial, composite. This plant is often taken for a species of dog fennel, but it is altogether distinct. It is indigenous to the south from Florida to North Carolina, which seems to be its northern limit. It is a yellow flowering weed that commences to bloom in August and keeps on till frost (Golden aster).

The goldenrod and the asters bloom till killed by frost. I esteem both these plants very highly for their honey-producing qualities. In some seasons I have hives filled with aster honey alone.

Germander, or wood sage.

GERANIUM.

The wild geranium, or cranesbill, is to be found in numerous species over a wide range of territory. The author has found it regarded as of some importance by beekeepers in the higher elevations of the Rocky Mountain region where the bees work it freely in mid-summer.

Although not often found in sufficient abundance to be important, some species of wild geranium is usually to be found in open woodlands and moist shady places from Canada to the Gulf of Mexico and from Nova Scotia to California.

GERMANDER or WOOD SAGE (Teucrium canadense).

The germander, also known as wood sage, is found in open woodlands and thickets from Nova Scotia to Nebraska, and south to Florida and Texas. It is common in the Central States, and is much sought by the bees.

The blooming period is long. In 1915 the bees were working on this plant in the writer's garden for nearly two months. Apparently this plant does not secrete nectar very freely, yet it is an excellent plant to keep the bees at work when they might otherwise be robbing. The writer does not recall ever having seen a locality where it was sufficiently abundant to amount to very much by itself, though it is a valuable addition to the other honey-producing flora.

Another species is *Teucrium lacinatum,* a desert mint common to the plains of southern Texas and westward. It is also found as far north as Kansas and Colorado. This is a low and spreading form recognized as of some value to the bees in Arizona. H. E. Weisner, of Tucson, states that the bees work it freely at times and that it is of principal value to stimulate brood rearing when little else is available. He regards it as rich in nectar from the eagerness with which the bees work it, and thinks that if it were sufficiently abundant it might even be the source of surplus. In the vicinity of Tucson, it is reported as blooming from April 15 to June 1st or longer.

Giant cactus or Sahuaro.

GIANT CACTUS

(Cereus giganteus) Sahuaro.

The giant cactus is confined to a rather limited range in Arizona and adjacent California and southward into Mexico. It is commonly known as sahuaro (also spelled saguaro) among the people of the region where it occurs.

The sahuaro is one of the most interesting and unusual plants to be found in America. It grows very slowly and lives to a considerable age, estimated at 200 years or more. It is supposed to require thirty years to reach a height of three feet and to be 150 years in attaining a height of thirty to thirty-five feet. It grows in a very dry region and the accordion like pleats expand to permit absorption of surplus water during a wet period and contract in like manner as the water is evaporated during a dry time.

In the region about Tucson it is the most characteristic plant and is present in abundance. The author found considerable difference of opinion among Arizona beekeepers as to the quantity of honey obtained from this source, but nearly all consulted reported that it is the source of surplus honey. A few reported heavy yields from it some years, although the honey is said to be mixed with mesquite as a usual thing. The honey is said to be of fine quality with little tendency to granulation. The body slightly heavier than mesquite, flavor mild, and less acid than the mesquite honey with which it is often mixed. The juice from the fruit is sometimes stored by the bees as honey.

GIANT HYSSOP, see Hyssop.

GILIA.

Jepson lists 18 species of gilia as common to California. The late Prof. A. J. Cook, writing of gilia in the American Bee Journal stated that all the gilias are good honey plants. He mentioned *Gilia capitata* as the source of quantities of sky-blue pollen. He mentions the fact that several species yield pollen of a blue color. It is probably a source of minor importance.

GILL-OVER-THE-GROUND (Nepeta glechoma) Ground Ivy.

Ground ivy, or gill, is a common creeping weed which was introduced from Europe and has become widely distributed about dwellings. The small blue flowers are attractive to the bees and continue blooming throughout spring and summer. Although often mentioned as a bee plant, it is not, apparently, as valuable as its near relative, the well-known catnip. In early editions of A. B. C. of Bee Culture, A. I. Root described it at length. The following quotation covers the important part of his description:

> "It transpired in later years that this plant yielded a great deal of honey, and in some localities favorable to its growth, such as the beds of streams where there is plenty of rich vegetable mould, it has furnished so much honey that it has been extracted in considerable quantity. * * * * The honey is rather dark and I believe a little strong, but if it is allowed to become perfectly ripened I think it will pass very well. Perhaps the greatest benefit to be derived from it, however, will be to bees uninterruptedly rearing brood until clover and locusts begin to furnish a supply."

GLADIOLUS.

Among the garden flowers few are more popular than the gladiolus. In a few places where large commercial plantations occur, it may be important to the bees, since it yields nectar freely.

Dr. O. W. Park writing in American Bee Journal, December, 1928, says:

> "The nectar seldom contains more than thirty per cent sugar, but is yielding in great abundance. The corolla tube is commonly filled with nectar to such a depth that, in spite of the fact that her tongue reach enables her to obtain only a fraction of the total amount present, the honeybee often obtains her load from a very few flowers.

Most of the bees gather nectar only, while a few gather only the pollen, which is of a beautiful orchid color; occasionally a bee gathers some of both.

As a rule, nectars which are secreted abundantly are not as rich in sugar as those secreted less abundantly. Gladiolus nectar is a typical example in that its yield is among the very highest and its sugar content among the lowest studied so far. It is sufficiently sweet, however, to induce the bees to gather it readily."

GLOBE THISTLE, see Chapman Honey Plant.
GOATWEED, see Croton.

GOLDEN ASTER (Chrysopsis).

There are more than twenty species of golden asters common to the southeastern states. They are mostly reported on sandy soil and bloom in summer and fall.

C. graminifolia is reported as the first of the fall bee pasturage from Georgia.

From Glennville, Georgia, W. C. Barnard writes that golden aster blooms in September and October and that the bees bring in large quantities of pollen and some nectar during its flowering time. The honey is yellow and of only fair quality but is usually mixed with Mexican clover which improves the product.

Chrysopsis villosa in our test gardens has failed to attract the bees at any time although under observation for several seasons.

GOLDEN GLOW, see Coneflower.

GOLDEN HONEY PLANT (Actinomeris alternifolia).

The golden honey plant is common from New York to Michigan and Nebraska, and southward to the Gulf States. This plant is closely related to the crownbeard and is sometimes classified as a verbesina. It grows in rich bottom lands and in the borders of woods and fields, reaching a height of 4 to 8 feet. The yellow flowers are very attractive to the bees and, where sufficiently common, are a valuable source of nectar.

GOLDENROD (Solidago).

Of the eighty species of goldenrod all but three or four belong to North America. It is one of our most widely distributed native plants. Some species seem adapted to nearly every condition from Canada to Mexico and from the Atlantic Coast to California. There is a wide difference, however, in the value of the different species to the beekeeper, and it is no easy task to get reliable information regarding the range of conditions under which it secretes nectar abundantly, nor is there much recorded information concerning the particular species which are most valuable for this purpose. It is a well-known fact that the secretion of nectar with any plant is greatly influenced by soil and climatic conditions. Some of our most valuable honey plants have been reported as producing no nectar when introduced into Australia.

It is very probable that when we have studied the matter carefully we will find that the same species of goldenrod varies as much in its nectar secretion under different conditions as we know to be the case with alfalfa.

Bushy goldenrod (*Solidago graminifolia*) is a fine honey plant in New England.

Lovell is of the opinion that all species of goldenrods secrete nectar in some localities. This is quite probable, although there is very little honey from goldenrod in Iowa. Along the upper Mississippi, in the northeastern counties, a few beekeepers report honey from goldenrod. In other sections of Iowa, beekeepers report that they have never seen a bee on the plant. Dr. L. H. Pammel, botanist at the State Experiment Station, reports nine species of goldenrods common to that State.

He lists *S. serotina, S. canadensis* and *S. graminifolia* as furnishing some honey here.

Richter lists only two species of this plant as important in California. The western goldenrod (*S. occidentalis*), he mentions as common in wet places such as marshes and river banks, from August to

Tall, hairy goldenrod (*Solidago rugosa*), one of the best for honey.

October, yielding an amber honey. *S. californica,* the common goldenrod of the coast, he describes as common on dry plains and hillsides or mountains throughout the State, from August to December. He lists it as a fair honey plant.

Scholl reports goldenrods as common to all parts of Texas, and states that the honey yield is good in favorable seasons when it is not too dry. He reports a long season, from April to November, but gives no list of the species furnishing nectar in that region.

Sladen reports finding eleven species of goldenrods about Ottawa. He finds that individually the *canadensis* group produce comparatively little nectar, but their great abundance makes them important collectively.

Sladen also notes the variation of the plant under different conditions and says that the nature of the land determines the presence and abundance of the best species. He reports that in the wet lands of Charlotte County, N. B., especially in the Honeydale district, they, together with asters, furnish the principal source of nectar, and that they are valuable generally as a source of surplus in coastal districts of New Brunswick and Nova Scotia. The same is said of eastern Manitoba. He places the yield at from 50 to 80 pounds per colony in localities where the best species of goldenrod and asters abound. The honey is usually of good quality, ranging in color in the different districts from white to dark amber. That gathered in swampy districts is usually bright golden. Evidently goldenrod honey is seldom stored separate from aster in localities from which these reports are made.

Mr. Sladen describes three types of locations in which the plants may be found in Canada:

1. Open swamp or bog, where *S. uliginosa* and *S. rugosa* are found. The former begins blooming in August, while the latter blooms until mid-September, so that there is more than a month of flow from these plants. Although the bogs are independent of rain during the honey flow, fine weather and moderate warmth are necessary to a crop.

2. Sandy or gravelly barrens or plains. On the coast, as well as inland, on such lands are found *S. puberula,* while inland are to be found in addition, *S. squarrosa* and the less important *S. hispida.* Good rains in early August, followed by fine and warm weather, bring best results.

3. A restricted area centering in Cumberland County, Nova Scotia, in which *S. Graminifolia* is a troublesome weed.

He further reports that the roadside goldenrods of old Ontario are not heavy producers of honey under ordinary conditions.

In an article on the Honey Flora of New England, which appeared in the April, 1916, American Bee Journal, Lovell states as follows:

> "If I were compelled to stake the existence of bee culture in New England on a single genus of plants I should select the goldenrods. There are many species, and they all yield nectar and pollen. They begin to bloom in midsummer and continue to bloom in October. They are very common and there are species adapted to the seashore, the fields, the rocks and the woods. I have never known the flow of nectar to fail, and a great quantity of heavy, yellow honey is stored annually."

Mr. Lovell has kindly sent me his field notes on these plants.

The notes include the study of six species, all of which produce some nectar in Maine. He describes the tall, hairy goldenrod *S. rugosa* as the latest to blossom and the most valuable as a honey plant. It is found in damp thickets and on moist land. While in bloom the bees work it very diligently and the honey is stored rapidly. The apiary is filled with a sour odor, which, in the evening, is noticeable at a distance.

Goldenrod honey, according to him, is deep golden yellow in color, thick and heavy, with a more decided flavor than white clover honey. When extracted it granulates in a month or two, but the bees winter on it perfectly.

Next in importance he places the bushy goldenrod (*S. graminifolia*). This is common in fields, open woodlands and hedgerows. The odor is faint, but the nectar is clearly visible in the flowers. He reports as many as six honeybees at work at one time on a single flower cluster. It will be noted that this is one of the species which Doctor Pammel

Common or Canada goldenrod (*Solidago canadensis*).

mentions as yielding nectar in Iowa. Sladen also cites it as important in Canada. Graenicher collected 135 different species of insects on this species in Wisconsin.

The cream-colored goldenrod, sometimes called white goldenrod (*S. bicolor*), is of special interest because of the fact that it is the only one of the group which is not yellow in color. Although it produces nectar, I can find no record which indicates that it is of much importance as a honey plant anywhere. Lovell says that it is of little value in Maine.

The early goldenrod (*S. juncea*), is the first to bloom in Maine and is very abundant in old fields. The bees visit it freely, but apparently do not get much honey from it. Graenicher states that he has collected 182 different species of insects on this plant, in Wisconsin, which indicates the presence of considerable nectar in that locality.

Early goldenrod (*Solidago juncea*).

In Illinois and Iowa, the author has seldom found the bees working on goldenrod. Although there are several species growing abundantly along the Mississippi River, some years of observation have failed to indicate any honey secured from this source here.

Edward Kretchmer who was an Iowa pioneer living in Montgomery County reported to the Iowa Beekeepers Association in 1914 that in the 1870's he depended mostly on wild flowers for his crop and that he had harvested honey from goldenrod, so thick that a table knife would stand upright in a bowl of it. This was probably from a prairie species that has since disappeared, possibly *Solidago rigida,* which is attractive to the bees.

On the author's "Sunset Ranch" in southern Holt County, Nebraska, there is still much open prairie where several species of goldenrod occur in abundance. There the bees gather some surplus honey of a bright amber color from goldenrod. Although they work to some extent on more than one species it seems that most of the honey comes from *S. nemoralis,* which is very common. The soil is light and sandy and the altitude is something like 1800 feet above sea level, which insures cool nights during the summer season.

In Journal of Economic Entomology for June, 1932 appears an extended article on "Honeyflow from Goldenrod in Relation to Temperature" by E. Oertel. Observations made at Donaldsonville, Louisiana in 1930 indicated "The corelations between average temperature, maximum temperature and daily gross gains are so small that they cannot be considered to show any relationship between these factors and colony gain."

From Indiana comes the following:—

> "At some yards where the land is low, as along wide river bottoms, goldenrod is the chief source of honey. This honey is of a bright yellow or golden color when unmixed with that from other sources and is of excellent quality. Because it is not white, being somewhat mixed with other honeys, it is quoted at a lower price, notwithstanding that its quality is superior to any sweet clover honey which we have been able to produce in this locality." E. S. Miller, Valpariso, Indiana, Gleanings page 488, August, 1933.

Concerning goldenrod, Robert M. Mead of Vermont wrote, (Am. Bee Journal Dec. 1936), as follows:

> "Good flows from goldenrod usually follow a summer when there has been plenty of rainfall and a very moderate crop from summer sources. The early goldenrods which begin to bloom as early as the first of August are worked by the bees at times but no noticeable amount of honey is gathered from them. The real flow comes from the first to the 21st of September. The best conditions for a goldenrod flow seem to be hot fall days—a cloudy but not a rainy, hot day giving the greatest flow. Cool nights should go with the warm days.
>
> "Goldenrod nectar apparently does not flow below 60 degrees F. and only moderately at 65 F., but as soon as the temperature reaches seventy degrees or higher the bees become very active. Changeable weather with cold north winds, driving showers or excessive dampness seem to spoil the chances of getting a crop from this plant.
>
> "Goldenrod honey when the flow is rapid, is a fair table honey when well ripened. It is yellow in color, smooth and mellow in taste but may have a characteristic sharp taste if too thin or green. When the flow is poor what honey is gathered is often dark in color and rather too sharply flavored for most tastes. A characteristic penetrating but pleasing goldenrod odor is very noticeable about the apiary during a good flow."

Along the Gulf Coast considerable surplus honey is secured from goldenrod. Dr. Oertel reports an average of 30 to 40 pounds per colony from *Solidago altissima* in the vicinity of Baton Rouge, Louisiana. It is valuable also as a source of pollen reserves for winter. Measure of the stores has shown as much as 1000 square inches of stored pollen from this source in a single hive.

Requires warm weather both day and night for heavy flow.

> "Two colonies of bees taken to a sandy plain forty miles north of Ottawa, August 25, each gathered, in three weeks, about 40 pounds of surplus honey from *S. puberula* and *S. squarrosa.* It is estimated that at least three-fourths of the honey came from *S. puberula,* which was much more abundant than *S. squarrosa.* The honey is of a light color and the flavor and aroma are pleasant and distinctly suggestive of goldenrod."—Sladen, in 36th report, Ontario, B. K. A.

GOLDEN WATTLE, see Acacia.

GOOSEBERRY (Ribes).

Gooseberries are native American shrubs with stems covered with sharp thorns or spines. There are several species widely distributed. They are very attractive to the bees and are of some importance, especially where grown in large plantings for market. The wild varieties are common in open woodlands in nearly every section of the country.

GOPHER APPLE (Chrysobalanus oblongifolius).

The gopher apple, sometimes called ground oak, is a low shrub which grows in patches and spreads by underground stems. It is found in sandy pine lands of Georgia, Alabama, Mississippi and Florida. In Alabama it is known as deer plum. The white flowers appear in spring and early summer and the fruit ripens in September. The ripe fruit is variously reported as yellowish-red and white, tinged with pink. It is about the size of a plum and very similar in appearance, fairly good when eaten fresh, but of little value otherwise. The honey from this source is reported by Florida beekeepers as a poor quality and the plant is not considered important as a source of nectar. It is regarded as of principal value for brood rearing.

The cocoa plum (*C. Icaco*), also called gopher plum, is a tree which reaches a height of 25 or 30 feet; common to peninsular Florida, Mexico, the West Indies and Central America. In Florida it is usually a shrub. The same, or a similar species, is also found in Africa.

Gorse or furze.

GOPHER PLUM, see Gopher Apple.

GOPHER WOOD, see Yellow Wood.

GORSE or FURZE (Ulex europaeus).

Gorse is a spiny evergreen shrub with yellow flowers that is common in Europe, where it is said to be used to some extent for fuel and fodder. There are few references to it in this country. In California it is said to bloom during all seasons, although much more freely in spring. Richter reports it as a very good honey plant on the hills of Marin County.

D. J. Hemming, in the British Bee Journal (Feb. 2, 1922), reports that in England it may be found in bloom all the year round. He reports it as a source of both pollen and nectar.

Bonnier lists it as a honey plant in France.

GRANJENO, see Hackberry.

GRAPE (Vitis).

The grape family is represented by wild species in all parts of the temperate regions of both hemispheres, and by cultivated species in

Wild grape in bloom.

nearly all parts of the world. There are about thirty species of wild grapes, and where sufficiently common they are very attractive to the bees. In many localities cultivated grapes are grown in large acreage. The nectar yield is not as abundant as with many plants, but is of some value where the vines are largely grown. Quantities of pollen are gathered from this source. At times honeydew is gathered from the leaves.

Scholl lists the mountain grape (*Vitis monticola*) as giving a fairly good honey yield and furnishing pollen for early brood rearing. Richter lists the California wild grape as yielding some honey.

Concerning wild grapes, S. W. Cole wrote in American Bee Journal, (page 30, Aug., 1869) as follows:

"Wild grapes bloomed May 24; ceased June 10. This is a splendid honey plant with us, not excelled even by the poplar. It is very abundant, growing everywhere. Indeed, it would be a very difficult matter to find a spot in any of the thick woods of west Tennessee that is not adorned by the rank foliage of *Vitis labrusca.*

In Mississippi the author was shown samples of honey said to be from wild grape.

Dr. Everett Oertel reports that bees work heavily on wild grapes on lowlands of Louisiana in vicinity of Baton Rouge.

Bees and Grapes

Of the disagreements between fruit growers and beekeepers, probably those growing out of the tendency of the bees to suck the juice from cracked grapes have been most serious. Many unfortunate misunderstandings have resulted from such circumstances, though the injury was as great to the beekeeper as to the fruit grower, in many cases.

This condition arises from a combination of circumstances which does not often occur in the average locality. In the first place, the bees do not seek the grapes when there is plenty of nectar in the field, and, beside, they are unable to reach the juice unless the grape has first cracked open through unfavorable weather conditions or has been injured by birds, wasps, or other agency. Grape growers, seeing the bees at work in the vineyards, have often accused the bees of injury to the fruit. The fact of the matter is that the bee is unable to puncture the fruit, and only sucks the juice from such fruits as have already been broken open and are already damaged.

Wet weather often causes ripening grapes to crack open to such an extent that they would be of little value, even though no insect touched them thereafter. In dry weather, also, birds sometimes pierce the skins, apparently in search of moisture from lack of an available water supply. Some authorities say that at such times a liberal supply of water in open vessels near the vineyard will stop the injury from birds. The English sparrow is accused of injury to grapes to a larger extent than most birds.

The grape-berry moth infests a great many grapes in some localities. In fact, entomologists state that in some localities as high as 50 per cent of the crop is injured by this insect alone. The fact that the honeybee sucks the juice from the berries which have already been opened by wet weather, grape-berry moths or other causes, does not greatly injure the grape grower, for such fruit is of little value.

The writer has visited the raisin districts of Southern California and discussed this condition with the raisin growers. The accompanying picture shows a bunch of raisin grapes that had been sucked dry by the bees. In that locality rains are very infrequent when the raisins are being dried. They are spread out in thin layers in crates and the crates left in the sun, or piled up one above another, till fully dry. Previous to the writer's visit there had been an unexpected rain, and some raisins were allowed to get wet. As a result they cracked open, and there being just then no available nectar for the bees, they swarmed over the raisins and sucked them dry, as shown in the pic-

Raisin grapes sucked dry by bees after the skins have been broken by rain.

ture. The grower admitted, however, that the raisins had been so badly damaged by the rain as to be of little value.

The thing which few grape growers seem to understand, is that it is unfortunate, indeed, for the beekeeper in northern regions, whose bees fill their hives with this grape juice. In the north there are long periods during the long winter months when the bees are unable to leave the hive. Since the bee is only able to void her excrement while on the wing, there is a large accumulation of feces during such long confinement. If the bees have only the best white honey for food the tax is severe at best. When they have fruit juice, honeydew or other food containing a large amount of waste matter, the intestines become so distended that the bees die for lack of opportunity of a cleansing flight. Grape growers will in many cases be surprised to learn that thousands of bees die from having filled their combs with fruit juice instead of honey. Of course, the wide-awake beekeeper will remove this material from the hive and give them good honey or sugar syrup instead, if it is possible to do so. This, however, involves a large amount of labor, and the gathering of the juice from the grapes, instead of being an advantage to the beekeeper, is a serious inconvenience to him. In southern California, where there is no winter confinement, there is no particular injury to the bees, other than spoiling the grade of any honey with which it may happen to be mixed.

That bees are valuable in securing the fertilization of the blossoms of some varieties of grapes, there is little question.

GRAPE FRUIT (Citrus decumana).

Grape fruit is recognized as a valuable source of honey, but not equal to orange and lemon. It is an evergreen tree similar in habit to the other citrus fruits, and the area devoted to its cultivation is being extended in various districts.

GRASSES.

There are many grasses which are attractive to the bees for pollen. Timothy or herd's grass produces pollen in great abundance and the bees are often observed gathering it. Occasionally some honeydew may be gathered from the grasses also, but no honey.

From Starkville, Mississippi, the author has received reports of honey-dew gathered from a fungus on *Paspalum dilatatum.*

In Beekeepers' Magazine for April 1885 appeared an extended account of honey from ergot on grasses, written by Prof. A. J. Cook. Samples of grass speckled with crystals of sugar had been received from several localities in Ohio and Michigan and was called "manna grass" by those sending it. The honeyed secretion appeared as characteristic of early stages of ergot. The ergotized grasses were reported as thronged with bees and in the vicinity of Quincey, Michigan, beekeepers were reported to have secured excellent honey from it.

GREASEWOOD.

The name "greasewood" is applied to several different shrubs common to the western states. At least two of these are of some importance to the beekeeper as sources of nectar. Creosote Bush (*Covillea glutinosa*) which is common to the mesas from western Texas to southern California is more frequently called greasewood by beekeepers than by its proper name. (See Creosote Bush).

Chamise (*Adenostoma fasiculatum*), an evergreen shrub common to the mountains of western California, is also often called "greasewood." (See Chamise).

GREEN BARK ACACIA, see Palo Verde.

GREENBRIAR (Smilax).

There are several species of greenbriar, some of which are common from Canada to Texas. They grow along watercourses and in open woodlands and the stems are covered with stiff and hard thorns. Scholl lists one species as yielding nectar in Texas. The author found reports to the effect that bees gather honey from greenbriar in McLennan County, Texas. At the same time it was stated that no honey was gathered from it in other nearby sections. The plant seems to be of doubtful value and is not sufficiently abundant to be important.

GREVILLEA.

There are a large number of species of *Grevillea* which are trees and shrubs, mostly Australian. Some species are noted as the source of honey.

G. theremanniana is an ornamental spreading shrub which is popular in California. R. M. Kellogg, of Artesia, writes that it is swarming with bees all the year and yields nectar freely.

GRINDELIA, see Gum Weed.
GROUND IVY, see Gill-Over-the-Ground.
GROUND LAUREL, see Arbutus.
GROUND OAK, see Gopher Apple.
GROUNDSEL, see Butterweed.
GUAIACUM, see Soapbush.
GUATEMOTE, see Baccharis.

GUAVA (Psidium guajava).

The guava is common to tropical America and is extensively cultivated in Brazil and northward to Mexico and the West Indies. It is grown to some extent in Florida. While the fruit is eaten raw it is best known in this country as the source of guava jelly. The fruit somewhat resembles the orange and varies in size from that of a plum to an apple.

Phillips lists guava as a source of nectar in Hawaii. (Bul. 75 pt. 5 Bureau of Ent.) He also mentions it as a source of honey in Porto Rico. (Bul. 15 Porto Rico Agr. Ex. Sta.).

GUAYACAN, see Soapbush.
GUM, see Tupelo, also Eucalyptus.

GUM-ELASTIC (Bumelia lanuginosa rigida).

Gum-elastic is the common name in south Texas for the Arizona buckthorn, a small tree occurring from western Texas to Arizona. It has a short stem with stiff and very spiny branches. It is frequently mentioned as a source of honey in southern Texas.

There are reports to the effect that it sometimes yields a surplus along the Trinity River and that the honey often sours in the combs after being sealed. (See Coma.)

GUM TREE, see Eucalyptus, also Tupelo.

GUM WEED or GUM PLANT (Grindelia squarrosa).

The gum weed, also called rosin weed, is a common plant from Wyoming and Colorado south to Arizona, New Mexico and Texas. It

Gumweed, or rosin wood, of the Great Plains.

occurs sparingly eastward to Minnesota and Missouri. The bright yellow flowers exude a milky resinous gum, which gives rise to its name, "gum weed." It is also widely known as rosin weed. It blooms in August and is much sought by the bees. The honey is yellow and of inferior flavor. It is often mixed with light honey in the super and the grade spoiled as a result. In Colorado it is well known by the name of "rosin weed." There the comb-honey producers complain that it often spoils the quality of their product through being mixed with the honey from alfalfa and sweet clover. Honey from gum weed candies very quickly. Comb honey which candies cannot readily be restored, hence in localities where gum weed is abundant the beekeeper may find it to his advantage to produce extracted honey. Even this does not solve the problem entirely, since the gum weed honey candies in the combs so readily as to make it difficult to extract. There are few reports of large surplus from this source. In most cases it is mixed with other honey and only in such quantity as to make it rather a nuisance than otherwise. The tendency to candy in the combs makes it undesirable for winter stores.

H

HACKBERRY (Celtis).

The hackberries are an important group of trees. There are about sixty species of trees and shrubs which are widely distributed in both the old world and America. There are about half a dozen species known to America, and of these, three are much valued by the beekeepers of Texas.

The granjeno (*Celtis pallida*) is a shrub common to the mesas and foothills of western and southern Texas. E. G. LeStourgeon, of San Antonio, regards it as the best of all the hackberries blooming after every good rain. The honey is pale amber in color and of good quality. He reports that honey from all hackberries is of good quality. When the author visited at Brownsville, in February, 1918, the bees were working on the above species very freely at that time.

The Southern hackberry or sugarberry (*Celtis mississippiensis*) is common all over the Gulf and South Atlantic States from Kentucky and southern Illinois to Florida, Arkansas and central Texas. LeStourgeon reports that this species blooms after the others, in March and April, and that while it yields some honey, it is not as good as the others.

The Northern hackberry (*Celtis occidentalis*) is also known as sugarberry, and is common from New England and Ontario to Minnesota, Nebraska, Colorado and southward to Georgia and Texas.

HAU TREE (Paritium tiliacium).

"The hau tree of Hawaii has nectaries on its leaves which secrete a honeydew. These are located on the veins of the leaves near the stem, and are one, two, three or five in number. Small drops of honeydew may frequently be seen on these spots. It is interesting to note that these extra-floral nectaries are present on the outside of the calyx of the flowers. There is apparently no true floral nectary. The hau tree is used extensively as a hedge, and grows from 20 to 30 feet high. It is doubtful whether this is the source of any great percentage of the honeydew honey."—E. F. Phillips, Bul. 75, Part 5, Bureau of Entomology.

HAW, see Hawthorne.

HAWKWEED (Hieracium aurantiacum) Devil's Paint Brush.

The orange hawkweed, or Devil's paint brush, also sometimes called missionary weed or red daisy, is common in fields and along roadsides from Ontario east to New Brunswick and south to Pennsylvania and New Jersey. It came originally from Europe, but is widely naturalized in the northeastern states.

Hawkweed is not important as a honey plant, although under favorable conditions it is reported as yielding some nectar. A Maine beekeeper states that when the weather is hot and the soil saturated with moisture it yields well. However, it is only an occasional season, when conditions are just right, that his bees find it attractive. He reports the honey from hawkweed to be of a light greenish color and of fair quality. A New York beekeeper states that the honey is dark amber and of poor quality. It is probable that in one or both cases the samples were mixed with honey from other sources. In New York the hawkweed is common over the state, but most abundant on acid soils of the plateau region of the southwestern part.

In a private letter to the author, under date of Nov. 3, 1934, Maurice N. Shutts of Merrill, New York says: "Our land is more or less run out so the fields are red with paint brush bloom. Paint Brush never fails to yield except in very wet weather. In fact it seems to do better during hot dry days."

HAWTHORN (Crataegus).

The picture of hawthorn in full bloom does not do justice to the masses of white flowers with which the tree was covered. This is an eastern species (*Crataegus punctata*), which occurs from Quebec to Ontario and south to Georgia. It was about the middle of June when this picture was taken and the bees were working on these trees everywhere we went. Clover had not begun to yield to any extent and the thorn was a great boost to the bees wherever it was plentiful.

There are many different species of hawthorn, or haw, some of which occur in Europe and Asia, as well as in North America. On this continent some species are common from Canada to Mexico and west to the treeless plains. Scholl reports the white thorn (*Crataegus spathulata*) as valuable for both honey and pollen in Texas, where it blooms in April. There are about 25 species of these trees within the United States, and all may be regarded as valuable sources of honey where they are sufficiently plentiful. In general, they may be regarded as similar to the tree-fruits in quality and quantity of nectar. Five species are known to occur in Ontario, where they are regarded as important.

Coleman lists the black-haw (*Crataegus douglassi*) as a source of white honey, in California, especially important for building up colonies early. This is the only species known to the Pacific Coast, although a varietal form is recognized.

The hawthorn blooms abundantly and yields nectar freely.

F. N. Howes in his Plants and Beekeeping, states that honey from hawthorn is of very high quality, very thick and of an appetizing rich flavor and dark amber in color.

HAZELNUT (Corylus americana).

The hazelnut is a slender growing shrub common in the borders of woodlands of the most of the temperate North America. It yields some pollen and is valuable where there is a scarcity of early pollen-bearing plants. The figure shows the male blossoms, which are more conspicuous than the fertile ones.

HEARTSEASE (Polygonum).

Polygonum is a large family with a variety of names. In some localities one name will apply, while in another the plant will be known by an entirely different one. Smartweed, knotweed, doorweed, persicaria, lady's thumb, water pepper, heartsease and several other names are applied to these plants. They are widely distributed, covering practically all of the United States and Canada, as well as much of Europe and Asia. *P. persicaria,* or lady's thumb, is most often called heartsease, and is also said to be the best honey producer. It is an introduced species, coming from Europe, and is still widely scattered through the sale of clover seed, the seed of this plant being commonly mixed with red clover seed.

The hazelnut is the source of some pollen.

The honey gathered from these plants varies greatly, both in quantity and quality. Some species do not seem to yield at all, at least not regularly, while others produce large quantities of nectar. The blooming period in the North is from midsummer until frost, and occasionally large yields are reported, an average of 200 pounds per colony not being the highest on record, from this source alone. Sometimes honey from these plants is of very good quality, while from other species it is very dark and of inferior grade. The better grade of honey is sometimes designated as heartsease honey, while the poorer grade is called smartweed honey.

These plants grow in moist fields everywhere, and frequently come up in grain fields and stubbles late in summer, after cultivation has ceased, thus offering plentiful forage for the bees, in fields where otherwise they would find nothing.

A section of comb honey sent to the author by F. W. Leubeck, Knox, Indiana, was of a light amber color, mild and pleasing flavor with cappings almost as white as clover honey. Concerning heartsease in his locality Mr. Leubeck wrote:—

> "Our amber or darker grades are gathered during August. Up to about the 25th we crowd our bees a little for room, then we put on another deep super and if the season is a rainy one, two supers. I have never seen any

Heartsease in bloom.

> dark honey in these last given supers, when the bees started to work on heartsease.
>
> "In 1925 we had a frost on July 16. All fall flowers were killed, no honey up to the first of September. Then heartsease came to be our salvation. The bees stored an average of 100 pounds surplus beside winter stores. A hive on scales in the home yard showed 14 pounds gain in one day and two days later 19 pounds in one day."

The Japanese knotweed or fleeceflower (*Polygonum cuspidatum*) frequently attracts the attention of beemen because of the large numbers of bees that swarm over the blossoms. Since the plant is only found in small plots of ornamentals it is of little importance.

Polygonum sachalinense, known as Chinese bamboo or Sachaline grows to a height of 12 feet and once established is difficult to eradicate. It spreads from strong rootstocks and is sometimes grown for rough forage. It has become established in a few localities in the Eastern states and is reported as swarming with bees during its period of bloom.

HEATHER (Calluna vulgaris).

Heather is a very important honey plant in Europe, but occurs in few places in America. Gray's botany lists it at the following places: "Low grounds, Massachusetts at Tewksbury and West Andover; Maine, at Cape Elizabeth; also Nova Scotia; Cape Breton; New Foundland, etc." It may in time become locally important in Eastern America, though it is doubtful whether it be so at present.

Writing in Gleanings, D. M. MacDonald has the following comments on heather honey in Scotland:

> "The product of heather is of a rich amber color, bright and sparkling, rather than dull and shady. It has a pronounced flavor, delicious to the palate when one has acquired a liking for it. The aroma is pungent and penetrating, making itself manifest in a room where heather honey is kept in a closed cupboard. Its consistency is so remarkable that it will not leave the comb by any amount of centrifugal force used in the extractor, and when desired in the liquid form, the combs have to be melted and pressed by heavy screw power in a specially constructed press. Most beekeepers in heather districts, therefore, work for sections only; but it pays well to press all defective combs preserved for the purpose, and thus renew the wax of the brood area periodically. On account of the profusion of the bloom, the flow is at times extraordinarily abundant; but as the lateness of the season frequently causes unfavorable weather conditions, the crop is an uncertain one.
>
> "Heather honey sells for about double the price obtained for any other kind in this country. While a great part of the flower, clover and lime honey brings the apiarist only 18 cents per pound, heather frequently fetches him 36 cents. While, too, the other kind drags on the market, heather honey sells readily and is often disposed of before it comes off the hives. Retail prices in warehouses in Edinburg and London are often as high as 48 to 60 cents per section."—October 1, 1910.

In November, 1913, C. P. Dadant, editor of the American Bee Journal, wrote as follows concerning heather in southern France, seen by him during his visit to Europe in 1913:

> "I had often heard of the 'Landes' of Gascony, but thought them low, sandy plains. They are rolling hills instead, and extend for scores of miles along the Gulf of Gascony.
>
> "The growth upon the 'Landes' is confined to numerous ferns, scrubby pines and cork oaks, with a very thick undergrowth of heather. Just now the heather is in its fullest bloom (September), and there are perhaps 20 different varieties, ranging from the palest pink to almost red and deep yellow in color. It is a mass of flowers upon which the bees work from June

Blossoms of heather.

until frost, which comes very late, usually not before November. So we may readily call this the eldorado of beekeeping. There is only one dark side to the picture—the heather honey is dark in color, a deep amber, strong in

> flavor and almost impossible to extract with the honey extractor. Here I ascertained positively what I already suspected, that when speaking of nectar containing 75 percent of water, we should confine ourselves to the nectar of our moist prairies. I am told that much of the nectar harvested from heather, in this dry, sandy soil, is too thick at the end of the first day to be thrown out readily."

In the northwest, from Alaska south to California, there are several species of the Ericaceae commonly known as heather or false heather, which are the source of some honey of an amber color and good quality, Of these false white heather *Cassiope mertensiana,* and *C. tetragona,* are low evergreen shrubs much branched, which are very common to the high mountain areas and often occur in large acreage.

Phyllodoce glanduliflora, and *P. empetriformis,* red false heather or red "fill-o-do-see" are also low evergreen undershrubs common to alpine regions and likewise known as heather.

All are apparently recognized as good honey plants by British Columbia beekeepers, and the honey from all is known as "heather honey." In an account of the British Columbia Honey Exhibit (Am. Bee Journal Dec. **1926**, p. **598**), "mountain honey from the Cascades, said to be chiefly heather was judged to be the finest honey present."

HEATHER, see also Wild Buckwheat.

HEATHERMINT (Elsholtzia stauntonia).

The heathermint is a hardy shrub which comes from North China. It is rather low growing, reaching a height of three to five feet and blooming profusely in September and October. The bees fairly swarm over the bloom and since it comes late after the harvest is mostly over it is a welcome addition to the autumn flora.

This is one shrub which should be included in all plantings for bee pasture.

HEDGE MUSTARD or JIM HILL MUSTARD, (Sisymbrium altissimum).

Hedge mustard is a weed introduced from Europe and established in numerous localities. It is a nuisance in grain fields in some places. It is a common roadside weed in Arizona and eastern Oregon where it is reported as the source of some nectar. Some pollen comes from this source also. Also called Tumble mustard.

Tansy mustard, (*S. pinnatum*) is reported by Scullen and Vansell as a source of nectar in Central Oregon.

HEDGE NETTLE or WOUNDWORT (Stachys).

The hedge nettles are herbs with flowers in terminal spikes. The corrollas are two-lipped and *Stachys agraria* is commonly known as "mint" in the Rio Grande Valley. At several places in the extreme south part of Texas the author heard reports of surplus honey from this source and from related species.

Richter lists three species, *Stachys ajugoides,* the white hedge nettle, *Stachys albens,* and *Stachys bullata,* as yielding honey in California.

Other species are common to the Eastern States from Ontario south to the Gulf States. See Pinkmint.

HELENIUM, see Bitterweed.

HELIANTHUS, see Sunflower.

HELIOTROPE (Heliotropium).

The cultivated heliotrope and also the wild heliotrope (*Heliotropium curassavicum*) are listed by Richter as sources of honey in California. There is also an occasional mention of heliotrope as a source of nectar in the bee magazines.

HELLEBORE (Helleborus viridis).

The green hellebore, or Christmas flower, is a native of Europe which has become established in a few localities in eastern America. It is now found in New York, New Jersey, Pennsylvania and West Virginia. The roots are used in medicine.

The plant is probably not sufficiently common to be of importance to American beekeepers, although in Europe it is regarded as of some value, as indicated by the following note from a German bee magazine:

> "On the hellebores (*H. viridis* and *foetidus*) which are practically the same color as the leaves except that the tint of the bloom is a shade lighter, I observed bees working hundreds of times, and as much for nectar as for pollen. Even where they grew or were planted in the wilderness I could count them as the most frequented of the spring flowers."—Wust, in Deutsche Illustrierte Bienenzeitung, September, 1921.

HERCULES CLUB or ANGELICA TREE (Aralia spinosa).

Hercules club or angelica tree is a shrub or small tree common to damp borders of the woods and river banks from Virginia to Missouri, and south to Florida and Louisiana. The flowers are white and appear in early summer. It is reported as yielding nectar abundantly, though not often as a source of surplus.

In western Pennsylvania it is reported to bloom with buckwheat and to yield best in seasons which are too dry for buckwheat to do well. Honey is said to be light in color, of good body and fair flavor.

The Hercules club, also called Devil's club, may readily be recognized by the large number of peculiar thorns.

HERON'S BILL, see Pine Clover.

HETEROMELES, see Christmas Berry.

HICKORY (Hicoria).

The hickories are an important group of forest trees of wide distribution. They are of special interest to the beekeepers as sources of large amounts of honeydew from aphis, which are frequently to be found on the leaves. These trees are important for pollen, also, in many localities.

HIMALAYA BERRY, see Blackberry.
HOARY VERBENA, see Vervain.
HOG GUM, see Poisonwood.

HOLLY (Ilex).

The holly family is a large one, with representatives in Europe, Asia and South America, as well as North America. There are more than a dozen species on this continent, most of them common to the Southeastern States. The common holly (*Ilex opaca*) is found from Massachusetts and New York south to Florida and west to Arkansas and Texas. It is reported as a valuable source of nectar in most of the Southern States. The Myrtle-leaf dahoon holly (*Ilex myrtifolia*) occurs from North Carolina to Florida and western Louisiana.

The holly trees bloom in May, and in Alabama, Georgia and Mississippi are reported as yielding rapidly for a short period of time, usually three to ten days.

The deciduous holly or privet (*Ilex decidua*), of the Southeast, is known in Texas as possum haw or bearberry. It ranges from southern Virginia to Florida and west to Missouri and Texas. It blooms early and is reported as yielding well for a short period.

The yaupon (*Ilex caroliniana*) is frequently reported as a source of honey by Texas beekeepers, though Scholl lists it as unimportant.

The gallberry (*Ilex glabra*) is probably the most important source of honey of the group, especially in the Gulf region. (See Gallberry).

From Florida the tall gallberry (*I. lucida*) and Krug Holly (*I. Krugiana*) are reported as sources of honey as well as the feverbush, black alder or winter-berry (*I. verticillata*). The feverbush is found in swamps and wet lands from Nova Scotia and Ontario to Wisconsin and Missouri and south to Florida.

In Arkansas *I. decidua* is known as buzzard's berry and is reported as valuable.

The common holly is reported as often yielding a surplus of fine table honey in Maryland according to George J. Abrams.

W. H. Hawkins of Camden, Arkansas, in a private letter writes as follows:—

"This honey is made from bloom of American Holly, (Ilex opaca). These blooms open from May 1 to 20 and are very fragrant. While the trees are in bloom the bees gather honey very fast. I have had hives store as much as 150 pounds in ten or twelve days. I think this honey is the finest in the state. Pure holly will not granulate while other honeys do."

The honey is of a rich amber color with a pleasant twang in its flavor.

HOLLYHOCK (Althaea rosea).

The hollyhock is one of the oldest cultivated garden flowers. It is a native of China and is to be found in every garden of old fashioned flowers. Wherever found, the bees seek it eagerly, and apparently it secretes nectar freely.

HONEYDEW.

Such material as the bees may store as honey, which is not secured from the nectaries of plants, is usually spoken of as honeydew. There are numerous exudations of plants which attract the bees and which can hardly be regarded as nectar, to which the term honeydew may well be applied.

The main source of honeydew, however, is from insects rather than from plants. Aphids, scale insects and leaf hoppers yield this sugar in great abundance. These sucking insects are often found on various trees or plants in large colonies, feeding on the sap; while bees and ants gather to feed upon their excretions. At times aphids are so abundant that they eject honeydew in such quantity as to cover the leaves on lower levels with the sticky substance. The bees gather honeydew readily in the absence of a natural honeyflow, carry it to the hive and seal it in their combs the same as honey.

The quality of most honeydew is inferior and it brings a low price in the markets, being in most demand for baking purposes. Since there is a much larger percentage of gums in honeydew than in honey it makes a poor food for winter stores. The excess matter clogs the intestines of the bees, and where they are confined on such stores for long periods without an opportunity for cleansing flight, a heavy mortality results.

There are hundreds of references to honeydew in the beekeeping literature. A few are given here to indicate the extent to which bees gather this insect product:

"The most copious flow of honeydew I ever saw was in 1897. It was from the pine. In early morning and late in the evening it could be seen dripping from the leaves, till all the leaves and even the bare ground beneath them were covered with the nectar. The bees swarmed over the trees and the hives were filled as I had never seen them before. The honey was light amber and of fine flavor, and gave my customers the best of satisfaction. While this flow was on, there was scarcely any honeydew to be found except on the pines, and every pine was dripping with it."—C. C. Parsons, Alabama. American Bee Journal, page 546, 1899.

"We have had the heaviest honeydew flow ever known in this part of the State. We have tons of the stuff."—Scholl. American Bee Journal, August, 1910.

"My bees stored a quantity of honeydew which granulated in the combs as fast as stored."—South Carolina. American Bee Journal, August, 1910.

"Of 250 colonies of bees in this town last fall there are not more than 20 left. It is not the winter that kills the bees, but poor honey. Honeydew is half an inch deep all over my honeyhouse floor; it soured and ran out of the combs where I packed up my hives. The bees will not touch the horrid stuff, nor can I get a swarm to go into a hive with one frame of it on one side, and good clean combs and frames of brood for the rest."—Vermont. American Bee Journal, page 458, 1904.

"Bees are working on honeydew, the trees just glistening with it. I have several hundred pounds of it in the supers. It is bad looking stuff, not fit to eat or sell."—Iowa. American Bee Journal, page 537, 1904.

Not all honeydew is of such bad color and flavor. There are numerous reports of honeydew of such flavor and quality that it finds a ready sale in competition with good honey.

HONEY LOCUST (Gleditsia triacanthos). THORNY LOCUST.

The honey locust is a well-known tree from Pennsylvania and Ontario to Florida and westward to Texas and Kansas. It prefers rich bottom lands, seldom being found on dry hillsides. The tree has spreading branches and very long, red brown thorns. The thorns are often branched and sometimes nearly a foot in length. At times they cover the trunk of the tree in great abundance. The flowers appear in May or June. The blooming period is too short for large yield.

While the honey locust is the source of considerable nectar, it is not equal to the black locust (*Robinia pseudo-acacia*). (See Locust.)

The two species are often confused, although the long thorns peculiar to the honey locust should distinguish it readily.

There are numerous reports to the effect that the bees work freely on honey locust, but it is seldom regarded as a source of surplus, while the black locust yields abundantly for a short period, under favorable conditions.

Dr. Everett Oertel reports surplus from this source in the vicinity of Baton Rouge, Louisiana where the trees are abundant in pastures.

Dr. J. Russell Smith of Columbia University has made a search for trees showing a high concentration of sugar in the seed pods. Some have been found whose pods contain more than thirty percent of sugar and these are propagated for planting in pastures to provide additional forage for livestock. It is argued that sugar might be profitably extracted from such a source.

HONEY POD, see Mesquite.

HONEYSUCKLE (Lonicera).

The honeysuckles are a widely distributed group in the temperate regions of the Northern Hemisphere. More than 150 species are known, of which about two dozen are native to America. They are very rich in nectar, as is indicated by the name. However, most of the species have corolla tubes so deep that the nectar is only accessible to humming birds and hawk moths. A few of the fly honeysuckles and the bush honeysuckle are much visited by the bees.

From Arkansas, W. C. Brass reports the winter honeysuckle (*L. morrowi*) as blooming in February and March, when it is valuable for simulating early brood rearing. This is a Japanese species which has escaped from cultivation and become naturalized in some localities, as far north as Massachusetts.

The black twinberry, or involucred fly-honeysuckle (*L. involucrata*) is found along streams and in canyons throughout the mountains of California and northward through Oregon, Washington and British Columbia. The fruit, a black berry, is very bitter. This species is reported as a source of surplus honey by British Columbia beekeepers near the coast, where it is very common.

It is probable that several species of the fly-honeysuckles are valuable locally. Some of these are common to the woodlands of the north from Nova Scotia and New Brunswick to Alaska.

The Tatarian honeysuckle, *L. tatarica,* is widely grown for ornamental purposes and is one of the most attractive to the bees. It is very hardy and is well known in western Canada. The author found it generally planted in Manitoba, Saskatchewan and Alberta. Even in the Peace River region it is a favorite.

The bush honeysuckle is also reported as valuable in a few places. (See Bush Honeysuckle).

HOP CLOVER, (Trifolium procumbens).

Hop clover is an annual common on sandy lands and along roadsides, which has been naturalized from Europe. It is not often recognized as a source of surplus honey, although the bees work upon it freely. It is reported to yield a rich golden honey of fine flavor and in a few localities is regarded highly. It is a low growing plant, not cultivated for forage and consequently is seldom present in a sufficient quantity to count for much with the beekeeper.

HOPS (Humulus Lupulus).

The common hop plant is too well known to need description. It is common from New England to British Columbia and southward. It is very generally cultivated for making yeast and for medicinal purposes. The small greenish flowers are wind-pollinated. It furnishes pollen in abundance, but no nectar.

HOP-TREE (Ptelea trifoliata), SHRUBBY TREFOIL.

The hop-tree occurs from New England and Ontario, south to Florida and westward to Michigan, Illinois and Missouri to central Texas. It is a shrub or small tree known also as ague-bark and quinine tree. The bitter fruit is sometimes used as a substitute for hops. The flowers have a disagreeable odor. Scholl lists the honey yield as good, and very good, in favorable seasons, where the shrub is abundant. The California hop-tree (*Ptelea crenulata*) occurs in the western foothills of the Sierra Nevada Mountains and in the Coast ranges. Its bloom has an agreeable aromatic odor. The author can find no reports that indicate it is regarded as important to the beekeepers of that region.

There are numerous reports to the effect that the hop-tree is a good source of nectar in the Eastern States.

HOREHOUND (Marrubium vulgare).

Horehound is a well-known plant, introduced from Europe, which has become naturalized from Canada south to the Gulf of Mexico. The plant is perennial, flowering from July to September. It occurs in waste places, along roadsides and near dwellings over a wide scope of country.

It is one of the chief sources of nectar in places in the Arkansas Valley in Kansas, also in portions of Texas. The honey is dark amber

Horehound.

and strong in flavor. Beekeepers report it as important at Seguin, Texas. Reports of honey from this source come from widely separated localities from the Eastern States to California.

Richter lists it as blooming in May and June in southern California, but at the usual period further north. He reports that Ventura and Los Angeles Counties produce horehound in considerable quantities, but that it is regarded with disfavor in the sage districts because a small quantity of this honey, mixed with the sage, impairs the color and flavor of the latter.

Wherever plentiful horehound may be expected to yield some honey, and in many places it yields surplus in quantity. Although not of the best quality, it is still an important honey plant.

HORSE-BEAN, see Jerusalem Thorn.

HORSE CHESTNUT, see Buckeye.

HORSEFLY-WEED, see Indigo Weed.

HORSEMINT (Monarda).

There are several species of horsemint, known also as bee balm, wild bergamot, etc. Some of the species are represented from New England to Texas. The corolla tubes of *M. fistulosa,* the wild bergamot of the North are so deep that, as a rule, the bees are unable to reach the nectar. In some cases it is reported as yielding freely and the author has seen times when the bees were apparently getting considerable nectar from this species. In parts of Wisconsin, *M. punctata,* according to Dr. L. H. Pammel, can be depended upon to yield an abundance of nectar every season. This is probably the most important species to the beekeeper. It is found more or less commonly on sandy soil from New York to Minnesota and south to Florida and Texas. In Texas, where it is called "sandy land horsemint" it is the source of very large quantities of surplus honey in seasons following wet winters and springs. The honey is a clear light amber with a decided minty flavor. It is one of the most important

Horsemint.

sources in Texas, where, together with *M. clinopodioides,* and *citriodora,* it is regarded very highly. In the Arkansas Valley of southern Kansas horsemint is also important, yielding as high as fifty pounds of surplus per colony. Honey from horsemint is of inferior quality and slow to granulate.

HOUND'S-TONGUE (Cynoglossum officinale).

Hound's-tongue is a weed introduced from Europe. It is now common in parts of Canada and the Northeastern States and occurs in Missouri and Arkansas, and from the Ohio Valley to the Carolinas and north Georgia. It is named from the shape and texture of the leaf. The illustration was photographed in open woods near Guelph, Ontario. There it is frequently mentioned as a good honey plant.

Wild bergamot (*monarda fistulosa*).

An Indiana beekeeper, writing in the Bee and Poultry Magazine for March, 1883, mentioned hound's tongue as a source of honey which never fails. From this article we quote:

> "There are few plants that yield more honey than it does. It comes between fruit bloom and white clover, extending into the season of the latter. Where it abounds as it does in this vicinity, bees neglect clover until it ceases to bloom. The honey has a reddish color and is not first class."

HUAJILLO (Acacia Berlandieri).

The beekeepers of the southwest boast of the quality of the huajillo honey. Huajillo is one of the acacias, but apparently not of very wide distribution. Coulter gives the range as "from the Nueces to the Rio Grande and west to Devil's River." It is also found in adjacent Mexico.

The honey is white and of mild flavor and in favorable seasons is

Hound's tongue.

stored in great quantity. Many carloads of this honey have been shipped from Uvalde, Texas, and nearby points. In the Uvalde region every beekeeper visited by the author spoke of huajillo as of first importance, although many spoke of it in connection with catsclaw and did not seem to know which was more important.

A rain while in bloom stops the flow and beekeepers report that the flow is more often cut short because of rain than for lack of it. With a little moisture present in the soil the desert plants bloom freely, and in this region it does not require much water to bring out the bloom. (See Acacia).

HUCKLEBERRY (Gaylussacia).

There are several species of huckleberry common to the Eastern States. The common huckleberry of the markets, the black huckle-

The huajillo is the source of high quality honey in southwest Texas.

berry (*Gaylussacia baccata*) is common from Eastern Canada to Minnesota and south to Georgia, in rocky woodlands and swamps. This species also occurs on the Pacific Coast, and W. J. Sheppard reports it is a honey plant in British Columbia.

Lovell reports that it is very abundant on Cape Cod and yields every year.

There are seven species of huckleberries found in the southeastern states, where they are an important addition to the nectar-bearing flora of early spring. The honey is reported by Southern beekeepers as thin in body with a pronounced flavor. See Blueberry also Farkleberry.

Black huckleberry.

HUISACHE, see Acacia.

HYDRANGEA

The hydrangea is much used as an ornamental and in some localities there are acres of them about large cities. The bees are very active on the blossoms and apparently secure both nectar and pollen.

Mr. George Gilbert, in a private letter to the author writes as follows concerning this source:

> "The bees tumble about in the pollen and suck the nectaries of the little flowerlets eagerly. They go away heavily laden. Where these shrubs are plentiful they must be of use, as they fit in between clover and buckwheat."

HYDROPHYLLUM, see Virginia Waterleaf.

HYSSOP (Agastache nepetoides).

Giant hyssop is a tall perennial herb with flowers crowded on a terminal spike, flowering in summer. It is common in woodland borders from New England to Minnesota and south to North Carolina and Texas. Lovell lists it as blooming about six weeks and much visited by honeybees. Probably nowhere important.

The fragrant giant hyssop (*Agastache foeniculum* or *anethiodora*) found on the prairies of Manitoba and Alberta south to Illinois, Nebraska and Colorado, has foliage with the scent of anise and is sometimes called anise-hyssop. In the seventies, when planting for honey was quite a fad, this species received some attention. Concerning it H. A. Terry wrote as follows in the Beekeepers' Journal, March, 1872:

"It produces honey in the greatest abundance, which possesses in a slight degree the same fragrance of the plant, which renders it exceedingly pleasant to the taste. It commences to bloom in May or early June and blooms incessantly until late in autumn. I firmly believe an acre of this plant well established would be ample pasturage for 100 colonies of bees.

"In manner of growth the plant somewhat resembles common catnip. On clear ground it may be sown broadcast, and when established will take care of itself, or it may be sown in drills and cultivated. The flowers are purple and it is well worthy of a place in the flower garden. The seeds may be sown in autumn or spring and when mature will self-sow so as to produce plants in greatest abundance. I find my bees work stronger on this plant than on any wild plant in this part of the country (Iowa)."

Anise-hyssop in flower.

The author found this species much visited by bees in the region north of Winnipeg, and there it is apparently the source of considerable honey.

In the bush country of western Canada from Winnipeg to Edmonton, the hyssop is an important source of surplus and in the vicinity of Edmonton the author found it reported as the source of from one third to one half the total crop. Honey said by local beekeepers to be from this plant was very light in color, of delicate minty flavor with delightful aroma and heavy body. The high quality and fine flavor would soon command a good demand once it became known in the markets.

Several years of test of the anise-hyssop in our garden at Atlantic, Iowa, have shown this plant to be unusually attractive to the bees. They visit the flowers from early morning until late evening from time the first flowers appear in June until the last are killed by frost in October. Not even sweet clover excels this plant in its attraction for them.

Prior to the coming of the white race, the Indians made a beverage from the leaves of the anise-hyssop and also used the leaves to flavor other food. In addition they extracted a remedy for colds from it. The full story of this observation is told in American Bee Journal for December, 1940, and December 1943.

In American Bee Journal, March 1933, George H. Vansell reported that *Agastache urticifolia* is abundant in the high Sierra Nevada mountains of northern California where it grows to a height of ten feet in thickets of ceanothus which protect it from sheep.

During July 1931 a surplus of 100 pounds per colony was harvested in Eldorado County. The honey was of fine quality with mint-like flavor. When first extracted it was water white and of good body. It showed no tendency toward granulation after several months.

I

ICE PLANT or SEA FIG (Mesembryanthemum aequilaterale).

The ice plant is recorded by Jepson as found on dunes and cliffs near the sea from Marin County, California, southward to San Diego. It is known as beach apple, also. The stems are several feet long, often forming extensive mats. The flowers are fragrant and showy, of bright rose-purple.

Coleman reports in Western Honeybee (April, 1921) that it is the source of considerable surplus in Santa Cruz and Monterey Counties and southward and especially in San Diego County. He states that the honey is white and of good flavor. Snow-on-the-mountain is also sometimes called ice plant, but is not to be confused with this.

IDAHO—Honey Sources of.

Situated in the inter-mountain region Idaho has a great variety of climatic conditions depending very much upon elevation. The altitude ranges from 738 feet above sea level at Lewiston to more than 12,000 feet in the highest mountains. Much of the state is mountainous with interspersed areas of plains. In the higher elevations the summers are short and cool, while in the lower valleys the summers are long and the winters mild.

The precipitation varies from less than 10 inches in the valley of the Snake River and adjacent plains to more than 40 inches on the slopes of some of the higher mountain ranges. In general, the parts of the state which are sufficiently level to encourage agriculture are deficient in rainfall. Although dry farming is successfully followed in some sections, it is precarious and agriculture only reaches its best development in the valleys under irrigation.

Because of the relatively low humidity, the nights are cool even in summer in all parts of the state, including the sheltered valleys. The climatic conditions are favorable to nectar secretion and large crops of honey are harvested.

Honey Sources

In the American Bee Journal for August, 1924, appeared an article on "Honey Production in Idaho" by Don B. Whelan which describes in some detail the conditions peculiar to this state. From this article, we learn that the sources of nectar are very similar to those of a large area along our northern border including all the Rocky Mountain region. Willows, poplars, elms and maples furnish early pollen with some nectar from the willows and maples. Following these, dandelions produce nectar and pollen in abundance to stimulate early brood rearing. In many of the valleys, fruit growing is extensively followed. Cherries, peaches, plums and apples are grown in large orchards and under favorable conditions furnish much early bee pasture. Raspberry grows wild in the forest regions and in some places yields abundant nectar. The cultivated raspberry is also valuable where grown. Black locust is commonly grown as a shade and ornamental tree as well as in small plantations for fence posts.

Wild mustard is a common weed in the grain fields, where it yields much poor quality honey. The surplus honey comes mostly from alfalfa, sweet clover (melilotus) and white and alsike clover. Both white and yellow sweet clover are found commonly along the roadsides and irrigation ditches. They are also grown to some extent for seed. White clover and alsike clover are found in the irrigated valleys where good honey crops are reported in the southern part of the state.

Among native plants may be mentioned snowberry, dogbane or Indian hemp, and fireweed or willow herb. These are of special importance in the northern part of the state. Goldenrods and asters are mentioned as of some value toward the end of the season.

ILLINOIS—Honey Sources of.

Illinois, situated in the heart of the Mississippi River Valley is 385 miles in length from north to south and 216 miles in width at the widest point. The total area is 56,665 square miles.

Because of the great distance between the northern and southern portions there is a marked difference in climate. This difference is due almost entirely to latitude as there are no important physical features which modify the climate to any great extent except in the vicinity of Lake Michigan, which the state borders for sixty miles on the northeast.

Like all the states in the interior of North America, Illinois has a continental climate with severe changes of temperature, especially in the winter months. The summer is marked by periods of high temperature and the winter by sudden changes and falling temperatures often fifty or more degrees within a few hours.

Soils

Northern Illinois is for the most part of glacial origin and is a rich loam which produces large crops of cereals, grasses and vegetables.

The soil in the northern section is lighter and better drained than farther south.

The central portion is also glacial and somewhat similar in character. The soils of the southern portion, however, are of a different type. Black loam is confined to small areas mostly in the river bottoms. The greater part of this region has a soil of a lighter color which is underlaid with hardpan. Most of this section was originally covered with timber.

Altitude

Northern Illinois is the highest portion of the state and, for the most part, is more than 800 feet above sea level. There are a few valleys along the streams where an elevation of less than 600 feet is to be found.

There is a general slope from north to south and the central portion of Illinois averages something like 200 feet lower than the northern third of the state. Most of this area is above the 600 foot level, with a few small areas above 800 feet.

The southern section is mostly below the 600 foot level, but the Ozark Hills extend for a distance of more than sixty miles in an easterly direction with an altitude of something like 1000 feet. This is but a narrow belt of a few miles in width.

Rainfall and Temperature

The annual precipitation for Northern Illinois averages about 33 inches, for Central Illinois about 36 inches and for the southern portion slightly above 41 inches. Rainfall is usually heavy in spring and ample for crop needs in summer and fall, although short periods of drouth sometimes occur.

Extremes of temperature recorded indicate a much greater range than often occurs. The highest recorded 112 degrees above zero at Ottawa in northern Illinois is much above the maximum of the average summer. The lowest 32 below zero at Ashton is likewise much below the usual winter temperature. Summer temperatures of 90 to 100 degrees and winter temperatures down to zero are not uncommon.

The average date of the last killing frost in spring is April 30 in the northern part of the state, April 20 in the central portion and April 13 in the southern portion. The average date of first killing frost in the fall is October 10 in the north, October 16 in the central section and October 21 in the south. The frost free growing season thus varies from 163 days in the north to 179 days in the central section and 191 days in the southern portion.

Honey Sources

Alsike and white clover are the principal sources of honey over the greater portion of the north half of Illinois. Sweet clover, however, is grown extensively in rotation of crops and also for seed. In the region where it is grown extensively, beekeeping is a prosperous business and large crops are the rule.

In the Illinois and Mississippi River bottoms much of the honey is gathered from fall flowers, including the bonesets, Spanish needles, heartsease, etc., with something from wild cucumber, button bush, ampelopsis, etc.

Most of the state has ample spring forage in the way of willows, maples, elms, dandelion, waterleaf (in woodlands), and fruit bloom. Indian currant (*symphoricarpos*) is abundant in the woodland pastures; basswood and black locust are sources of surplus in a few places. Commercial orcharding is general in the southern section and the bees are sought to pollinate the blossoms.

INCENSE CEDAR (Libocedrus decurrens).

The incense cedar is an important forest tree in the higher elevations of the mountains of California and Oregon from which comes honeydew in commercial quantity.

The tree grows to a height of 70 to 100 feet or more and to a diameter of trunk of from two to six feet. According to Sudworth (Forest Trees of the Pacific Slope) incense cedar is found in the Mountains of Southern Oregon, Sierras and Coast Ranges of California, western edge of Nevada and northern Lower California. It is found generally at elevations of 2500 to 6000 feet although in some localities its range varies from as low as 1200 feet to 8000 feet.

It is most abundant and largest on west slope of Sierras. Usually found with sugar and yellow pine, also with white fir in lower part of the latters vertical range.

Due to the presence of a scale insect honeydew is produced in great abundance at times and beekeepers secure large crops in favorable situations, from 100 to 300 pounds per colony being reported. Because of the wide area occupied by this tree it bids fair to become an important source of honeydew honey when roads are opened into the forests to make it available.

According to F. E. Todd the honey is of heavy body, non-granulating and of indifferent or neutral flavor. In the market it brings a fair price and in view of the large yields is a promising source to beemen within reach of it.

> "The incense cedars in certain areas, at varying elevation from 2000 to 7000 feet, are infested with a cypress scale (*Xylococculus macrocarpae*). This insect beginning in early summer 1931, produced large quantities of a very gummy, amber-colored honeydew, which was freely gathered by the hive bees. The color of the exudate remained constant until fall, after the bees clustered for winter. A visit to the mountains on November 3 revealed a most interesting fact. The honeydew which was being thrown off at that time was crystal white. The large beads of it sparkled beautifully like dewdrops or diamonds. After drops became large they dripped away to the soil.

"The white cedar honey, however, is reported to come to the hives in July, during certain years."—George H. Vansell, American Bee Journal, Sept. 1932.

INDIANA—Honey Sources of.

The map prepared by E. G. Baldwin, gives a good idea of the bounds of the principal natural honey regions of Indiana. In the American Bee Journal for February, 1921, he gave a detailed description of each of these regions. He described region 2 along the Wabash River from Cass to Wells and Adams Counties as the best beekeeping territory in the entire state, with white clover furnishing the surplus. To the south of this, in region 3, is found a somewhat similar area where white clover is the principal source of nectar, with alsike and sweet clover also found to some extent. Because of lighter soils, less rainfall and more southern latitude the flows are less dependable. His region 4, along the eastern border of the state, overlaps both the above mentioned areas and is rather arbitrary division to indicate the range where sweet clover is of special importance. In the southeastern part of the state, to the north of the Ohio River, locust is of first importance as a source of nectar. Bordering the streams in the southwestern section the bluevine, or climbing milkweed yields heavily. This area overlaps the heartsease region to some extent. The territory included in the circle is the poorest for honey production in the state, with no important source of nectar present in sufficient quantity to insure dependable honey flows.

The Kankakee Marsh region, in the northwestern section is subject to overflow and the honey crop is largely gathered from fall flowers, including buckwheat, button-bush, Spanish needles, milkweed, and the bonesets or thoroughworts.

The following list of honey plants of the state is taken from the second annual report of the Inspector of Apiaries, by G. S. Demuth:

Crowfoot, April-May.
Dandelion, April-September.
Elm, March-April.
Box elder, March-April.
Silver maple, March-April.
Fruit trees, April-May.
Currant and gooseberry, May.
Grapevine, May.
Honey locust, May-June.
Raspberry, May-June.
White clover, May-June-July.
Red clover, June-July.
Basswood, June or July.
Button bush, July.
Teasel, July.
Sugar maple, April-May.
Judas tree, April-May.
Willow, March-April.
Red maple, March-April.
Skunk cabbage, March.
Ground Ivy, May-August.
Buckeye, May.
Black locust, May.
Poplar (tulip-tree), May-June.
Blackberry, May-June.
Alsike, June-July.
Sweet clover, June till frost.
Mustard, June till frost.
Asters, August till frost.
Catnip, July and August.

Sumac, August.
Figwort, July till frost.
Ironweed, July till frost.
Marsh milkweed, July-August.
Smartweed, August-September.
Marsh sunflower, August till frost.
Spanish needle (*Coreopsis*), August.
Cucumber, melons, etc., July-August.
Boneset, July till frost.
Buckwheat, August.
Beggarticks, August.
Jewelweed, August-September.
Goldenrod, August till frost.

INDIAN BEAN, see Catalpa.
INDIAN CHERRY, see Serviceberry—also Buckthorn.

INDIAN CORN or MAIZE (Zea mays).

Indian corn is native to North America and was cultivated by the Indians at the time of the discovery of the continent. It has become one of the most important grain crops and is grown in large acreage. It furnishes an abundance of pollen in mid-summer and the bees work upon the tassels so freely as to give rise to a very general impression that it furnishes honey in abundance. At times, aphides are to be found upon the stalk and the bees also get a sweet substance from the axils of the leaves. While on infrequent occasions the bees may get honeydew from Indian corn, its principal value to the beekeeper is as a source of pollen.

> "If the weather is favorable for the reproduction of plant lice, we may always expect them to attack the tassel, making the top leaves sticky and discolored. I have seen the bees pile on the tassel until you could scarcely see anything but the bees gathering this honeydew. The honey thus obtained is dark, but of very fair flavor."—Wm. R. Howard, White Rock, Texas. American Bee Journal, page 225. May, 1880.

W. K. Morrison, writing in Gleanings, states as follows regarding corn:

> "Corn is not generally set down as a yielder of the nectar sublime, but in tropical countries it is a very valuable honey plant, showing the importance of locality, showing also that corn is a native of the tropics." (Aug. 1, 1905.)

On page 599 of Gleanings for 1882, H. M. Morris of Rantoul, Illinois, reports the bees working heavily on the corn plant, getting honey at the joint at the base of every leaf. He stated that the bees worked on corn for about two weeks and that he secured 500 pounds of extracted honey beside some in sections. The editor comments that the sample sent to him would rank with the best in both color and flavor. There are so many similar reports from different localities that the fact that the bees sometimes store surplus from corn can hardly be questioned.

> "The smooth gland or plate, situated at the base of the leaf, or rather at the junction of the blade with stalk, usually furnishes honey, and, at

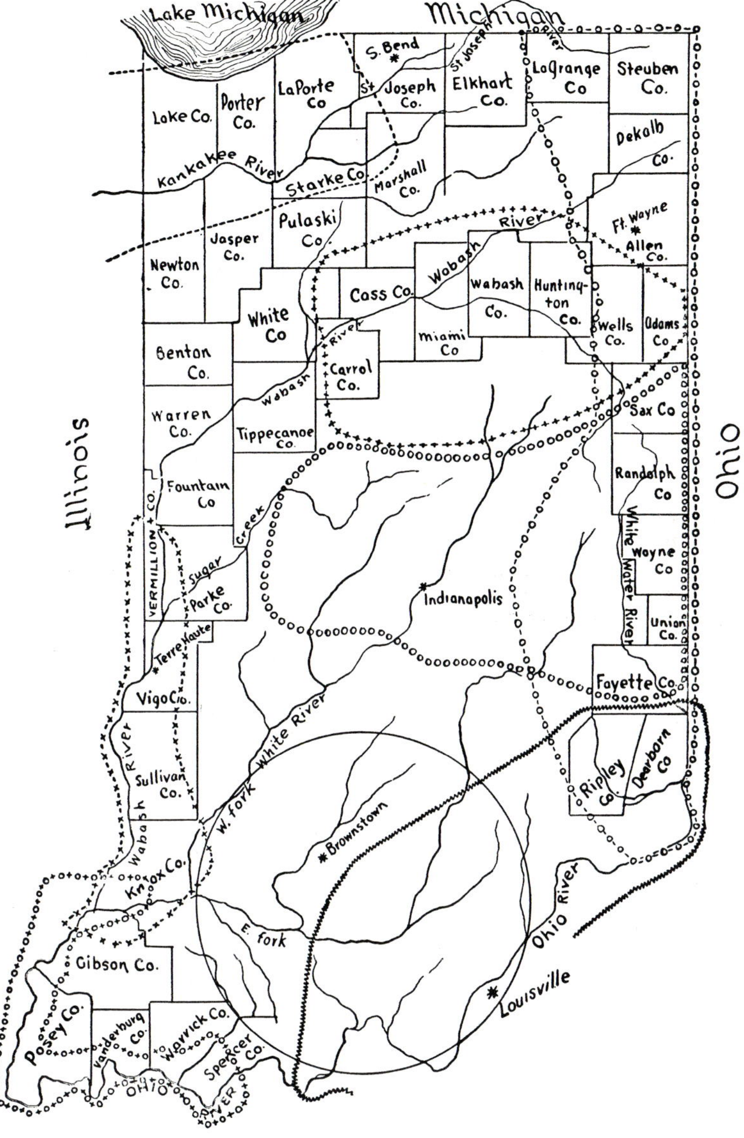

Map showing range of principal sources of honey in Indiana.

Region 1. .... Kankakee River Marsh lands.
Region 2. xxxxxxxxxxx Best white clover and basswood.
Region 3. oooooooooooooo Fair white clover, with alsike and sweet clover increasing.
Region 4. -o-o-o-o-o-o-o-o Best sweet clover.
Region 5. vvvvvvvvvvv Locust.
Region 6. xoxoxoxoxox Bluevine or climbing milkweed.
Region 7. -x-x-x-x-x-x-x Heartsease.
Circle—Poorest region for beekeeping.

certain climatic changes the flow from this source is quite heavy."—R. B. Robbins, Gleanings April 1882.

In American Bee Journal, February 1936, Alfred H. Pering, wrote of instances when corn was growing rapidly the stalks had split and permitted the sap to run out. At such times the bees gather the sap.

From Gleanings in Bee Culture, page 123, March 1882, the following by Elisha Gallup, which indicates that at times corn may secrete some nectar, is taken:

> "One season in Iowa, my bees gathered honey very freely from corn. They gathered large quantities of pollen from the tassels, but the honey was gathered from the silk. The silk fairly glistened with the sweet—so much so that you could taste it with your tongue, and also see it with the naked eye. I succeeded in getting four kegs of 150 lbs. each, almost pure corn honey; and when it granulated it was the coarsest grained honey I ever saw. It was quite yellow and had a peculiar taste. I could detect somewhat the flavor of corn silk. Never, in all my experience before or since, have I seen the bees work so freely on corn as they did that season. The atmosphere was quite humid and hot, night and day, during said flow of honey."

INDIAN CURRANT (Symphoricarpos orbiculatus).

The Indian currant, also known as coral berry or buckbrush, is a widely distributed shrub that furnishes considerable nectar in late summer. It may be found in the woodland borders and open forest from New York, west to the Dakotas, south to Missouri and Arkansas, and from New Jersey south along the mountains to Georgia and Alabama.

Indian currant in bloom.

The blossoms are very small and inconspicuous, but where the plant is abundant it is much sought by the bees. In southeastern Iowa, the season of 1914 was a very poor one for the bees, and many colonies required feeding to get them through the winter. In a few localities, where buckbrush abounds, they not only were well prepared for winter, but stored some surplus. The blooming season is July and August in most Northern States, so that the clover harvest is usually nearly over when it comes on. The bunches of red berries that hang on the bushes after the leaves have withered and dropped, will be instantly recognized by anyone familiar with the plant. These berries are often about the only winter food available for small birds when the ground is covered with snow.

The snowberry (*S. racemosus*) is a related species with white berries. It occurs from New England west to Nebraska and Dakota, also on the North Pacific Coast. The plant is quite similar to the red-berried species in habit and growth and is also often called buckbrush.

INDIAN FIG, see Prickly Pear.

INDIAN HEMP, see Dogbane.

INDIAN PEAR, see Serviceberry.

INDIGO BUSH (Dalea spinosa).

Indigo bush is a spiny shrub common to the desert region of southern California and adjacent Mexico. It sometimes reaches the proportions of a small tree about fifteen feet in height, but is crooked and much branched in appearance. The flowers are small and resemble those of the pea family. Because of the deep blue color the name Indigo Bush is applied. The flowers appear in spring.

Coleman lists this species as the source of both pollen and nectar. There are numerous related species of dalea which are valuable also. (See Dalea).

Indian currant or buck brush, showing fruit.

INDIGO-WEED (Baptisia tinctoria). Yellow Indigo or Clover Broom.

The yellow indigo is known by a number of local names, among which may be mentioned shoofly, rattlebush, wild indigo, clover broom, horsefly weed and bugle-weed.

It is common on dry and sandy soils from New England to Minnesota and south to Florida and Louisiana. The showy yellow flowers appear in mid-summer.

E. M. Barteau, of Brookhaven, Long Island, writes that it grows there on waste, sandy lands and most luxuriantly on burned-over areas in woodlands. It grows about two feet high, with an extensive root system penetrating three or four feet into the soil. It yields nectar freely under almost any kind of weather conditions. Immediately after rain or on hot, dry days, there is little difference apparent in the yield.

The honey is light amber with a characteristic flavor, neither mild nor strong. Coming at a time when otherwise there would be a dearth, it is very valuable. In 1920 Mr. Barteau had a heavy flow from yellow indigo, which grew in the greatest profusion in a burned-over woodland. From July 20 to August 5, his hive on scales showed a net gain of from 3 to 18 pounds per day, with the exception of only two days. The total net gain from this source in 17 days was 157½ pounds.

There are several other species of *Baptisia,* some of which have white or blue flowers that may be expected to yield more or less nectar.

Baptisia australis native to the prairies from Pennsylvania to

Seed pods of *Baptisia australis.*

Georgia and west to Arkansas and Nebraska has proved very attractive to the bees in the author's test garden. Indications are that where abundant this species may provide good bee pasture.

INKBERRY, see Gallberry.

IOWA—Honey Sources of.

There is no other state of equal area in our country which has as little waste land or as little diversity of physical conditions as has Iowa. There are no mountains, no large areas of non-productive land nor any large lakes within her borders. Since Iowa is bounded on the east by the Mississippi River and on the west by the Missouri, it lies in the broad valleys of these two great streams.

There are three factors to be considered in comparing the different parts of Iowa from a beekeeping standpoint: altitude, soil and honey plants. There is not sufficient difference in the temperature or the rainfall of Iowa localities to greatly influence the honey crops.

Surface

Iowa contains 55,475 square miles of territory carved out of the heart of the continent. The richest portion of each of the six surrounding states in every case adjoins Iowa. The surface was originally prairie, interspersed with belts of timber along the streams. The streams are sufficiently numerous to afford excellent drainage. With settlement, groves were planted on nearly every prairie farm and the timber areas were largely cleared.

The altitude ranges from about 575 feet at the top of the bluff, at Keokuk, in the extreme southeast corner, to a little more than 1,600 feet at Ocheyedan, in Osceola County. There is, therefore, only a range of about 1,100 feet in altitude for the entire state. The average altitude for most of the Mississippi drainage basin for northeastern Iowa is below 1,000 feet, while the greater portion of the Des Moines drainage basin and western Iowa is above 1,000 feet. Only a small area in Osceola, O'Brien and nearby counties in northwestern Iowa is above 1,500 feet in elevation.

Climate

Publications of the United States Weather Bureau, from which much of the data contained in this paper were secured, are quoted as follows, concerning the Iowa climate:

> "Iowa has all the essential factors of climate to make it the most productive State in the Union. The cold of winter, while not as severe as in States farther north, is sufficient to disintegrate the soil; and the rains and heat of the summers, while not as intense as in the states farther south, are sufficient to insure bounteous crops every season. Situated near the geographical center of the United States, the climate of Iowa is continental in type. This implies a wide range in temperature, winters of considerable severity, summers of almost tropical heat, and a large percentage of sunshine. There being but little range in elevation in the various sections, the climate is similar, but with such variation of temperature and rainfall as result from latitude and location with reference to the pathway of the cyclones which pass eastward over the continent."

The last killing frost in spring usually comes from the middle of April to the first week in May. It has occurred as late as May 21, at Dubuque, and May 28 in western Iowa, although frost is rare later than the 2nd or 3rd of May. The first killing frost in autumn usually comes in October, in the eastern part of the state, and more often in September in the western part. For the state as a whole, the frostless period is usually about five months.

Rainfall

In the southeast part of the state the average annual rainfall is about 37 inches. There is a gradual decrease toward the northwest, where the average is 27 inches, with an average for the entire state of nearly 32 inches. Seventy-one per cent of the precipitation occurs during the six crop months from April to September. More than half of the entire rainfall comes during the four summer months from May to August when it is most helpful to growing crops.

The precipitation is heavier in the eastern part of the state during the winter months and heavier in the west in summer. Occasionally there is a severe drought for a few weeks during the growing season, but entire failure of crops is unknown.

Soils

Iowa soil is glacial in origin, and for the most part is very rich. The beekeeping possibilities seem to be best in the western and northwestern part of the state on the Mississippi loess and the Wisconsin drift. The southern loess in the southern portion of the state, offers the poorest prospect to the beekeeper. In general, the north half of the state furnishes far more certain honey flows than does the south half, although the sources of nectar are similar.

The Honey Plants

White clover was long the main source of surplus in nearly every county. In many of the counties there is no surplus in the years when white clover fails. The flow from white clover increases in intensity and becomes more dependable from south to north. From the author's long experience at Atlantic, in Cass County, which is in the edge of the Missouri loess soil, about 70 miles from the southern boundary of the state, it seems safe to say that there will be one big crop, two fair crops, one light crop and one failure in every five. It sometimes happens that two big years will come together, or two very poor ones.

Alsike clover is important in some Iowa localities, but there are many counties where it is little grown. There are many localities where the beekeeping possibilities are very limited because of the fact that little clover of any kind except the red clover is grown. The corolla tubes of the red clover are too deep for the bees, and except under unusual conditions it is not of much value to the beekeeper. In the dairy districts, where the white clover pastures are extensive and where alsike meadows are common and where sweet clover is generally grown we find the best bee pastures.

Sweet clover is a farm crop. It has long been present along the rail and wagon roads, but the area was not sufficient to add much to the beekeepers' product. Of late it is coming into popularity on the farms. It is more dependable as a source of nectar under Iowa conditions than is white clover, and with the larger acreage is doing much to encourage honey production. In one county in northwestern Iowa, there are estimated to be more than 5,000 acres of sweet clover on the farms.

Without the clovers, beekeeping in Iowa could hardly be taken seriously, while, with them, it is becoming more and more a recognized industry.

Basswood was once important in the fringes of woodland along the streams, but is rapidly disappearing. There are still a few places along the Mississippi and Missouri Rivers and some of the small streams where enough basswood still remains to yield some surplus, but the area is constantly growing less.

In the river bottoms and in the swampy areas in the north central and northwest parts of the state, heartsease furnishes an important yield of fall honey. A few localities also have sufficient Spanish needle or boneset to be of some value. In the woodlands, also, buckbrush (*Symphoricarpos orbiculatus* and *S. racemosus*) yield some surplus in midsummer. Asters are also of some value in late autumn, but rarely yield surplus.

In early spring the willows, maples, fruit trees and dandelion yield nectar and pollen for brood rearing and prepare the bees for the later clover flows. Dandelion is especially abundant over most of the state, and a surplus of dark honey is sometimes secured from it.

There is a long list of minor plants which add something and which are of some importance to keep the bees busy between flows. Iowa, however, has a short list of important honey plants.

IRONWEED (Vernonia).

There are many species of the ironweeds to be found in many countries. They are common in Asia and Africa, as well as North America. They are common from New England south to Florida and west to Dakota and Texas. The picture shows the flowers of the western ironweed (*Vernonia fasciculata*). In the middle west they grow very commonly in pastures, and the purple blossoms are very conspicuous in late summer. At times the bees work them very eagerly, but it is doubtful whether they are often of much value as a source of surplus.

Blossoms of western ironweed.

IRONWOOD (Olneya Tesota).

The Mexican or Sonora Ironwood is a common tree along desert watercourses and dry washes in southern Arizona and adjacent California and Mexico. Although the name ironwood has been applied to numerous eastern trees before this species was known, to none is the name so appropriate. The wood is very heavy with crooked grain and dark color. The grain is too uneven and checked for use except as fire-

Ironwood tree blossoms.

wood, although small specimens are sometimes used as canes. It commonly grows to a height of 12 to 15 feet, although the author has seen a few specimens of much larger size. Like most desert trees it takes a bushy form with a short, thick stem or trunk. The leaves remain on the tree over winter, falling when the new ones appear. They are compound, similar to the acacias and of a dusty greyish green color. The flowers are purple and resemble small pea blossoms. The blooming period varies somewhat depending upon available moisture from spring to early summer. Old residents of the region state that the tree does not bloom every year.

Beekeepers generally, regard it as an important source of honey although, like other desert trees, somewhat uncertain in its yield. The honey is said to be light amber in color and of good quality. In favorable locations, a super of honey is reported from this source.

Since the tree occurs in company with mesquite and palo verde, the honey is likely to be more or less mixed with that secured from these trees as well as desert shrubs.

Most plants in this region are equally well known by Mexican names and the Mexican Ironwood is also called Arbol de Hierro and Palo de Hierro.

The name ironwood has been applied to so many different trees that much confusion is the result. Several are important sources of nectar. The reader is referred to the following additional species which are locally called ironwood: ti-ti, coma, mesquite.

IRONWOOD, see Coma, also Titi.

IVY (Hedera Helix).

The ivy, often called English ivy, is a root climbing evergreen vine commonly cultivated for covering brick or stone walls. It is native to Europe and Northern Africa but is very generally cultivated in temperate regions. Individual plants are known to live for long periods. One at Montpellier is mentioned by Ferd. Von Mueller as known to be 440 years old. He also mentions it as a bee plant.

This species is very variable and according to Bailey (Cyclopaedia of Horticulture) there are more than sixty varieties cultivated in European gardens. It is very commonly found covering the walls of ancient buildings of the old world and factories, stone fences and brick or stone houses in America.

In the Feb. 1926, issue of American Bee Journal, Foloppe Brothers of Nice, France, have an extended article concerning this plant from which the following is quoted:

> "The bees are not quiet they are working actively and come home heavily laden. We are now in front of the hives and the odor is almost excessive, resembling that of syringa or tuberose. Toward evening the odor becomes still stronger.
>
> "A very plain buzzing drew attention to a steep hillside covered with ivy in full bloom. This was the blossom that gave the penetrating perfume for the delicate little yellow blossoms were covering the entire abrupt face of the stone.
>
> "Notwithstanding the lateness of the season we managed to secure some twenty pounds of the nectar, selecting the unsealed combs, so as to avoid as nearly as possible mixing it with other honey, especially buckwheat honey.
>
> "This honey was comparatively thick, perhaps owing to the low temperature: its aroma was very pronounced and even if there was some mixture of buckwheat it was sufficiently *sui generis* to cover the strong aroma of buckwheat honey.
>
> "Later and in spite of the great humidity of the atmosphere, it granulated readily; its strong aroma was noticeably weakened. * * * * it does not have the bitter flavor of the leaves of that plant, neither does it seem to

possess the perniciousness that might be expected from a plant with very purgative properties. The honey may be classed as aromatic. Its grain is very fine, in granulation."

IVYWOOD, see Laurel.

J

JACKASS CLOVER or STINKWEED (Wislizenia refracta).

Jackass clover or stinkweed is a rank scented annual plant with yellow flowers, growing two to six feet high. It is common from Sacramento to Lathrop and southward in the San Joaquin Valley, according to Jepson.—Flora of Middle Western California.

It is an important honey plant in the interior valleys, where it is reported as blooming freely only every other year. The blooming period is from August to December. C. R. Snyder, of Selma, regarded it as a main source. He reported as high as 100 pounds of surplus per colony. He usually extracted two or three times from this source in September, and has extracted as late as December. A heavy rain or a frost will stop the flow. The honey is light and of good quality. A sample presented to the author is a light amber with a peculiar flavor, unlike our eastern honey. The flavor reminds one somewhat of butter-scotch candy. It is rather strong, but agreeable.

Richter comments on this plant as follows:

> "Honey water white, mild in flavor and of good body; granulates in three to six months, when it resembles a paste made from powdered sugar. During the fall of 1909 a Fresno beekeeper reported that he extracted thirty pounds per colony each week for six weeks from this source. Another beekeeper of the San Joaquin Valley relates that during the jackass clover flow the noise was terrific, and that homecoming bees flew so slowly that they could be picked out of the air. It was Henry T. Christman, of Colinga, who first became aware of its value as a honey plant and gave it its present name."

In Stanislaus and San Joaquin counties it is called alkali mustard.

Commenting on the above quotation the California "Beetimes," January, 1929, quotes Mr. Christman as disclaiming credit for giving the plant its name. It is stated that he called the plant Golden Clover and that C. I. Graham was responsible for the name jackass clover which has come into common use. It is said that Graham lost a team of burros on which he depended for moving his outfit, and that they were hidden from sight among these plants which grew to a height of four feet or more. After hunting for the lost animals for a week and losing a honeycrop from failure to move to a new field, he called the plant "Jackass Clover." The name seems to have struck a popular fancy for it is the one by which it is commonly called.

The plant is found also in Arizona, New Mexico and in Texas as far east as San Angelo and as far north as the Red River. There it is reported as visited freely by the bees.

JAMAICA DOGWOOD (Ichthyomenthia piscipula).

The Jamaica dogwood of southern Florida and the Keys is a tropical tree with scaly bark of a reddish brown color. The branches are stout and irregular and the flowers are reddish white.

It is frequently reported as a source of surplus honey.

JASMINE, see Yellow Jessamine.
JERUSALEM ARTICHOKE, see Sunflower.

JERUSALEM THORN (Parkinsonia). Horse-Bean.

There are several species of *Parkinsonia* common to the southwest. They are shrubs or small trees with smooth bark, armed with thorns. The flowers are yellow or white in short racemes.

The horse-bean (*P. aculeata*) is found as far east as Florida and as far west as California. In Texas, where it is regarded as a valuable source of nectar it is known as "Retama."

The Jerusalem thorn (*P. microphylla*) is also known by the Mexican name of Palo Verde, which is properly applied to a related plant. (See Palo Verde). It is also known as "Desert Bush." It is found in southern Arizona, California and south into Mexico.

Probably all the species of *Parkinsonia* are valuable to the beekeepers. In many desert regions they are among the most important sources of nectar.

JEWEL-WEED or Touch-Me-Not (Impatiens).

The jewel-weed is common in wet places and along small streams in shady situations. Some are to be found from Alaska to New England and south to Florida and Louisiana. The plant has an odd hanging blossom, as shown in the picture. It is often called touch-me-not, from the sudden bursting of the seed pod when touched.

Jewel weed, or touch-me-not.

It is usually regarded as a bumblebee flower rather than a bee flower, but is reported as a source of honey in both Michigan and Wisconsin, where it is quite common in places.

Chas. Bessey included two species of jewel-weed in his list of Nebraska honey plants published by the Experiment Station in 1895.

JIMSON WEED, see Datura.
JOE-PYE WEED, see Boneset.
JOINTED CHARLOCK, see Radish.
JUDAS-TREE, see Red-Bud.
JUNEBERRY, see Serviceberry.
JUNE-BUD, see Red-Bud.
JUNIPER, see Red Cedar.

K

KANSAS—Honey Sources of.

Kansas has a great variety of climatic and soil conditions and the honeyflows vary greatly in different parts of the state. There is great variation in the rainfall and extreme changes in temperature. In the

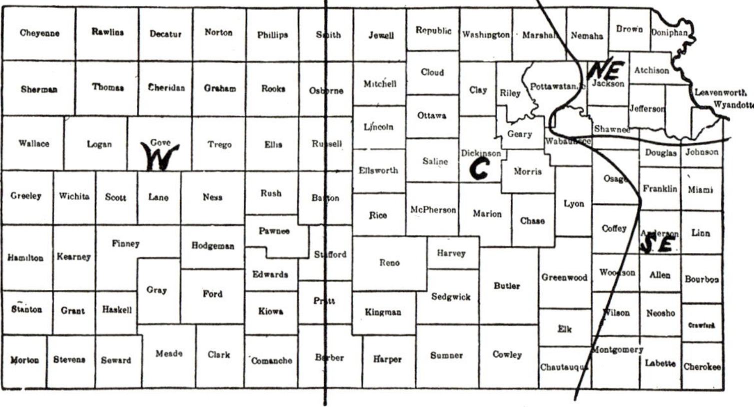

Kansas is divided into four natural beekeeping regions, as shown by this map.

southeastern portion the altitude is less than 1,000 feet above sea level, while the high plateaus of the western part are above 3,000 feet. The extremes are about 750 feet at the mouth of the Kansas River to 4,000 feet at the west line of the state.

The rainfall decreases from 42 inches in the southeast to about 15 inches in the west. Three-fourths of the annual precipitation comes in the crop-growing months from April to September.

The northeast corner of the state has a glacial soil and in this region white clover is of some importance. Southeastern Kansas soils are formed from shale and leguminous plants are not important in this section. The central portion has a mixture of limestone in its formation and sweet clover and alfalfa succeed well there. The western part of the state is largely composed of wind blown and outwashed soils rich in lime.

In the January (1922) issue of American Bee Journal, Dr. J. H. Merrill, of the Kansas College of Agriculture, divides the state into four general regions, as shown on the accompanying map.

The rainfall is heaviest in the northeastern and southeastern regions. In the northeastern region white clover, sweet clover, heartsease, Spanish needle, dandelion, fruit bloom, basswood and forest trees furnish the principal nectar sources.

In the southeastern region Spanish needle, persimmon, sumac, goldenrod, heartsease and boneset are the important sources.

The Central Region

The altitude of the central region ranges from 1,000 feet on the east to 2,000 feet on the west. The rainfall varies from 31 inches east to 21 inches west. Most of the commercial beekeepers of Kansas live in this part and the apiaries, for the most part, are located in the valleys of the Arkansas, Blue, Kaw, Verdigris and Solomon Rivers. Maple, elm and dandelion furnish early nectar and pollen and the principal surplus comes from sweet clover and alfalfa. Yellow sweet clover is found in many places, thus furnishing a good flow in advance of that from the white sweet clover and alfalfa. Heartsease also furnishes considerable surplus in this region.

The West End Region

This part of Kansas, commonly called the short grass country, varies from 2,000 to above 3,000 feet in altitude. Most of the beekeeping is found in the valleys where large acreage of alfalfa is grown under irrigation. In addition to alfalfa, most of the surplus comes from sweet clover. In some sections the Rocky Mountain bee plant grows abundantly, also. The principal drawback lies in the lack of early pollen and nectar for spring brood rearing.

Nectar and Pollen Sources

Taking the state as a whole, Merrill lists the plants of importance to the beekeeper in the following order:

Alfalfa, yielding surplus in all parts of the state above 1,000 feet, where soil and moisture conditions will permit, blooming about June 1st.

White sweet clover, yields through June and July and is regarded as of first importance in many parts of the state. On the whole it crowds alfalfa closely for first place.

Yellow sweet clover, blooms before the white variety and at a time when little else is within reach.

White clover, blooms in May and June and is of value only on the glacial soils of the northeast portion of the state. In favorable seasons large yields are secured.

Heartsease, blooms from August until frost. It is found on suitable situations throughout the eastern half of Kansas. It is usually the source of a good crop unless cut short by frost. The quality varies greatly in different sections, due perhaps to the different species of the plant predominating in different localities.

Alsike clover does well in northeast Kansas, blooming in July. The acreage is restricted but will be important in this region if largely grown.

Dandelion is the source of large quantities of pollen in April in the northeast and north central sections. Nectar for early brood rearing is also secured from dandelion.

Basswood is found only along a few streams in the northeast section. Because of limited distribution, it is only locally valuable.

Goldenrod is reported as important in a few southeastern counties.

Spanish needle is important along the entire eastern border.

Fruit bloom is an important source of both nectar and pollen in the orchard districts of the northeast section and the Arkansas Valley and other valleys of the central section. While of principal value for early brood rearing, considerable surplus is secured in the vicinity of the large apple orchards.

Maple, elm and box-elder are valuable along the streams and in the vicinity of towns where these trees are grown for shade and street trees.

Persimmon is valuable in Cherokee County. Aster, black locust and sumac are locally valuable in eastern Kansas and canteloupes in some sections in western Kansas.

KENTUCKY—Honey Sources of.

Conditions in Kentucky, on the whole, closely resemble those of Tennessee. Beekeeping as a commercial enterprise is but little developed and, while small apiaries are numerous, there is little to encourage large scale honey production except in a few localities.

There is a long list of minor sources to encourage brood rearing over a long period, but few places offer sufficient flows for storage of good crops of surplus on a scale to justify a large expenditure of time and money.

Spring opens with a supply of pollen and nectar from such sources as elm, maple, willow, fruit trees and bush fruits. White and alsike clover are found abundantly in some neighborhoods and sparingly in others, but climatic conditions are such that dependable yields are not secured. Sweet clover is largely grown in some counties and in favorable seasons yields well. Some beekeepers report crops from buckwheat. Among the trees which are important may be mentioned basswood, sourwood, tulip-poplar, yellowwood, and black locust. Good crops are often secured in small apiaries from these sources.

Blackberries are abundant in fence corners and waste places and give some honey. Honeydew is harvested in large quantities from forest trees at rather frequent intervals. Goldenrod and boneset are reported as yielding well in late summer. The asters are among the most dependable honey plants in this area and seldom fail to give something in the way of surplus. Crown-beard in some localities is valuable.

KIDNEY-BEAN TREE, see Wisteria.

KINNIKINNIK (Rhus virens).

Kinnikinnik is a sumac which grows from the Colorado River to the Rio Grande and westward. The Indians and Mexicans mix the leaves with tobacco and smoke them. It grows in large quantity in the hills some distance north of Uvalde, Texas, and blooms from September till frost, with sufficient rain. Local beekeepers report that as high as sixty pounds of honey per colony is sometimes secured from this source. The honey is said to be green in color, with a rank, strong taste, and does not granulate. The combs are capped very white when the bees are working on kinnikinnik. This should not be confused with the species of dogwood which is also commonly called by the same name. There are several species of sumac which are valuable honey plants. (See sumac).

H. B. Parks states that the name kinnikinnik, often pronounced with an extra syllable, kin-ni-nick-i-nick, is also applied to *Rhus microphylla,* the Shawnee-haw, which grows in company with *R. virens,* and that the nectar from the two plants is so mixed that honey from either is seldom secured by itself.

In the Pacific northwest the name kinnikinnik is often applied to the bearberry (*Arctostaphylos Uva-ursi*). This prostrate or trailing shrub with red bark forms dense mats in the open woods of Washington and British Columbia. It is reported as one of the most important native sources of honey in that region early in the season. (See Manzanita). It is found as far north as Alaska and in several of the western mountain states.

KNAPWEED, see Star Thistle.
KNOTWEED, see Heartsease.

L

LABRADOR TEA (Ledum groenlandicum).

Labrador tea is a small shrub with entire leaves which are woolly underneath. The flowers are small and white and appear in terminal clusters. The blooming period is May and June. It is commonly found in swampy places in the north, from Greenland and Labrador to British Columbia and south to New Jersey and Minnesota. It is also found in the sphagnum bogs of Washington, while a related species, *L. glandulosum,* is found in higher altitudes of the Sierra Nevada Mountains of California.

Sladen reports the first named species as the source of nectar in northern Ontario. The last named, together with *L. latifolium,* is listed for western Washington in the first report of the Division of Apiculture for that state.

Two species are cultivated to some extent for their flowers and the attractive evergreen foliage. *L. groenlandicum* is said to have been used as a substitute for tea during the American war for independence, hence the name "Labrador tea."

Some species are said to be poisonous.

LACE FLOWER, see Carrot.
LADIES' EAR DROPS, see Brunnichia.

LADINO CLOVER (Trifolium repens latum).

A giant form of white clover has been developed in Italy and for years cultivated in the irrigated lands of Lombardy. The name comes from Lodi from whence it came. According to an article in Country Gentleman which attracted wide attention, in its Italian home it sometimes outyields alfalfa, producing five or more tons per acre. It is a true perennial and on moist lands will stand for a long time unless grazed too closely. An acre in Idaho was reported to have carried six head of stock for the whole season, to the latter part of August when a good margin remained.

Ladino clover appears to be suited especially to irrigated lands in arid climates where it makes a growth up to twelve inches in height. It is reported principally from California, Idaho and British Columbia where it is reported as a good honey plant. From available reports it appears that it is similar to white clover in yield and quality of honey. The flowers are larger than those of white clover.

Cross pollination is necessary to seed production and honeybees are effective pollinating agents.

LADY'S THUMB, see Heartsease.
LAMBKILL, see Laurel.
LAMIUM, see Dead Nettle.

LANTANA.

There are several species of Lantana common to the Gulf States. Most of the species common to this country are shrubs. Some species are cultivated as ornamentals.

Lantana is reported as a weedy pest on cattle ranges in Hawaii and is found in all parts of Porto Rico. Phillips reports it as a source of nectar in both places. (Bul. 75 Bu. of Ent. and Bul. 15 Porto Rico Ex. Sta.) Probably of little importance.

LARKSPUR (Delphinium).

The genus delphinium is a large one with many species and also many horticultural varieties. The late W. J. Sheppard of British Columbia mentioned these plants as attractive to the bees but they are seldom mentioned in beekeeping literature.

Mr. Henry J. Moore, in his book, "The Culture of Flowers," states:

> "The Larkspur is an excellent bee plant, and for this reason may be sown broadcast on areas near apiaries, where it will not only improve greatly the summer appearance of the surroundings, but will also prove a source of revenue in the fine and fragrant honey which the bees will produce from its nectar."

LAUREL (Kalmia).

The mountain laurel (*Kalmia latifolia*), also known as calico-bush or spoon-wood and in the Southern States as poison ivy, is a common shrub occurring in the higher altitudes from New England and Ontario south to the Gulf States. It is widely credited as being the source of poisonous honey. (See Poisonous Honey.) The sheep laurel or lambkill is a closely related species which occurs from Newfoundland and west to Michigan and south to north Georgia. They are shrubs with showy flowers which are not often reported as important honey plants. In some places, hillsides are covered with the mountain laurel, which makes a pleasing sight when in bloom. These plants are so well known that a great variety of common names have become known in different localities.

Sheep laurel, or lambkill.

In Gleanings for January 1875, Dr. J. Grammer who was a surgeon in the confederate army reported having been poisoned by eating laurel honey. He states that it has a highly poisonous effect being an extremely distressing narcotic, and states that while campaigning he had numerous opportunities for observing its effects.

J. J. Wilder reports *Kalmia hirsuta,* which is called wickey in many southern localities, as the source of a surplus of white honey from May to July in south Georgia.

According to Sladen, the sheep laurel (*Kalmia angustifolia*) is one of the important sources of honey in Nova Scotia. See also Coral Bean.

LAUREL-TREE, see Red Bay.

LAVENDER (Lavandula).

Lavender is the source of the well-known oil of lavender, which is much used in perfume. There are several species native to Europe and North Africa. They are little known in America, although cultivated

Lavender is the source of rich honey.

to a limited extent in California. In some sections of England lavender is extensively cultivated. In Italy it grows spontaneously on dry hillsides at an altitude of 1,600 to 5,000 feet. The plant is well adapted to California conditions and is said to thrive in the black belt of Alabama, also.

The lavenders belong to the mint family, which furnishes some of the finest honey plants. In Italy, *L. officinalis,* shown in the illustration, is regarded as extremely valuable to the beekeeper.

Since the essential oil derived from lavender is a well known article of commerce, it might be profitable to grow it on a large scale in suitable locations. One species is said to grow as far north as Norway. The quality of the oil differs somewhat according to the locality in which the plants are grown, but several species are listed as the source of oil.

An Australian publication (Select Extra-Tropical Plants, by Von Mueller) estimates that a ton of the finest flavored honey can be obtained annually from an acre of *L. stoechas.* This species is recommended to stay sand.

A sample of lavender honey received from France granulated with a very smooth grain, almost equal to butter and almost as hard as candy when fully granulated. The flavor is peculiar and unlike any honey with which the author is familiar, but of good quality. It is amber in color and has a slight tinge of the lavender aroma.

LEAD PLANT, see False Indigo.

LEATHERLEAF (Chamaedaphne calyculata) Cassandra.

Leatherleaf is a low and much branched shrub, with nearly evergreen leaves that grows in bogs from Labrador to Georgia. The blooming time is April and May. H. C. Merriam of Bar Mills, Maine, calls attention to the fact that it is a good source of early stimulation around the peat bogs of New England and helps to get the bees ready for the clover flow.

LEATHERWOOD, see Ti-ti.

LEMON (Citrus limon).

The lemon is a valuable source of honey in southern California, though much less is heard of it than of orange. It is cultivated principally in the coast region, and Richter suggests that the proximity to the ocean of the principal lemon groves may account for the fact that it does not yield as well as orange. Unlike other fruit trees, it blooms more or less continuously throughout the year. This again would make its real value less apparent, since a plant which yields a little nectar for a long period of time may give a total greater than one which gives a heavy yield for a short period. It is noted, also, that oranges yield less freely along the coast than in the interior valleys.

LEMON WEED, see Psoralea.

LESPEDEZA or BUSH CLOVER.

The lespedezas are relatively new in American agriculture and much remains to be learned concerning their behavior under different environmental conditions. Indications are that the species already cultivated offer little promise of becoming of major importance as the source of surplus honey. There are many other species, however, which have not yet come into cultivation that are under investigation in various localities and one may yet be found which is an important bee plant.

So far twenty species have been planted in the American Bee Journal honey plant test garden at Atlantic, Iowa. Of these some have failed entirely and others have not been under observation long enough to enable us to judge their value. Of those observed only three appear to offer much attraction to the bees under our conditions and all three of these are very woody in their habit of growth.

Only a few reports of others who have observed the lespedezas in relation to the bees have so far appeared so any report must be regarded as preliminary and by no means conclusive. Many plants which are decidedly disappointing on some soils are extremely valuable under other conditions. Until we have reports from many localities which represent various conditions of both soil and climate we must reserve final judgment concerning these plants.

An annual lespedeza commonly called Japan Clover, (*Lespedeza striata*) was brought to this country many years ago and has become widely naturalized throughout the southern states. This species was common prior to the war between the states and was probably scattered in many new localities by the movement of the armies. This is a warm climate species which will not stand frost and its range is limited to areas with a season long enough to permit it to mature seed.

Another annual form of much later introduction is the Korean lespedeza, (*Lespedeza stipulácea*). It came from Korea in 1919 and was introduced by the United States Department of Agriculture. It blooms earlier and matures seed nearly a month before the Japan clover so will thrive much farther north. Kobe, Harbin and Tennessee are special strains of these two annual forms.

Reports of bees working on these annual forms are very rare. If they could be regarded as important honey plants certainly more reports would be available.

In American Bee Journal for July 1937, Edgar Abernathy reported that the bees work Korean lespedeza vigorously in North Carolina but do not visit the Kobe. He reports that in three years of observation the bees gathered no honey from lespedeza in one but that in each of the other two seasons his best colonies stored about twenty pounds of surplus from this source. The honeyflow lasted about two weeks in late August and early September. The bees worked the flowers freely in the forenoons and on damp days to some extent in the afternoon before the flowers withered. He described the honey as bright golden in color and of delicate flavor unequalled by any other local honey. He expressed the opinion that large yields will never be secured except under very favorable circumstances. In a letter to the author in Feb. 1943 he stated that he had not secured surplus from lespedeza after 1937 although there had been no reduction in acreage.

The bees have not worked the annual forms for us and a wide search of available reports brings to light no other as favorable as that of Abernathy. Since only his best colonies stored surplus while most of the others showed no gain in the supers we have no ground for regarding the annual lespedezas as important bee pasture.

Perennial Lespedezas

Lespedeza sericea is the only one of the perennial forms which has as yet attracted sufficient notice to result in extensive cultivation. Although this species was tried at the North Carolina Experiment Station as long ago as 1896 it was not recognized as of value until 1924. In that year the Bureau of Plant Industry secured seed from Japan. Later seed was distributed for trial and it has become established as a farm crop much sooner than is usual with plants of foreign origin. It is now grown in large acreage in many localities notably in Missouri.

We have had a plot in the test garden for eight years and it has shown no indication of winter injury although it had been assumed that we are near the northern limit of its range. This species is promising for soils where the clovers and alfalfa refuse to thrive because of acidity. While the stems are woody and would not be looked upon with favor by one who could grow alfalfa successfully, the plant does produce a large amount for forage if cut before the stems become too coarse.

We have been disappointed in its attraction for the bees. In two of the eight seasons we have failed to find the bees working the plant. In 1941 the bees worked the flowers vigorously for a few hours and gave every indication of finding a satisfactory harvest. Just what change in the weather was responsible for this attraction we have been unable to determine. In 1942 they worked the blossoms with equal vigor for a somewhat longer time but only visited the flowers for a few days.

Lespedeza sericea is somewhat slow in getting fully established and it may do somewhat better as it grows older but certainly it would have to do a great deal better to be of value to a beekeeper here. Frank Van Haltern reports that at times the bees work this species freely at the Georgia station but did not indicate that surplus is obtained from it. (American Bee Journal Nov. 1936.)

In American Bee Journal, May 1938, W. E. Watson of Tennessee reported that observation revealed that it was not yielding in some fields while the bees were getting a heavy flow from it in others. Thus while we have evidence that at times *Lespedeza sericea* does yield some honey, there are few reports of important surplus from this source.

A somewhat similar species is *Lespedeza daurica* a species of late introduction from Manchuria by the Bureau of Plant Industry. When it first came into bloom in our test plots the bees found the flowers at once and worked them so vigorously that we hoped at last to find the one which yields honey copiously. However, the flowering period proved too short to permit much of a harvest even though it might yield heavily. The bloom was about all over within a week from time it opened and that is too short a time to provide a profitable harvest. In other seasons the bees failed to show the same enthusiasm.

Thus far three species, *Lespedeza bicolor, Lespedeza cyrtobotra* and *Lespedeza thungbergii,* are the only species which offer much promise for the bees and these are all three very coarse and woody in their habit of growth. They grow to a height of from six to eight feet with stems a half inch in diameter at the base, far too coarse for satisfactory forage. It is reported that when cut quite young hay of fair quality can be secured but that, of course, eliminates any chance for bloom and thus for bee pasture.

From Japan comes a report that *Lespedeza bicolor* is the source of much of the surplus honey from some areas in that country. It is also reported as occurring in thickets where forests have been cut in Siberia.

The shrubby lespedezas yield a dark honey of fine quality.

In our test plots it was rather slow in getting established, requiring two years to reach a normal blooming stage. The bees work it freely and indications are that if it was present in sufficient acreage we might also get surplus honey from it here. It has rather a long blooming period in late summer. It blooms luxuriantly and makes a very fine showing as a garden ornamental. The bright pink flowers and attractive light green foliage combine effectively to make an attractive shrub for the perennial border.

Lespedeza cyrtobotra has not been reported as a honey plant anywhere so far as we are able to find. Yet it is by far the best of them all for the bees in our garden. The bees work it more freely and over a longer period than is the case with any of the others. While they desert *bicolor* at certain times they are nearly always to be found on *cyrtobotra* when the flowers are open. It blooms a little later than *bicolor* and the flowers are often caught by October frost before the seed has matured. Further south this species should be desirable for control of soil erosion and for browse in permanent pastures.

Lespedeza thungbergii is quite similar to the last two species and blooms at about the same time. The bees visit the flowers freely and apparently get considerable honey. Its woody stems limit its value to the farmer but it has attracted some attention on the part of nurserymen who are offering it as a desirable ornamental.

One native species, *Lespedeza virginica,* is reported by Frank Van Haltern, in the article already mentioned as attracting the bees in Georgia. In the wild state it is found from New England to Florida, Kansas and Texas. It may have some value as a cultivated crop but as yet has not been put to the test for that purpose to any extent. This species thrives in our garden, as might be expected in the case of a native plant, but thus far the bees have not manifested any great attraction for its blossoms.

Lespedezas are likely to grow in importance because of their adaptability to soils where so few legumes will succeed and it is important to the beekeeper that varieties be found which will yield nectar freely if there are any such. Thus far our search, except for the coarse stemmed species has not been promising.

In the Southeastern States the shrubby lespedezas have been planted in considerable areas for game cover and erosion control. From Clemson, South Carolina, Prof. David Dunavan sent to the author a sample of honey from this source. It was very dark in color but with a delicious flavor that should find ready demand in the markets.

LETTUCE (Lactuca floridana).

Wild blue lettuce is common in the woodland borders, in rich soil, from Pennsylvania to Iowa and south to Florida and Texas. The plant produces hundreds of blue flowers in late summer and early fall. Although it is of no special importance as a honey plant, the bees visit it frequently and apparently get some nectar from its blossoms. In the South it is reported as blooming in May and June. The writer does not recall having seen it in bloom in Iowa earlier than August, while it blooms into September. There are a considerable number of species of wild lettuce, some of which, like the prickly lettuce, become very troublesome weeds. The writer has not observed the bees working to any extent on any except the blue fall lettuce above described.

LEONURUS, see Motherwort.

LIGNUM-VITAE, see Soapbush.

LIGUSTRUM, see Privet.

LILAC, see Mountain Lilac.

LIME (Citrus acida).

The lime is a fruit very similar to the lemon, which has been introduced to some extent into the citrus districts. Since it is not generally grown it is not very important, though it yields honey freely.

LIMETREE, see Basswood.
LINDEN, see Basswood.
LION'S TAIL, see Motherwort.
LIPPIA GRASS, see Carpet Grass.

LIQUORICE (Glycyrrhiza lepidota).

The wild liquorice is a plant of the pulse family occurring from Minnesota to Missouri and westward to Washington and Arizona. It is found in Colorado and Wyoming at 4,000 to 8,000 feet altitude. It is reported as quite abundant in places in Wyoming, and some years as being a great yielder of nectar.

LIVE OAK, see Oak.

LIZARD'S TAIL (Saururus cernuus).

The lizard's tail, also called swamp-lily or breastweed, occurs in swamps and on muddy banks from New England, Ontario and Minnesota, south to Florida and Tennessee. It is a marsh herb with slender root stock, jointed stem and the small white flowers are borne in dense spikes opposite the ovate leaves. It grows from one to five feet high.

A sample received from Jes Dalton, of Louisiana, was accompanied by the statement that it began yielding in 1920 before the close of the spring flow and carried the bees through a period which was usually one of dearth till the summer flow in July. This overlap provided a continuous flow from the time spring opened till the close of the summer flow. He described the honey as fine flavored and yellow colored, but of a different shade of yellow than any other he has seen. The flow was said to be heavy. Although there were hundreds of acres of the plant within reach of his bees he had not previously known a honey flow from it. He stated that it would be of inestimable value if it could be depended upon to yield as it did that season.

Dr. Everett Oertel reports that while the plant is abundant in Louisiana no other reports of honey from this source have been received.

LOBELIA.

The Cardinal flower, best known of the Lobelias is a humming bird flower and unsuited to the honeybee. Bumblebees and moths also visit it freely. Honeybees are only found as chance visitors and under normal conditions are unable to secure the nectar.

Charles Robertson, (Botanical Gazette March 1891) states that in his neighborhood, (Carlinville, Illinois), *Lobelia leptostachys* is visited by the honeybee for nectar.

Bessey in his list of Nebraska honey plants lists Great Lobelia *Lobelia syphilitica,* as yielding honey.

These plants can hardly be regarded as important to the beeman.

LOCALITY.

The plants which yield honey in surplus quantity are comparatively few in number. In the average locality there will be only two or three important sources of nectar, with a considerable number of minor ones which furnish the bees with a partial living between the main flows.

As an example, a typical Iowa locality may be mentioned. There sweet clover is the principal source of surplus honey. Now and then some surplus will be gathered in the fall, also, from heartsease. Yet the list of plants which yield nectar in that particular neighborhood is a rather long one. Were it not for these minor sources, to provide for the bees before and after the clover flow, beekeeping would hardly be possible there. To begin the season, fruit trees, apples, cherries and plums, furnish a liberal supply of nectar. These trees bloom early in spring when the bees are still weak from the long winter. If it were possible for the beekeeper to so conserve the strength of his colonies that they would come through the winter in as good condition as they are in mid-summer, a good crop of surplus honey would be gathered from fruit bloom in favorable seasons. Following fruit bloom, the dandelions come. There are several days of bloom from dandelion and the bees make the most of it. Considerable nectar and an abundance of pollen are gathered from dandelion, only to be used in brood-rearing. It takes a large amount of honey to rear a big force of bees, and without strong colonies of bees, profitable crops of honey cannot be harvested. It takes honey, then, to make bees, and it takes bees again to gather a big crop of honey.

The dandelion continues to bloom, in some localities, until within a short time of the opening of the white clover flow. In favorable seasons, when the clover is abundant and weather conditions favorable, a liberal crop may be expected from the clover. Following the clover there is a long period when but little honey is coming from the field. In one neighborhood the bees may be entirely idle, while a short distance away there may be sufficient forage to support the colony. In the one location the surplus gathered from clover will be consumed, in part, to support the colony until the fall flow from heartsease; in the other the bees will gather enough from minor plants to support them. The one locality thus becomes a good one for beekeeping, while the other is poor, and perhaps they are but a few miles apart.

This is a fair example of general conditions. The locality may not be in the clover region, but the presence or absence of the minor plants is extremely important to the beekeeper who would support his family from the products of his apiary.

There are many locations where the presence of plants which yield pollen abundantly are second in importance only to the plants which furnish the main honeyflow. As an example, there are places along the Appalachicola River, in Florida, where enormous yields of surplus honey are sometimes secured from tupelo, but where there is so little pollen for the rest of the summer that the bees suffer seriously for the lack of it. In some places the beekeepers find it necessary to move their bees to other locations in order to maintain the strength of their colonies, following the flow from tupelo. Thousands of colonies of bees have died in locations where wonderful yields of honey have been harvested, because no pollen was available to enable them to continue brood-rearing.

It often happens that the crop will be a failure in one location, while but a few miles away it would be possible to gather a surplus by moving the bees. As an example, at Hamilton, Illinois, in 1919, the clover crop was a failure and the bees were on the point of requiring feed at the time when the clover harvest should be coming in. By moving their bees from fifteen to forty miles, the Dadants were able to harvest about forty thousand pounds of honey, instead of feeding to carry their bees through the winter.

It has become the practice of many California beekeepers to secure more than one crop by moving their bees several times during the season. Following the harvest from one source, the bees are moved by truck to other fields. Thus it is sometimes possible to use the same apiaries to harvest two or three good crops in a single season.

In like manner a well-known California queen breeder has found it possible to greatly increase his output by moving his queen-rearing yards to new fields when the nectar supply had ceased in his own location.

The variation in the supply of nectar in different localities is extremely great. A few locations will support several hundred colonies in one yard. Other locations will hardly support twenty colonies profitably. A careful study of the flora within reach is most important to the beekeeper. One who fully understands his location can adapt his system to his conditions and succeed where failure would otherwise result. John W. Cash, of north Georgia, had about eight hundred colonies of bees in a section where thirty colonies would overstock any single location. By establishing a large number of apiaries in widely scattered situations, he was able to secure a surprising uniformity of yield. For a period of several years he never secured less than 56 pounds average of surplus per colony. At the same time his highest average was 86 pounds per colony.

The combination of plants which yield surplus honey with those which yield pollen and those which furnish some nectar during the periods between the flows, together with climatic conditions, determine the value of a location for beekeeping. It does not require much skill to secure a crop of honey in a location where every condition is favorable, but the man who is fully acquainted with the flora will discount the crops of the hit-and-miss apiarist even there. In the poor locality it takes a thorough knowledge of the sources of honey and pollen, together with expert beekeeping to succeed.

LOCO WEED (Astragalus). Also called Rattle Weed or Buffalo Bean.

There are several species of *Astragalus,* common to the Rocky Mountain region, from Manitoba south to Texas, and west to California. The loco weeds are herbs with odd-pinnate leaves, and spikes or racemes of purple, white, or pale yellow flowers.

Several varieties are poisonous and are the source of heavy losses among stockmen of the plains and mountain regions. Honey from loco is reported principally from Colorado, where it is mentioned as blooming with horsemint, in May and June. It is to be found mostly in the foothills and the yield is not dependable. The honey is reported as of light color and good flavor.

According to an article by Vansell and Watkins in the May 1934 issue of Bees and Honey, the spotted loco, (*A. lentiginosus*) is the

Blossoms of black locust.

cause of the death of large numbers of bees in Nevada, particularly in Lyon County. Apparently only the adult bees are poisoned and brood suffers for lack of attention because of the diminishing numbers of adult bees. (See Milk Vetch.)

LOCUST (Robinia pseudo-acacia).

The black locust, or false acacia, is a native tree from Pennsylvania to Iowa and southward. However, it has been widely introduced into other States, thus greatly extending its range. It is now to be found in many places from New England and Canada southward. It is reported as producing a surplus of honey in parts of California, and is listed among the honey plants of Texas.

The wood is desirable for posts, railroad ties and other purposes requiring durability. Large plantations are often set for utility purposes, so that in some localities the beekeeper may readily expect a surplus from this source. Borers are a serious menace to the life of this tree, and whole plantations of locusts are sometimes injured by the insects, which kill the branches and sometimes the bodies of the trees, causing them to sprout again from the root.

The flowering time is very short thus limiting the value as a source of surplus honey.

The honey is water-white, of heavy body and mild flavor. The picture shows the blossoms and leaves. The flowers, it will be noted, much resemble those of the garden pea.

In some localities the tree is known as white or yellow locust. (See Honey Locust.)

LOGWOOD (Haematoxylon campechianum).

The logwood is an important tree common to the West Indies and Central America, but probably does not occur within the United States. It is the principal source of vegetable dyes, most of which have been displaced by chemical products.

It grows in dense forests over large areas and in Jamaica is regarded as the principal source of honey. When conditions are favorable enormous crops of honey are harvested, single colonies sometimes gathering several hundred pounds. Since the forests are often miles in extent, large apiaries can be supported in a single location. The plant usually blooms twice during the year, once in November and the second time near the holiday period. That weather conditions affect the flow from logwood as readily as that of other plants will be seen by the following quotations from American Bee Journal of June 8, 1905:

> "I noticed unmistakable evidences of an almost universal bloom, and about ten days later it came out in all its glory. It was truly a magnificent sight, and although the house was about 500 feet from the apiary, the roar of the bees passing to and fro was a sound to make glad the heart of any beekeeper. I went down to the apiary one morning about 6 o'clock, and if

I live to be 100 years old, I never expect to see a more stirring scene in any apiary than I looked upon in that yard of 250 colonies. * * * They kept up this pace for four days; but, alas, it rained that Saturday night, and the next morning the logwood blossoms were as brown as though they had been burnt, and the flow was over. Six thousand pounds for the four days was the record."

(See also Brazil.)

Ordetx in his Plantas Melliferas de Cuba, writes concerning logwood:

"It is one of the plants most affected by the humidity of the soil; if it does not rain at the opportune time, the flow of nectar can be insignificant, although flowering may be great; and in some cases of extreme drought the buds can wither and fall off without opening. On the contrary, ε hard

The loquat.

rain at the proper time can bring about a profuse flowering for several days with an intense flow of nectar.

"The honey of this plant, of pearl white color, and delicious taste is comparable in quality with that from aguinaldo. (Campanilla)

LOMBARDY STAR THISTLE, see Centaurea.

LOQUAT (Eriobotrya japonica).

The loquat is a Japanese fruit of evergreen habit and fragrant white flowers which has been introduced into parts of California. The tree is cultivated for ornament and for its edible fruit, which resembles a small yellow pear. It blooms in winter in California and is said to be an excellent honey plant.

From a commercial report on that country we learn that the loquat is one of the principal sources of surplus honey in China.

In an article in the China Journal, March, 1937, Claude R. Kellogg mentions loquat as a source of surplus honey in the Fukian province of China in the month of November.

LOOSESTRIFE, see Purple Loosestrife.

LOTIBUSH or TEXAS BUCKTHORN (Zizyphus obtusifolia).

The lotibush or lote-bush, also called Texas buckthorn, is a common chaparral bush on the plains and prairies of Texas, New Mexico, Arizona and adjacent Mexico. Coulter (Botany of Western Texas) describes it as one of the most widespread and abundant shrubs in western and southern Texas, on gravelly mesas, slopes and bluffs. A related species, *Z. lycioides,* occurs in extreme western Texas and westward. This apparently is the more common species in New Mexico.

According to E. G. LeStourgeon, lotibush is often confused with brazil. (See Brazil). He states that it is a valuable honey plant, blooming after every rain. (See also Buckthorn).

LOUISIANA—Honey Sources of.

The northern portion of Louisiana is composed of uplands with elevations of from 100 to 200 feet above sea level, with a few ridges reaching 400 feet. For the most part the soil is fertile although there are some areas of forest land poorly suited to cultivation.

The southern part of the state is low and marshy with numerous bayous. The greater part of this area is sixty feet or less above sea level with a few limited areas rising to 200 feet.

The climate of the state is moderate with mild winters and severe freezing temperatures seldom occur. At Shreveport according to Government weather reports, there are only 22 days on an average each year when temperatures drop below freezing. The summers are warm with temperatures rising to 100 degrees at some time nearly every year.

The rainfall is abundant, usually fifty to sixty inches annually over the state.

The breeding season is long, the bees getting nectar and pollen from maples in late winter. Sometimes a small surplus is harvested from this source. There are immense areas of willows in the swamps and willow honey is a common product. White clover is abundant and usually blooms from February to July. While honey is harvested from this source its yield is not so dependable as in the northern states. Tupelo, pepper-vine, heartsease, thoroughwort, goldenrod, holly, locust, cotton, blackberry, and Spanish needle are among the plants which yield surplus.

Since the greater part of the commercial beekeeping is in the vicinity of swamps the honey is a blend from many sources, some of which are hard to determine. Climbing boneset, Virginia creeper, and morning glory are among the vines which add to the wealth of the beekeeper. When visiting the state in 1920, the author found one report of 27 colonies which produced an average of 150 pounds per colony from heartsease. Another beekeeper reported a crop of 135 barrels of honey, mostly from pepper-vine. A crop of 210 pounds of honey gathered by a single colony from willow was another report.

Louisiana has a very rich flora and an unusual variety of honey plants. Buttonbush, vervain, gallberry, sweet clover, alfalfa, horsemint, orange, sumac, basswood, magnolia, wild cherry and many others might be mentioned as of local or secondary importance.

Willow yields honey in such abundance that much honey is stored from it at times. White clover is also important. Senna, horsemint, goldenrod and Spanish needle yield some honey in the fall.

LUCERNE, see Alfalfa.

LUDWIGIA, False Loosestrife or Water Purslane.

There are several species of Ludwigia, mostly confined to low, wet lands of the South. Two or three species are also found in swamps of the Eastern States.

The indications are that Ludwigia may be very valuable as a source of nectar, but because of their habit of growing in the swamps have been generally overlooked by the beekeepers. The author first had his attention called to the group by finding the bees busily working on *Ludwigia pilosa* near Gulfport, Mississippi, in August, 1920. From the eager manner in which the bees sought the blossoms it was readily apparent that they were finding good pasture, but local beekeepers were unable to give information concerning the importance of the plant.

H. B. Parks has written the author that *Ludwigia natans* is very common everywhere there is water in south Texas. He reports that the bees work upon it very freely. One Brazoria County beekeeper reports that the honey is light amber, but without particular flavor, being merely sweet.

L. palustris is known as water purslane in California. It is found in moist situations similar to those of other species.

LUPINE (Lupinus).

There are many species of lupines which are common, especially in the plains region and west to the Pacific Coast. Some are of no value to the bees, or yield pollen only. Richter lists *Lupinus affinis* as a source of nectar in California.

The blue lupine or bluebonnet (*Lupinus subcarnosus*) is widely distributed over southern and western Texas, fairly covering large areas when in bloom. The blooming period comes in March or April. Some claim it yields only pollen. Scholl lists it as a source of honey.

In Colorado, beemen regard some lupines as good honey plants, also.

Von Mueller in his "Select Extra-Tropical Plants" lists lupines among the most important sources of nectar and states that some, if not all, lupines can be counted among honey plants.

LYCIUM.

There are some species of *Lycium* native to Africa and Asia which are used for hedges (box-thorn). Some of these are mentioned as important bee plants. Being grown in this country only for ornament they are of little importance. Most species are tender and their use is confined to the southern states.

In the desert regions of the southwestern states are to be found several native species mostly shrubs, some of which are of considerable importance as sources of early nectar. One species common to southern Arizona, commonly known as "Squawbush" or "Squawberry," blooms in winter or very early spring and provides for winter brood rearing. In moist winters it begins blooming early and furnishes some

Blossoms of desert matrimony.

nectar for a long period during the winter months. The author found that beekeepers in the Salt River Valley generally regarded it highly as a means of stimulating winter and early spring brood rearing.

There are at least three species of "desert matrimony" found in New Mexico. They are also known as "Tomatilla." In that section they are likewise reported as important for spring stimulation, but not as the source of surplus.

Most of the desert species are dormant during long periods of dry weather, but may bloom at intervals following rains. Both leaves and flowers may appear within a few days after rain; if the weather then continues dry the leaves may be shed again within a few weeks and the plant return to the dormant condition.

One species *L. Carolinianum* is found through southern Texas and eastward to South Carolina and Florida. Some of the desert species extend their range as far north as Colorado and Utah.

Matrimony vine (*Lycium halimifolium*), is a low, spiny shrub, with very long, lithe and almost climbing branches. It came originally from the Mediterranean region and was very generally planted for ornament about American homes a generation ago. It is very persistent, and has run wild about many deserted home sites. The small flowers clustered in the axils of the leaves are very attractive to the bees, and, where sufficiently common, the plant is probably of considerable importance.

LYTHRUM, see Purple Loosestrife.

M

MADRONA (Arbutus menziesii). THE ARBUTE TREE.

Jepson describes the madrona tree as evergreen with glossy, leathery leaves, widely branching, 20 to 125 feet high; bark polished, crimson or terra cotta, on old trunks dark brown, and fissured into small scales. Coast ranges of California, oak hills, etc. Grows on high ridges, mountain slopes and in gravelly valleys.

According to Richter it yields both nectar and pollen from the flowers.

Stephen J. Harmeling, in first report of Division of Apiculture, of Washington, states that it reaches its highest perfection in the Puget Sound region. It blooms in May and the flowers resemble those of the lily of the valley. Large nectaries are located at the base of the urn-shaped corollos. The berries are scarlet and edible.

Coleman, in Western Honeybee (Oct. 1921), writes as follows concerning madrona:

> "When in full bloom in May and June, the great crown laden with its honey-cups full to overflowing, around which the honeybees, bumblebees and

other nectar-loving insects, gather as at a feast, is a sight to make the heart of any nature lover glad. The honey is light golden amber, with a very heavy body, and a delicious woodsy flavor, which easily distinguishes it from all others."

Dr. C. E. Ehinger stated that the hummingbirds are attracted to the madrona trees in considerable numbers and that he has seen fifteen to twenty of these little birds humming about the blossoms at one time near Chico, Washington.

MAGNOLIA or BULL BAY TREE (Magnolia grandiflora).

The magnolia is native to moist soils from North Carolina to Florida and west to Texas. It is the largest leaved tree of the evergreens. It is a magnificent tree when uncrowded and is a favorite shade tree in the South. The thick, leathery leaves persist over winter until the new ones have appeared. The flowers are large and showy and the blooming period is from spring till midsummer.

Bloom of the madrona tree.

The sweet bay (*Magnolia virginiana*), known also as laurel magnolia, swamp laurel and beaver tree, grows in swamps from Massachusetts to southern Florida and west to Arkansas and Texas. In Texas it is reported as yielding very dark honey of poor flavor.

Magnolia is not often mentioned as important, though an occasional report of surplus in the South is received. One beekeeper reports that in Mississippi the honey is so dark and strong as to be unpalatable.

"We had three days of cool, damp, cloudy weather last year in August. During that time I visited my magnolia apiary, and on approaching it I heard the heavy roar of bees. I first thought that wholesale robbing was in full force, and I soon saw that they were gathering honey, and on looking at the alighting boards I saw particles of magnolia blooms. This told the tale. I went a few rods into the swamp, which was decorated with the large, rich magnolia blooms. I examined a bloom and there was the nectar visible, and all the bees had to do was to alight, fill themselves and return. The weather soon cleared off and the magnolia honey was no more. Those three days of damp, cool and cloudy weather saved many old style gums from being turned bottom up, in the spring, in this section."—J. J. Wilder, Cordele, Georgia. American Bee Journal, Feb. 15, 1906.

Geo. H. Rea reports that cucumber tree (*Magnolia acuminata*) yields a heavy flow of honeydew in western Pennsylvania, also that it yields nectar in May and early June. The cucumber tree is common in many sections of the Eastern States from New York and Ontario south to Alabama and Mississippi. It is probably an important source where the tree is abundant.

MAGUEY, see Century Plant.

MAHALA MATS (Ceanothus prostatus).

Mahala mats belongs to a well-known group of shrubs which yield nectar freely. (See Mountain Lilac and Buckthorn). The branches of this species are prostrate or trailing and take root freely, thus forming mats. In some localities it is known as squaw carpets. The leaves are small with sharp-toothed edges. The flowers, appearing in May and June, are blue, followed by red berries.

Mahala mats is native to the Pacific Coast, where it forms large patches on poor, gravelly soils. In the pine woods of the Sierra Nevada Mountains and the north coast ranges of California, it is abundant. It was reported as an important source of nectar in western Washington by Stephen Harmeling in the first report of the Division of Apiculture of that state.

MAHOGANY GUM, see Eucalyptus.
MAHOGANY SUMAC, see Sumac.

MAINE—Honey Sources of.

Maine covers an area of about thirty thousand square miles and is the most northerly and easterly portion of the United States. The greater portion of the state is composed of forest interspersed with lakes and streams.

It is a region of severe winters with long continued cold and heavy snowfall. Snow falls as early in October and as late as April, with short and cool summers. Maine is a favorite resort for the heated summer period. The average precipitation is above forty inches, much of which is in the form of snow.

Killing frosts occur as late as June and as early as August. According to publications of the U. S. Weather Bureau, the frost free season averages about 145 days in the southern portion and 105 days in the northern region.

The agriculture of the state is confined to comparatively small areas and beekeeping is relatively unimportant.

The most important honey sources are white and alsike clover, fireweed, goldenrod, raspberry, fruit bloom and buckwheat. Asters are abundant and of some value, boneset, buttonbush, Canada thistle, carrot, dandelion and milkweed add to the production of the hive. Maples, horse chestnut, black locust, mustard, willow and sumac are generally appreciated by the beekeepers, although of minor importance in most cases.

MAIZE, see Indian Corn.

MALLOW (Malva).

Several lists of honey plants contain some mention of the mallows. The bees visit the flowers for both nectar and pollen, but the author can find no record to indicate that they are anywhere important.

The common mallow (*Malva rotundifolia*) is found in barnyards and waste places over a large portion of the American continent. It is a common weed, introduced from Europe and generally naturalized in this country. It has a number of common names, as Dutch cheese, doll cheese or fairy cheese (from the shape of the small seed-pod), blue mallow, or country mallow. It blooms through a long period and is still in flower late in fall, when there is little to attract the bees.

A specimen has been sent to the author from Quebec with the statement that it blooms there in September, and in mild seasons October, and is very attractive to the bees. While the bees work upon the blossoms to some extent, it is probably at times when there is little else to attract them. The author regards it as of little value.

The author has found the bees eagerly visiting the blossoms of the Turk's Cap Mallow in the lower Rio Grande Valley in winter. These plants are commonly grown for ornament in the warmer parts of the country.

E. G. Carr reports the rose mallow or swamp mallow as the source of surplus in New Jersey in late summer.

MANCHINEEL (Hippomane mancinella).

The manchineel is an evergreen tree common to the beaches and marshes of southern Florida and the Keys; also found in the Bahamas and Tropical America. It is very poisonous, and Britton credits it with being the most poisonous of our American trees. He states that the milky juice was used by the Caribs to poison their arrows.

It yields nectar abundantly some seasons and is the source of surplus honey. According to E. G. Baldwin, it blooms with pigeon cherry and with dogwood, and the late O. O. Poppleton was the only man to attempt to harvest a crop from the tree, growing as they do on the Keys, where not easily accessible. He credits Poppleton with a yield of 28,000 pounds from the three together in 1910.—Gleanings, April 1, 1911.

MANGO, (Mangifera indica).

The mango is a well known tropical fruit which is now grown, to a limited extent, in Florida. The tree is large and spreading in habit and attains a height of eighty to ninety feet. The better varieties are ranked as best among tropical fruits.

It has a long blooming period and is listed by Phillips as a source of nectar in Porto Rico. (Bul. 15 Porto Rico Ex. Sta.) He also quotes W. E. Hess of that station to the effect that the bees are fond of the overripe fruit.

J. W. Barney of Bradenton, Florida reported that mango yields with citrus, that the honey is darker in color, heavier and more aromatic. Bees work it freely when in bloom according to his report.

Ordetx in his "Plantas Melliferas de Cuba" writes:—

> "When weather conditions are favorable, the mango secretes a considerable quantity of nectar, early in the morning; during the rest of the day, the flow almost ceases except in exceptional cases.
>
> "The honey from the mango tree is of amber color, very dense, and of delicious taste."

MANGROVE, see Black Mangrove, also White Mangrove.

MANITOBA—Honey Sources of.

The possibilities of honey production in Manitoba until recently have not been properly appreciated. While wintering is a serious problem because of the long-continued cold weather, the province offers exceptional opportunities for commercial beekeeping. This is especially true in light bush or park country, and also in that part of the prairie region where sweet clover is grown extensively.

A considerable portion of the province is included in the northern coniferous area where the principal trees are pines and spruce. It is hardly to be expected that beekeeping would be important in this region. The altitude of this region is about 400 feet. The central region is covered with mixed woods of poplar, birch, spruce, tamarac, willow, etc. In this area the wild raspberry is abundant. Saskatoon or billberry, wolfberry, fireweed, several mints, goldenrod and asters, and numerous other natives furnish nectar freely. The highbush cranberry is valuable in some sections. The willows furnish both nectar and pollen and cottonwoods and poplars furnish early pollen in quantity. Along the Red River north of Winnipeg are to be found locations supporting 200 colonies or more in a single yard. In this section is to be found some clover, as well as such introduced weeds as sow thistle, dandelion and Canada thistle. Most of the honey produced is amber in color and of pronounced flavor. The altitude of the Central region averages about 800 feet.

The transition region of open prairie with patches of poplar and dogwood thickets begins a little to the southwest of Winnipeg and extends south to the international boundary. The altitude is approximately 1,000 feet. This is probably the best beekeeping territory in the province, although there are large areas given over exclusively to the cultivation of grain, where little forage is available to the bees.

On the prairie the crocus or pasque flower (*Anemone patens*) blooms in great abundance early in spring. It often blooms before the snow has entirely disappeared. It is common over large areas and furnishes pollen in great abundance for early brood rearing. Dandelion also is common in spring. Along the streams and in wet places the pussy willow is common and adds to the early pasture. Elm is also common along the streams. Next comes the Siberian pea-tree or carragana, which is an introduced shrub commonly planted for hedges and windbreaks. Considerable quantities of fine quality white honey is harvested from this source. In the wheat fields and on roadsides the Canada thistle and sow thistle are extremely abundant over a large area. Gumweed and snowberry or wolfberry, commonly called badger brush, are widely distributed and both are important sources of nectar.

It is from alfalfa and sweet clover, however, that the biggest crops are harvested. Where either of these is grown in sufficient acreage the beekeeper has a very dependable crop. One beekeeper not far from Winnipeg, who has the bush flora as well as alfalfa within reach, reports a ten-year average of 150 pounds per colony. Since the tendency is toward mixed farming, a larger acreage of these crops will constantly be grown and the bee pasture will accordingly improve as the country grows older.

In the southwest corner of the province there is a small area of short grass plains with an altitude of about 1,500 feet. The native flora is not favorable for beekeeping, but the introduction of alfalfa and sweet clover provides some good locations.

In the central and northern regions there are considerable areas of muskeg or swamp where beekeeping is not practical.

MANZANITA or BEARBERRY (Arctostaphylos).

The manzanita is seldom heard of as a honey plant east of California. The following information is copied from Richter's "Honey Plants of California":

> "*Arctostaphylos,* manzanita, bearberry. Throughout coast ranges, Sierra Nevada foothills and San Bernardino Mountains (2,000 to 9,000 feet), November to February.
>
> "The honey is amber and of excellent flavor, much like manzanita itself (Colusa County); pollen. San Diego County reports a white honey from the manzanita. One of the most important honey plants to induce bees to early breeding. In some parts of Monterey, Colusa and Eldorado Counties, a 20 to 40 pound surplus is obtained, and on very warm days (Monterey County) nectar can be shaken from the bloom. A beekeeper from Applegate reports it to be his best honey yielder."

The *Arctostaphylos uva ursi,* bearberry or bear grape, according to Gray, occurs on the rocks and bare hills from New Jersey and Pennsylvania to Missouri and far north and westward. It is also said to be common in Europe and Asia. It is recorded in the local lists of plants of Connecticut and Ontario, although, probably because nowhere abundant, it is not known as a source of honey. Although Richter's list does not give the species from which their honey is secured, it is probably *A. manzanita* or *A. tomentosa,* or other species peculiar to the West Coast.

The leaves of the eastern species are much used in medicine. It is said to be an astringent tonic, used in diseases of the liver.

Manzanita.

Coleman describes four California species of manzanita as sources of nectar in Western Honeybee (Nov. 1921). Of these he describes the common manzanita (*A. manzanita*), as most abundant and widely distributed.

It blooms for six weeks to two months, beginning in February. He describes the honey as white or light amber and of delicious flavor.

A. tomentosa (hairy manzanita) he mentions as next in abundance to the above and also furnishing honey of fine flavor, light amber in color. The third species on his list is big-berried manzanita (*A. glauca*), common on the inner coast ranges and in some parts of the Sierra Nevada foothills. This species blooms during the winter months, beginning in December. It produces nectar abundantly and strong colonies gather considerable surplus of light color and good quality. The dwarf manzanita (*A. numularia*) he mentions as common in the upper Sierra Nevada timber belt where it blooms in mid-summer. The honey from the dwarf species is described as similar to that from the others.

Another species, *A. nevadensis,* is common in the Sierra Nevada and Cascade Mountains at 8,000 to 10,000 foot elevations. It is found from Washington to the King's River region, according to Smiley's Flora of the Sierra Nevada.

Scullen and Vansell (Bul. 412 Oregon Ex. Sta.) report several species as common to Oregon of special value for building up in spring. They list three species, *A. viscida, A. patula* and *A. columbiana* as important in the Rogue River Valley.

They mention manzanita honey as having a faint bitter taste that disappears after extracting.

MAPLE (Acer).

The maples are mostly large trees and are common to nearly all sections of the United States and Canada. More than twenty species are native to America, several of which are commonly planted for shade and ornament. Probably all yield nectar.

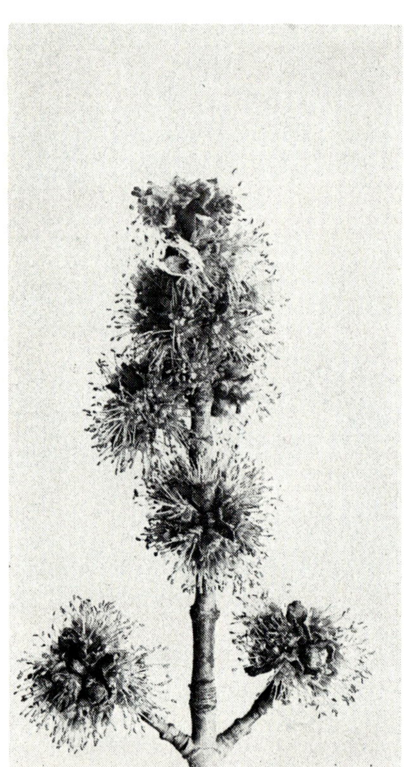

Blossoms of the soft maple.

The sugar maple or hard maple, (*Acer saccharum*) is a beautiful and long lived tree. The soft maple, also known as silver maple, and white maple, (*Acer saccharinum*), grows much more rapidly and has been very commonly planted for shade and windbreak in the central west.

The blossoms open early, when especially valuable for building up colonies for the main honeyflow. If the bees were as numerous as later, the nectar stored from maple bloom would make a creditable yield. Mr. C. L. Pinney of Iowa, reports that one year his scale hive showed a gain of from one to two pounds daily from soft maple, when the ground was still covered with snow.

The red or swamp maple, (*Acer rubrum*) is widely distributed in swamps and wet grounds from New Brunswick to Florida and westward. It blooms early and is regarded as of considerable importance as a source of nectar.

The big-leaf maple of the Pacific Coast (*Acer macrophyllum*), also known as Oregon maple, California maple, water maple or white maple, is reported as of special importance in Oregon and British Columbia, its yield is usually cut short by rains. It is found in the

Sierra Nevada and coast ranges of mountains in California, north to southern Alaska. (See also Box-elder).

In British Columbia, Washington, Oregon and Northern California the vine maple (*Acer circinatum*) is a sprawling, vine-like tree, with crooked limbs. It is reported as an important source of nectar in many localities in the states mentioned.

While most of the maples bloom so early in spring that the bees are not yet in condition to make the most of them, there is one, the mountain maple, commonly called water maple or swamp maple, (*Acer spicatum*) which blooms in June and should be of far more value. It is a small shrubby tree which always grows in the shade of larger trees in moist places from New Foundland and Labrador to Hudson's Bay, south to Michigan and New York. It is also found in higher altitudes farther south. In Northern Michigan and similar locations it is the source of considerable nectar.

Hubbard Brothers of Boyne Falls, Michigan reported that the honey is amber and of rather poor flavor. With them the bees stored fifteen or twenty pounds surplus in favorable weather but they used it again for brood rearing. In that locality the flow comes when there is little else to be had and it is valuable to sustain brood rearing for the later flows.

MARIGOLD (Gaillardia pulchella).

There are several species of *Gaillardia* common to a wide scope of country. This particular species ranges from Louisiana west to New Mexico and Arizona. It is the source of large quantities of yellow honey in Texas, where it is highly regarded as a honey plant. It is reported as yielding surplus occasionally as far south as Brownsville, though not frequently.

Scholl lists it as "a main source of surplus; honey dark amber and of good quality. May and June." (See also Spanish needle.)

The garden marigolds of the genus *Tagetes* are strong scented annuals with yellow or orange flowers. Most of them are natives of Mexico. They are attractive to the bees which visit them in large numbers in late summer when but little forage is available. Much public interest was manifested in the breeding work of the Burpee Seed Company when developing new varieties of marigolds. About five acres of seedlings were grown with numerous hives of bees placed in the field to insure crossing of the different types. Small flowered varieties and large flowered ones were planted in alternate rows to insure the bees spreading the pollen properly.

MARJORAM (Origanum vulgare).

The marjoram is a European plant, cultivated in gardens, which has become naturalized in the Eastern States. It is a favorite of the honeybees and in a few localities in New York and Vermont has become sufficiently common to enable the bees to secure considerable honey from it. It is common to Europe, North Africa and Northern and Middle Asia. The plant is perennial and prefers limestone soils. From its wide distribution in the old world, it may be expected to establish itself over much wider areas in America and its importance will probably increase with the years.

Marjoram is the source of surplus honey in some parts of New England where it has become well established.

Honey from marjoram is reported to be of fine quality, equal to that from clover, but with a distinct flavor.

It is sometimes called winter sweet or organy.

MARRUBIUM, see Hoarhound.

MARSH FLEABANE (Pluchea).

There are several species of marsh fleabane common to the southeastern states. They are found in marshes and on wet lands and sometimes in woods. Specimens of *P. petiolata* have reached the author from Louisiana with the statement that this plant is of some value to the bees in that region. The blooming period is late summer and fall.

L. A. Schott of Benton, Missouri, writes that in his locality *P. petiolata,* is called "skunk-weed" and that it is almost a sure crop. It blooms in September and the bees work it eagerly. He states that in 1922, O. M. Headlee of Morehouse secured an average of 160 pounds per colony from this source in about two weeks. He describes the honey as a bright amber in color with a disagreeable odor and taste somewhat similar to that emanating from the plant. The honey is of good body, but candies very quickly after it is extracted. The plant is found in abundance on black lands where swamps have been drained. There are large areas of this kind in southeast Missouri.

MARYLAND—Honey Sources of.

G. H. Cale divides Maryland into five natural beekeeping regions. (American Bee Journal, August, 1921). The altitude ranges from sea level in the swampy regions of the eastern shore to 3,000 feet in the mountains in the western part of the state. Being near the coast, Maryland has an abundant rainfall and a rich flora.

In the mountain region there is much land unsuited to cultivation and forests occupy the larger part of the area. Buckwheat is an important source of nectar. This is a mining region and farming is not extensive.

In the upper midland there is a variety of soils, sands and sandy loams predominating in the west. In the valleys of Middletown and Hagerstown clay loams are found and farming is profitable. Fruit growing is extensively followed in some sections.

In the lower midland the soils are loams and clays with small strips of barrens. This is a highly developed agricultural country with grain and vegetables grown in large acreage. There are considerable areas of forest interspersed with the farms in both the upper and lower midland.

In the western shore district there is much land formerly cultivated to tobacco and other crops which has become so badly run down that it is little used at present. Nearly half of this region is in forest. The sandy loam is naturally fertile, but the single crop system of farming has resulted disastrously, as it always does where carried on for a sufficient period of time. Vegetable gardening is carried on extensively in the region south of Baltimore.

The eastern shore is largely composed of sandy soils in the northern and southeastern counties, with clay soils in the counties along the bay. Large quantities of garden truck are grown in this region. Fruit growing is also important. In the northern part there are considerable areas of deciduous forest and large pine areas in the south.

Nectar Sources

White clover and tulip-poplar, together with alsike clover, are the source of most of the surplus honey produced in Maryland. In the upper and lower midland, where farming and dairying are carried on extensively, both white and alsike clover are abundant. The lack of lime limits the growth of clover in the shore regions and the climatic conditions are not always favorable to the secretion of nectar from clover in the lower altitudes. Cale estimates that good crops of clover are secured about once in three years in the region where the plant is abundant.

The tulip-tree is common to the forests from the swampy lands of the east to the mountain slopes of the west, being more abundant on the dry flood plains of the higher lands. Tulip-tree yields abundantly in May before the opening of the clover flow.

In addition to the above the following are mentioned in the above quoted article:

Sugar maple, red maple, black locust, basswood, sumac, chestnut, sweet clover, blue thistle, buckwheat, goldenrod, aster, Joe-Pye weed, sweet pepper bush, chickweed, dandelion, crimson clover, raspberry, black gum, hairy vetch, mountain mint and muskmelon.

MASSACHUSETTS—Honey Flora of.

Honey production is relatively unimportant in Massachusetts. The demand for bees for pollination purposes by the cranberry growers and by those who grow cucumbers under glass rather overshadows the production of honey.

There are a few localities where commercial honey production might be followed with fair returns but generally those interested in bees have little investment and regard them as a sideline of minor importance.

Taking the State as a whole, the soil is poor and there are few large areas where honey plants of major importance are found. Winters are rather long, and temperature changes are frequent and severe. The variety of nectar sources is varied and local markets are favorable. In the localities where there are good pastures, beekeeping might be followed much more extensively than it is.

The following list of honey plants by Burton N. Gates is from Bulletin 129, Mass. Ag. Ex. Sta.:—

White clover, found in nearly all quarters of the State.

Alsike clover. Under favorable conditions it yields not only a good quality of nectar but large quantities of it.

Red clover. The second flowering is somewhat accessible to bees.

Sweet clover. Two species, neither abundant in Massachusetts.

Goldenrod and asters. Rank close to clovers in nectar production.

Fruit bloom. Apple, pear, cherry, plum, peach, etc., source of early stores, upon which the colonies build up for the clover harvest. The body is heavy, the color clear and light and the flow comes with a rush, which insures handsome sections; but best of all is the exquisite aroma of the apple blossom, which places fruit bloom honey in a class by itself.

Linden or basswood. Doubtless the most valuable tree honey plant in Massachusetts.

Buckwheat. Reported from all counties in Massachusetts.

Wild raspberry and blackberry. The nectar flow is of long duration, beginning after apple bloom has ceased.

Several species of sumac are important honey sources which are greatly underestimated.

Locust. A valuable forage for bees. Reported as sporadic in yield.

Maple. Probably of less importance as a honey plant than the mints, strawberry and milkweed.

Clethra. Known also as black alder and sweet pepper bush. A valuable honey secreting plant, largely confined to a belt paralleling the eastern coast.

Milkweed. Where milkweed occurs in large quantities it is a valuable honey plant. Reported as important from Berkshire County.

Wild cherry, knotweed, dandelion, strawberry, chestnut, mints, gill-over-the-ground and mustard also reported occasionally as valuable.

Willow and skunk cabbage valuable for early pollen as well as some nectar.

MAT GRASS, see Carpet Grass.

MAYPOPS, see Passion Flower.

MAYWEED (Anthemis Cotula).

The mayweed or dog fennel is an old-world weed extensively naturalized from Canada to the Gulf of Mexico and west to Texas. It is common along roadsides, in barn lots and waste places generally. Lovell credits it with yielding a honey that is light yellow in color and very bitter. It is seldom mentioned as a source of nectar.

Richter states that because of its blooming between spring and summer it is of considerable value to many Sacramento Valley beekeepers.

MEADOW FOAM (Limnanthes Douglassii).

Meadow foam is a spreading herb with attractive white or yellow flowers about one inch in diameter. It forms large patches on low ground in or near shallow water in the valleys of the Coast Ranges of

California. It is also cultivated as a garden annual to some extent elsewhere, growing to about one foot in height.

The fragrant flowers are attractive to the bees, and although it has attracted little attention among American beekeepers, it is cultivated as a honey plant in England and Scotland. It is probable that it is of some importance in the California valleys.

MEDICAGO, see Alfalfa.

MELEZITOSE

A special demand for melezitose honey brings it to the attention of beekeepers on rather infrequent occasions. Melezitose is a rare trisaccharide which sometimes occurs in the honeydews of certain pines and other trees. It is used in the study of certain microbic organisms and their metabolism on various sugar media. Such use is of course extremely limited and no large quantity of melezitose is needed for this purpose.

C. E. Burnside reported in American Bee Journal, November 1929, (Occurrence of Melezitose in Honey), that on several occasions much of this product was stored from *Pinus virginiana,* the scrub pine in Pennsylvania, Maryland, Virginia and nearby regions.

It has also been found on the Douglas fir of British Columbia. (See Douglas Fir.) It is probable that it frequently may be gathered by the bees in other localities and not be recognized as Melezitose honey.

MELIA, see China Tree.
MELILOTUS, see Sweet Clover.
MELISSA, see Bee Balm.

MELONS (Cucumis melo).

Melons are valuable sources of honey. In locations where they are grown in large acreage, as in the Rocky Ford district of Colorado, considerable quantities of honey are stored from them. There are numerous varieties of muskmelons, canteloupes, etc., but apparently there is not much difference in the value of the various sorts to the beekeeper.

In the Imperial Valley of California canteloupes are raised in large acreage. In 1922 more than five hundred carloads were shipped from the valley each day for some time. The total was enormous. More than thirty thousand acres of the crop were grown that year. A full train load of canteloupes left the valley every two hours of the 24 during the height of the season.

Dr. P. Holt, of El Centro, wrote to the author that the bloom of canteloupes furnished a fine source of nectar and that in his opinion one-third of the honey produced in the entire valley came from it in 1922. The honey there is largely blended with alfalfa.

In some southeast Missouri counties the beekeepers report melons as furnishing honey in surplus quantity, also.

MENTZELIA.

Mentzelia speciosa is a conspicuous plant along roadsides and waste places in Colorado. Its large white blossoms are closed during the heat of the day, but beekeepers report that the bees work upon it freely in early morning and late evening. It is probably not important.

Mesquite is an important source of good quality honey in the southwest.

In Oregon the stickleaf (*Mentzelia albicaulis*) grows freely on dry, stony land. It has a spreading habit and its pale orange colored flowers are very attractive to the bees.

MESCAL, see Century Plant.
MESOSPHAERUM (Swamp Basil). See Purple-Flowered Mint.

MESQUITE (Prosopis).

Most of the group represented in the mesquites are natives of tropical countries and are only found in the dryer sections of North America. They are small, shrubby trees, much branched and of slow growth. The wood is hard and durable, but because of the small size and rugged form of the tree is not of much commercial importance.

Some species are found in dry regions of Asia, extending from India to Afghanistan, Persia and Syria. It is primarily a tree of the desert or semi-arid regions.

The most common form in the southwest is *P. glandulosa,* which is occasionally found as far north as southern Kansas, Colorado and Utah, but is more abundant southward.

The mesquite, or mezquit, is the most important plant of the arid

The mesquite tree.

regions from central Texas to New Mexico and Eastern California. At a distance it has much the appearance of an aged peach tree. The tree is much branched and spreading in habit and is the source of fuel for the inhabitants of the region. Live stock also are fond of the leaves and pods and the Indians of the Southwest eat the seeds, first grinding them into meal and baking them.

The same or a closely related species is common in Hawaii, where it is known as algaroba. There it is one of the most important sources of honey. This species is sometimes called honey-pod or honey-locust, although the true honey-locust is a very different tree.

In the desert regions of California it is an important source of large quantities of honey, as it is in Arizona, New Mexico and Texas. In southwest Texas mesquite is the principal source. The honey is light amber and of good quality. Beekeepers report that it is lighter in color some years than others, even though nothing is blooming at the same season, so the difference cannot be laid to a mixture of honey from different sources. The quality of mesquite honey is good. It is reported to yield more regularly on sandy land than on heavy soil.

There are two blooming periods, the first in spring, usually in April, and the second in July. If there has been plenty of moisture previously the mesquite blooms profusely, due to the fact that it roots very deeply and can reach any moisture that is available in the soil.

P. odorata is a similar tree which is found in western Texas, southern New Mexico, Arizona, California and adjacent Mexico. It is locally called tornillo.

Concerning mesquite as a source of honey in Texas, H. B. Parks writes as follows:

> "Mesquite is not a reliable honey plant, as there are so many factors governing its nectar flows. In 1914, 1918 and 1921 there were heavy flows, but in the years between there was little or no flow. The ideal conditions seem to be plenty of moisture up to April and then hot weather till the flow is over. When moisture and heat conditions are right, it is no uncommon thing to see five or six sets of different aged beans on the same tree. Trees standing alone, especially those in yards or roads, bloom nearly every year, while those in the chaparral do not.
>
> "The honey is light amber, well flavored, and granulates rapidly. The flow comes on rapidly and is very heavy. Surplus up to 200 pounds on individual colonies is recorded. Mesquite and horsemint are rivals for first place in honey production. When mesquite yields it is far ahead, but when it fails, horsemint holds first place. As with alfalfa, a species of thrips often reduces the mesquite flow. While the heavy flows are restricted to central and southwestern Texas, its rapid spread gives hopes of increased yields from this excellent honey plant." See screwbean.

MEXICAN CLOVER (Richardia scabra).

The name clover is a misnomer, for this plant does not belong to the clovers but to an entirely different group. It is a luxuriant annual weed growing to a height of two feet or more. The bees are reported as working upon it quite late in the season.

The plant was introduced from the tropics and has become naturalized in Georgia, Florida, Alabama and Mississippi.

Although recognized as a weed, it is often cut for hay and is regarded by many farmers in the Gulf region as a valuable forage plant. It comes up in the fields after cultivation has ceased and covers the ground with a dense carpet. It begins to bloom about the first of September and yields nectar freely for several weeks.

J. J. Wilder (Dixie Beekeeper, Dec., 1920) reports that it yields an average of ten to fifty pounds of honey per colony. The honey is light amber in color and of good flavor. It granulates readily when extracted.

In Gleanings, (December, 1922) W. C. Barnard says that Mexican Clover is the most important summer and fall plant to south Georgia beekeepers. He gives the blooming period as from May 15 until frost, furnishing surplus during August, September and October. It is usually

Mexican clover is a common weed in the cultivated fields in the Gulf Coast region.

blended with cotton and velvet bean nectar and is secured in its purity only late in the season.

MEXICAN LAVENDER, see Vitex.

MICHIGAN—Honey Sources of.

While Michigan is within the clover belt, there are a greater variety of sources of surplus honey than in most nearby States. In spring, willows, maples, fruit bloom and dandelion have the usual importance. Both white and alsike clover yield surplus in Michigan. To these may be added wild raspberry, fireweed or willow herb and milkweeds, all of which are important sources of surplus honey in the northern part of the State. Basswood was formerly important and still produces surplus in some sections, though many of the basswood forests have been cut. Buckwheat is also a source of surplus in some parts of Michigan, though according to E. D. Townsend surplus from buckwheat is only to be expected on "a rather poor quality of sandy soil." Townsend says, further, that alsike is worth all other sources put together in the southern two-thirds of the lower peninsula, and that aster yields surplus in Sanilac County.—Gleanings, page 1184, 1908.

In American Bee Journal (December, 1922) B. F. Kindig stated that "It is hardly possible to divide Michigan into clearly defined areas which are characterized by the production of a single flavor of honey." He mentioned the fact that alsike and white clover are found growing together in all parts of the state and that while white clover may be found in greater abundance in old pastures, alsike will be found dominant in the lower lying lands in the same vicinity. Thus, except in some sections of the northern part of the state, the credit for honey from clover must be about equally divided between white and alsike. On the clay soils clover thus becomes the source of greatest importance.

Wild raspberry is valuable principally in the northern regions on the cut-over lands. In such situations it is at times so dense as to cover the ground, and yields heavy flows of high quality honey. For a few years following the forest fires, which run over the country at frequent periods, fireweed is abundant and yields heavily, but is gradually displaced by the re-established undergrowth. Fireweed yields when the weather is very cool and conditions are unfavorable for some other sources. Milkweed is of special importance on the lighter soils of the northwestern part of the peninsula. (See milkweed).

Kindig also mentions Canada thistle as important in some seasons, being particularly noticeable in seasons when the more important plants fail to yield normal crops.

Among the minor sources may be mentioned asters which yield surplus some seasons. Kindig reports as high as 200 pounds of surplus from aster from the upper peninsula. Verbenas are of some value in the pasture lands of southern Michigan. Surplus honey is reported from several counties. (Bul. 4 Dept. of Agr.) Blackberries cover large areas of the cut over lands in the northern region. Cucumbers are grown commercially in several localities where they yield some surplus honey. Goldenrod is considered the most important source of fall honey taking the state as a whole. Spanish needle and boneset furnish important yields in the swampy areas of Michigan, the former being more important in the southern section. Fruit bloom is important, especially in the fruit belt near Lake Michigan.

MIGNONETTE (Reseda odorata). COMMON MIGNONETTE.

The mignonette of our gardens is a native of Egypt which came to America by way of Europe. It is frequently mentioned as a honey plant, especially in older literature. Some extravagant claims have been made for it, but perhaps it has never been given a fair test on a sufficiently large scale to demonstrate fully its value. The following extract from an article which appeared in the American Bee Journal, Page 47, 1878, is of interest in this connection:

> "After being started under diligent cultivation it was astonishing to see the rapid progress which they made. The plants soon covered the ground, where soil was good, and were out in blossom in a short time, and from that time forward the bees were working on them by the thousand from morning till late afternoon. I have seen them thick on it by 8 o'clock. It yields pollen as well as honey. * * *
>
> "I have found that on account of the spikes of the blossom being so much longer, the bees must work on the larger varieties. I have some sorts which stand two feet high, the spikes being from eight to ten inches long. * * *
>
> "A correspondent from California stated that he thought an acre of mignonette would be adequate for a hundred colonies. * * *
>
> "When you give them this in addition to what they would otherwise have, it will certainly secure an immense addition to the honey produced. There is no plant within the range of our knowledge as valuable for bee forage as mignonette. It will bloom year after year if not disturbed by frost and gives a longer period of bloom than any other plant. It gives more blossoms in a given space and more forage than any plant we have ever seen. Honey from this plant has the most delicious fragrance of any we have ever tasted."—William Thompson.

It was later tested at the Michigan Agricultural College, but failed to bear out the above claims.

> "I expected great things of this plant, as the bee papers were very high in their praise of its qualities. June 23 it began to blossom, and it was not till the 27th that the bees began their work upon it. They did not seem to take it very readily, for on every occasion that I made observations, I found very few bees present. With us it proved a failure. Others have corroborated this statement. * * * It is rather a delicate plant for this climate."—Page 83, A. B. J., 1878.

It is listed by Richter in Bulletin 217, "Honey Plants of California," with the statement that it is "very much visited by bees whenever in bloom."

MILK VETCH (Astragalus).

The milk vetch is a close relative of the loco (see Loco), and like the loco is visited by the bees. Some species are of some value for honey, especially in the plains region west of the Missouri River.

Frank H. Drexel of Crawford, Colorado, reports a species of milk vetch locally known as "Bee Weed." This is probably *Astragalus haydenianus.* There the bees work it freely and a considerable yield of honey is secured. Although he is unable to determine with certainty the quality of honey he is of the opinion that it is not the best. Coming as it does between dandelion and the clovers it is very welcome.

From Yasuo Hiratsuka we learn that the principal source of honey in his part of Japan is "Genge," (*Astragalus sinicus*) and that the water white honey of high quality is harvested in May.

C. T. McKnight of Shreveport, Louisiana, reports the bees working freely on *Astragalus obcordata* in early spring.

MILKWEED (Asclepias).

The milkweeds are a large family of plants common to the temperate and tropical regions of many parts of the world. North America alone has 55 recognized species. These plants are also known as butterfly weeds and silkweeds. The blossoms are borne in large ball-shaped clusters. The seeds are attached to silken parachutes, on which they are carried by the wind. It is these silky attachments that give rise to the name "silkweed." Remarkable yields of honey are sometimes secured from *A. syriaca*. An average yield of 100 pounds per colony from this source is occasionally reported through the bee magazines.

Blossoms of the milkweed.

Much has been written about the entangling of bees in the pollen masses of milkweed. It frequently happens that bees thus entangled are unable to free themselves and die as a result. Some species of milkweed is included in nearly every list of honey plants which the author has consulted. Apparently it may be regarded as of some value almost everywhere. The honey is said to be light in color and of good quality.

It is of special importance in northern Michigan, where it grows in great abundance. In some locations beekeepers report an average per colony production of 50 pounds, year after year, from milkweed. (See also Dogbane and Snow-on-the-Mountain.)

Scullen and Vansell, (Bul. 412 Oregon Ex. Sta.) report some honey from two species, *A. speciosa* and *A. mexicana* in Oregon.

"C. A. Jamieson, Bee Division, Ottawa, in speaking on Nectar Secretion of the Milkweed Plant, said that (1) it depends solely on insects for pollination, (2) it secretes a larger amount of nectar in the forenoon with lower

In northern Michigan milkweeds grow abundantly over large areas.

sugar content and a smaller amount in the afternoon, with greater sugar content, (3) as humidity increased concentration decreased, and (4) it requires approximately 311 milkweed plants to produce enough nectar for one pound of honey with 17% moisture content."—Gleanings, pages 51, Jan. 1940.

Concerning pleurisy root, (*A. tuberosa*) which was formerly very common on the prairies of the middle west, James Heddon wrote in American Apiculturist, (P. 294 Nov. 1887).

"If there is any plant to the growing of which good land may be exclusively devoted for the sole purpose of honey production, I think it is this. I would rather have one acre of it than three of sweet clover. It blossoms through July and the half of August, and the bees never desert pleurisy root for basswood or anything else. The blossoms always look bright and fresh, and yield honey continuously in wet and in dry weather. Bees work on it in the rain and during the excessive drought of the past season, it did not cease to secrete nectar in abundance."

"It is the best honey yielding plant with which I am acquainted, white clover and basswood not excepted. It is eminently adapted to light sandy soil, doing well upon land too poor to produce ordinary farm crops. * * * The honey is very light colored and of excellent flavor.

In the same magazine, January 1888, John Haskins of Missouri wrote concerning pleurisy root:

> "The bees work on it from morn till night whether the weather is wet or dry. This year has been the dryest season for twenty-seven years; and still pleurisy root keeps yielding honey right along when nearly everything else that blossoms has dried up. * * * I never knew a year that it did not yield honey in abundance."

MILKWEED VINE, see Bluevine.

MINNESOTA—Honey Sources of.

Minnesota, in common with several states along the international boundary, has some splendid locations for honey production. The beekeeper who is well located in this state need hardly look farther.

The state is divided into three natural divisions, as will be seen by the map shown herewith. This map, which originally appeared in connection with an article on Minnesota by Prof. Francis Jager, in the American Bee Journal (Feb., 1923), shows also the average annual rainfall for different parts of the state.

The greater portion of Minnesota ranges from 1,000 to 1,500 feet above sea level. There is a small area along the streams of both the eastern and western borders below 1,000 feet of elevation, and a few scattered areas with an altitude above 1,500 feet.

The Coniferous Region

About one-half of the total area, including all of the northeast part of the state was originally one vast white pine forest. This immense area has largely been logged off and is now known as the cut-over region. Fire has run over a large portion of it, killing the young growth, which was following the original pine. The pine stumps are hard to remove and do not rot readily, making the land extremely hard to clear. Under present conditions much of this area is of little value for any purpose, although the soil is rich in lime and clover grows luxuriantly when given an opportunity. This section is sparsely settled, but in the portions where dairying is extensively carried on it offers some of the finest bee range. Alsike is found at its best here and grows wild along railroads, wagon roads and among the stumps. For two or three years following fires, the fireweed or *Epilobium* covers the burned districts and yields great crops of nectar. Raspberry gradually replaces the fireweed and this, together with alsike, holds sway until the underbrush finally deprives them of the necessary sunlight. With raspberry, fireweed and alsike furnishing white honey of the finest quality, it is not uncommon for the beekeeper to harvest an average of 200 pounds of white honey per colony.

The Park Region

From southeast to northwest there stretches diagonally across the state an area known as the hardwood or park region. This region is most extensive in the southeastern part and gradually narrows to a point some distance south of the northern border. The largest cities of

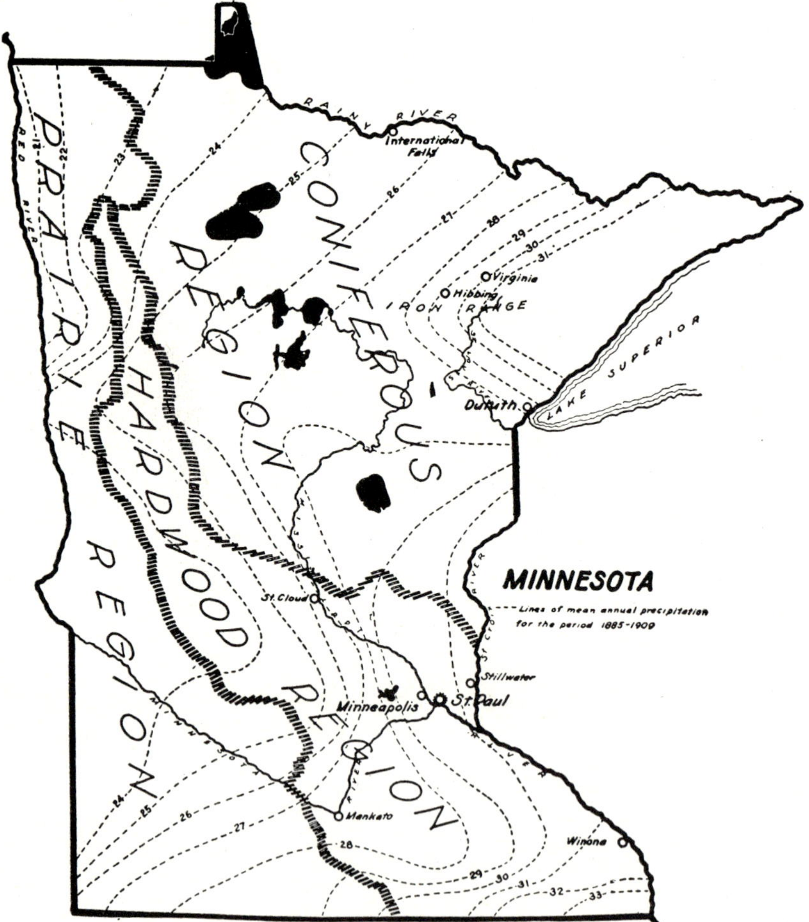

Map showing natural regions of Minnesota.

the state are found in this section and it has largely been cleared and occupied by prosperous farms and small towns. There are several valuable sources of nectar in this section. White and alsike clover, sweet clover, basswood and asters furnish the greater part of the surplus. Fruit bloom, dandelion, elm, maples, willows and numerous minor sources furnish both nectar and pollen in abundance. According to Jager, the honeyflow in this region varies from three to eight weeks in duration. In some seasons the flow is very intense, single colonies storing as high as 20 pounds per day. In other seasons the maximum is not above five pounds per day. Good crops are the rule.

The Prairie Region

Along the western border is the prairie region of Minnesota. This area is largely composed of stock and grain farms, with the clovers as the principal source of honey. High winds sweep across the prairie country and windbreaks are necessary for the comfort of the farmers as well as the bees. The soil is rich and land values are high. While alsike and white clover yield well in this region, the best locations for beekeeping are in the Red River Valley, where sweet clover is largely grown.

Minor Sources

In addition to the more important sources already mentioned, the following are common to some portions of the state. Snowberry, vervain, ironweed, jewel weed, goldenrod, wild plum, milkweed, dogbane, Virginia waterleaf, borage, hound's tongue, boneset, several mints, figwort, buttonbush, bush honeysuckle, Canada thistle, wild sunflowers, chicory, sowthistle, wood-sage, etc. Minnesota offers a wide range of conditions and a long list of sources of nectar.

MINT (Mentha).

Richter lists the spearmint, (*Mentha spicata*), as yielding a great abundance of amber colored honey in Sacramento County, California and southward. Spearmint was introduced from Europe and has escaped from cultivation and become naturalized in many places, both east and west.

He also lists the tule mint (*Mentha canadensis*) as yielding honey from July to October. The mints belong to an important group of honey plants and where sufficiently common are valuable. They should not be confused with hedge nettle, which is locally called mint in many places. (See Hedge Nettle).

Peppermint *Mentha piperita* is grown extensively as a field crop on the muck soils of southern Michigan and northern Indiana. It is estimated that a hundred tons of mint oil is produced annually in that region. Much of the crop is sold through a cooperative marketing organization. In times past the oil has sold as low as $2.50 or less per pound while it has also ranged as high as $26.00. In a good season the mint fields yield 20 pounds of mint oil per acre or more.

Beekeepers living in the vicinity of these fields of peppermint are able to secure considerable crops of mint honey.

MISSIONARY WEED, see Hawkweed.

MISSISSIPPI—Honey Sources of.

Because of its mild climate Mississippi is capable of supporting an intensive and profitable agriculture. There are large areas not yet in cultivation, most of which are in timber. The clay soils of the interior require fertilization for best results. The Delta of the Yazoo and Mississippi Rivers in the northwestern part, is a very rich region with an area of nearly six thousand square miles where cotton, corn and other staple crops are extensively cultivated. The southern part of the state near the Gulf has a sandy soil well adapted to the growing of vegetables. The winters are short and mild and the summers long and hot with oppressive humidity at times. The greater part of the state is composed of level plains or gently rolling hills.

The long season is favorable for the breeding of bees for the northern package trade and for the rearing of queens, although unfavorable weather conditions in early spring at times interfere with breeding operations. There are numerous locations where commercial honey production can be profitably followed.

Along the eastern border of the State there is a large area where sweet clover is widely disseminated. In this section large yields of surplus honey from this source are reported. An average yield of 140 pounds per colony of surplus from 700 colonies in one yard near Prairie Point has been reported to the author by a prominent Mississippi apiarist.

Willow, maple, fruit bloom and elm are reported as important for early pollen and nectar. White clover yields some honey in the northern part of the State. Persimmon is given as an important source of dark honey. Bitterweed yields freely, but the quality is so poor as to be of little value. Tupelo and cotton are valuable in some sections of the State. There is some fall honey from asters and goldenrod.

The coast region offers a vast area of gallberry and there is much titi along the streams.

In addition to the above important sources there are a large number which yield some nectar over a large area and a few which are of considerable importance in limited localities. The climbing boneset is abundant in the river bottoms in parts of the state. Butterweed yields nectar on the prairie lands over the state and in some places yields surplus. Mexican clover is valuable along the coast, buttonbush is valuable on the low lands, gaura, heartsease, holly, Chinaberry, gum, privet, French mulberry and basswood are to be considered in limited areas. Honeydew from several trees, including red cedar, is common. Scrub palmetto is found near the coast and Chickasaw plum yields early nectar over the state. Red sumac is not abundant but is sometimes reported as valuable locally. Wild grapes in the swampy sections are considered important. The juice from the fruit of the wild blackberry is often stored, making a fine fruit honey-syrup. Some honey is reported from partridge pea and Spanish needle.

MISSOURI CURRANT, **see Buffalo Currant.**

MISSOURI—Honey Sources of.

Beekeeping is a minor industry in Missouri and in few localities is honey production followed on a commercial scale. White clover is the principal source of surplus over the state. In a few counties sweet clover is grown extensively and these offer better opportunity than elsewhere. The white clover crop fails frequently from drouth in midsummer. Along the streams where heartsease, Spanish needle, boneset and aster insure favorable pasture good crops are the rule. The Missouri and Mississippi River valleys furnish numerous locations where clover is within reach on the hills and the fall flowers in the bottom lands. Away from the streams the honeyflows are not dependable over much of Missouri.

In the southeastern section there are several counties where canteloupes, watermelons and cow-peas are raised in large acreage. Here good crops are to be expected. White clover begins to bloom in early May, with melons following about June 1st. The first planting of cowpeas comes on soon after, overlapping the melon flow. The second crop of cow-peas comes in August, and this again is followed by boneset, heartsease, Spanish needle, Joe Pye weed, etc. Persimmon is common in the entire south half of the state and is a valuable source of honey in some localities.

Indian currant or coral berry is common over the northern counties and is important, especially in the northeastern part of the state. Bluevine (*Gonolobus laevis*) yields surplus in a few localities, and bitterweed is common in the Ozark region and other southern counties. The honey from this source often spoils the quality of the good honey stored earlier in the season, when the beekeeper fails to remove it before the bitterweed comes into bloom.

Missouri has a very large number of the minor sources of nectar common in the Mississippi Valley. There are large commercial orchards in some sections and fruit blossoms, followed by dandelion, furnish ideal spring pasture.

MISTLETOE (Phoradendron flavescens).

The mistletoe occurs as a parasite on trees from southern New Jersey and Missouri south to Florida and Texas. It is a yellowish-green shrub, much branched, which grows on the branches of the trees to which it attaches itself. It flowers early, usually in February and March, and is frequently mentioned as a honey plant in the Southern States. Scholl lists it as the first source of nectar and pollen in Texas, blooming there in January and February. Many Texas beekeepers regard it as valuable for spring stimulation.

Texas beekeepers reported a flow from mistletoe in January, 1923, and strong colonies were said to store as high as 15 pounds from this source.

Coleman lists three species of mistletoe (Western Honeybee, Dec.,

Mistletoe, first source of pollen in Texas.

1921) as sources of pollen in California. He states that some nectar is secured from one species.

The author found the bees working freely on mistletoe near San Antonio, Texas, on January 6, 1925. They were also found to be visiting a leafless species in southern Arizona very freely during the months of January and February. The flowers are minute three cornered cups set in yellow bands at short intervals along the spikes. They develop into one seeded berries which ripen at about the time the blossoms of the following year appear. The honey is reported to be light in color, very sweet and of a sticky, gluey texture hard to extract.

MONARDA, see Horsemint.

MONTANA—Honey Sources of.

Montana is a state of high mountains, dry plains and great extremes of temperature. In summer the thermometer often registers 100 degrees fahr., and in winter the temperature occasionally drops to 40 degrees below zero. The lowest degree recorded for the United States was 65 below near Miles City in January, 1888.

The precipitation is light and there is a low humidity which mitigates the extremes of temperature to a great extent. The rainfall averages between 13 and 15 inches in southeastern Montana to about 30 inches in some mountain localities. While the soil is rich and productive, the agriculture is limited because of the available moisture.

Montana is a high and dry state, the altitude ranging from slightly

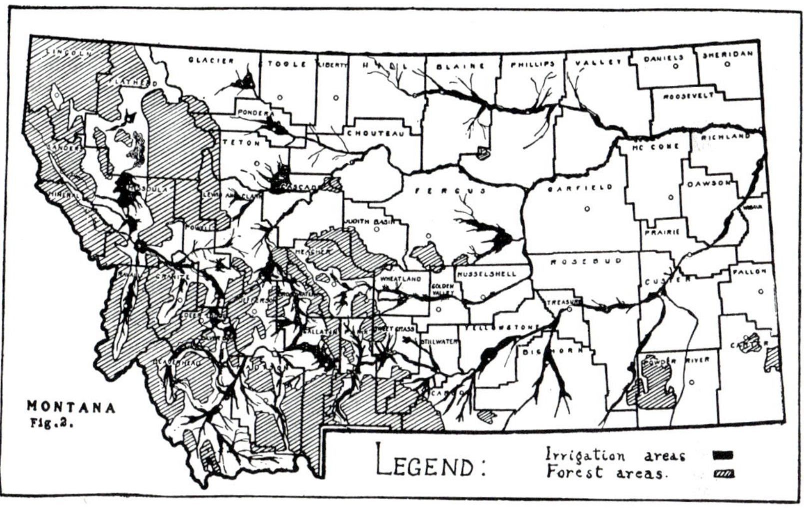

Map showing irrigated and forest areas of Montana.

below 2000 feet in northeast to spots above 10,000 feet in the mountains. In the valleys where irrigation is practiced splendid crops of alfalfa, vegetables and small grains are grown. While dry farming is practiced successfully in some sections, the grazing of livestock is the principal agricultural industry outside the irrigated valleys.

Conditions for honey production are very favorable in the irrigated valleys and also in some dry farming neighborhoods where sweet clover is extensively grown. Most of the honey which goes to market from this state is from alfalfa and sweet clover. Large average yields are harvested and the honey is of the best quality.

In an article on Beekeeping Conditions in Montana in American Bee Journal, (May, 1924), O. A. Sipple stated that the best honey producing regions are situated between the 3500 foot and the 5000 foot elevations. He estimates the average yields in a commercial apiary to be between 200 and 300 pounds per colony.

Honey Plants

As already stated, the principal surplus honey comes from alfalfa and sweet clover (melilotus) but minor sources are important to the man who seeks a permanent location. Sippel lists the following as common to the state:

The prairie crocus (anemone) is the first source of spring pollen followed by pussy-willow. Later comes dandelion and in Bitterroot Valley fruit bloom from such trees as plums, apples, etc. Alsike and white clover are important and in some localities furnish the chief surplus, as high as 200 pounds per colony having been harvested from these clovers. Box elder, wild raspberry, choke cherry, wild currant and blueberry are widely distributed. Canada thistle has been introduced in many neighborhoods and, while a serious weed pest, is of much value to the bees. Bear-berry or kinnikinick, snowberry and buckbrush are common native shrubs which yield nectar freely.

The gumweed, (*Grindelia*) is common on the plains and yields a honey of poor quality which granulates quickly. Some honey is reported from goldenrod in late summer. Rabbit brush is present in the west.

MONUMENT PLANT (Frasera).

The monument plant is variously known as columbo, deer's ears (a translation of the Navaho name), and Frasera. It grows in high altitudes from 7,000 to 10,000 feet, from South Dakota westward to Montana and Oregon and south to New Mexico and California.

It has large, creamy blossoms, about two inches in diameter, on tall flower stalks growing to a height of three feet. It is common in the Rocky Mountains and is much sought by the bees. According to Wesley Foster it is an important source of honey in May. It is a striking plant which once seen is not likely to be forgotten.

MORNING GLORY (Ipomoea).

There are several species of native morning glory common to nearly all parts of America east of the Rocky Mountains, which are classified as *Ipomoea*. These are especially abundant in the southeastern states, where surplus honey is often reported. A species introduced from the tropics for cultivation for ornament has escaped and established itself in waste places over a wide area.

Richter reports morning glory (*Convolvulus arvensis*) in the list of plants occasionally yielding surplus in California. This is the small bindweed which is so common over almost the entire country from Nova Scotia to California. It is native to Europe and Asia.

The monument plant (*Frasera speciosa*).

Scholl lists *Ipomoea Caroliniana* as furnishing a light, yield of honey and pollen from June to November in Texas. There are nearly twenty species found in Texas and it is probable that several are of some importance. *I. trifida* is commonly known as tievine and is the source of surplus in some localities. At Victoria, Texas, M. B. Talley reported to the writer that he secured surplus from this source in September and that the quality was very good.

Concerning the morning glory as a source of honey, in the American Bee Journal for Nov. 10, 1892, W. S. Douglas of Lexington, Texas, wrote as follows:

> "On September 1, the "shower" came in earnest from this wonderful honey plant. There was at least 3000 acres of it in reach of my bees. One colony stored 80 pounds of surplus from it alone. This was my best colony. The honey is of the finest quality, as clear as spring water, and the flavor is just splendid."

Morning glory in bloom.

Oertel reports morning glory as a minor source of honey in Louisiana. (See Campanilla.)

MOTHERWORT (Leonurus Cardiaca).

The common motherwort is a weed introduced from Europe and Northern Asia. It is now quite generally naturalized from Canada to Florida and west to Louisiana. It is an excellent source of nectar. Reports of bees working on this plant very freely, when most other plants fail to yield anything, are frequent.

The motherwort grows in clumps in waste places in old barn lots, along railroads, in factory grounds, etc. It grows from two to six feet high, with small flower clusters in the axils of the leaves. It is a relative of the catnip, and apparently equally attractive to the bees. This plant was formerly used to some extent in medicine, especially for diseases of women. It is also known as lion's tail.

Clump of motherwort in barnyard.

"Motherwort is one of the best honey plants that I have ever seen. It begins blooming here in the mountains early in May and today, Nov. 5, you can still find my bees working on it. * * * Drouth has no effect upon it. The honey from it is a light orange color." Otis A. Griffith, Scholten, Missouri, Gleanings page 40, 1911.

MOUNTAIN BALM, see Yerba Santa.
MOUNTAIN LAUREL, see Laurel.

MOUNTAIN LILAC (Ceanothus).

There are several species of mountain lilac common to California and the Western States. They are closely related to the buckthorns.

(See Buckthorn). New Jersey tea belongs to this group. (See New Jersey tea).

Jepson lists fourteen species of *Ceanothus* common to California. They are shrubs or small trees with small but showy flowers, borne in umbels or panicles. The California species are mostly evergreen.

They are known by various names, as deer-brush, California lilac, mountain lilac, etc. One species is called buckbrush.

Some species are common to the Mountain States from Wyoming and Colorado to Arizona and west to Texas. Both honey and pollen are secured from this source, though there are probably not many places where surplus yields may be expected.

Motherwort in bloom.

MOUNTAIN MAHOGANY (Cercocarpus).

There are several species of *Cercocarpus* common to the Rocky Mountain region. They are shrubs or small trees with very hard and tough wood of a reddish brown color, sometimes used for tool handles and similar purposes.

They grow usually on dry mountain slopes in poor and rocky soils in regions of light rainfall and are accordingly of very slow growth, although generally thought to be long lived. The leaves persist over winter, falling when the new leaves appear or in some cases lasting through the second season. Twigs of the low growing forms often furnish an important browse for cattle grazing on the ranges. Young twigs have a slightly aromatic flavor.

The long tail-like attachment to the fruit is a common means of identification featured in most descriptions of this group. This serves to carry the ripened seed with the wind or sometimes becomes attached to the hair of passing animals, thus assisting in its distribution.

C. montanus or as sometimes classified *C. parvifolius,* is reported as an important source of surplus honey at Flagstaff, Arizona by S. M. Campbell. There it occurs in abundance on the steep mountain sides and, according to his report, yields a light colored honey of good quality. He states that his bees have averaged fifty pounds of bulk comb honey per colony from this source some years and that it blooms in June and sometimes again in September.

In southern California the mahogany sumac (*Rhus integrifolia*) is often called mountain mahogany. It is also a source of nectar. (See sumac.)

MOUNTAIN MINT, see Basil.

MOUNTAIN MISERY (Chamaebatia foliolosa).

Mountain misery, also called bear clover or bear-mat is reported by Prof. G. H. Vansell as the source of good crops of good quality honey in the high altitudes of Tulare County, California This plant is sometimes called tarweed.

MUNG BEAN (Phaseolus aureus).

The mung bean is an Asiatic plant that has been cultivated since ancient times. It is known by numerous common names as Oregon pea, Jerusalem pea, and Chop Suey bean. It is a legume which has been cultivated to some extent in southern states, although some varieties mature in about eighty days so could be used anywhere in the corn belt.

W. C. Barnard of Glennville, Georgia, reports that bees work the blossoms freely over the entire day for a period of about three weeks and he regards mung beans as superior to cowpeas for honey. George W. Bohne of Luling, Louisiana, after extensive trial reported that the seed shattered badly and that the hay was less attractive than that from peavines. His bees worked a 400 acre field as well as cowpeas but the honey had a "Beany" flavor which he regarded as inferior.

In American Bee Journal honey plant garden in Iowa the bees ignored it completely although it bloomed freely and yielded beans abundantly.

MOUNTAIN QUEEN, see Yucca.
MULE FAT, see Baccharis.

MUSTARD (Brassica campestris).

The Figure shows the common yellow mustard (*Brassica campestris*), which is common all over North America and in Europe. The black mustard (*B. nigra*) also has a very wide distribution in Europe and America. There are about 50 species, including the closely related cultivated varieties of cabbage, turnips, rutabagas and mustard. All produce some nectar, and in some localities they are an important source of honey. In parts of California, notably the Lompoc Valley, mustard is grown commercially for seed; much honey is stored from this source. The honey is said to be light in color and mild in flavor. Apparently there is much variation in the amount of nectar, according to climatic conditions. In any locality where it is sufficiently abundant, mustard can be expected to add something to the product of the apiary.

The late J. S. Harbison wrote, in his "Beekeepers' Directory,":

Mustard in bloom.

"Mustard affords a larger amount of valuable pasturage to the acre than almost any other plant. It blooms throughout the month of May, and part of June. During this time, bees increase in numbers, and store from it large quantities of honey of a clear yellowish color, but partaking slightly of the taste of the plant."

In Bees and Honey, March 1934, G. H. Vansell states that in the Sacramento Valley in California, black mustard, *B. nigra* is a fickle source, sometimes yielding many tons of honey and at other times pollen only. The white mustard, *B. alba,* grown commercially in the Lompoc Valley he reports as a consistent source of honey. He also states that *B. napus* from Napa Valley is a reliable source of honey in early spring.

Scullen and Vansell, (Bul. 412 Oregon Ex. Sta.) report that common mustard frequently attracts bees away from fruit trees in the Willamette Valley and yields considerable nectar there.

MYRTLE, see Wax Myrtle, also California Laurel.

N

NAMA (Hydrolea).

Nama is a water-loving plant found on the banks of streams or ponds. *H. ovata,* the ovate-leaf nama, is found from southern Missouri to Louisiana and Texas and south to the tropics. It has been reported as valuable only from Texas. H. B. Parks writes that it is found with every pond in the eastern part of that state and is worked heavily by the bees from May to September.

NAPA THISTLE (Centaurea melitensis).

The Napa thistle, or tocalote, (*Centaurea melitensis*) is known in California as Napa thistle because it was first introduced into that state at Napa and distributed from there. It belongs to a group of plants known as star thistles.

This species yields a light amber honey of good quality. (See star thistle.)

NEBRASKA—Honey Sources of.

Nebraska may be divided roughly into two divisions for honey production. The eastern or white clover region follows the loam soil known as loess. This is found in a southwesterly direction through southwestern Antelope, southern Holt, northwestern Boone, southern Wheeler, Garfield and Loup, northwestern Custer, eastern Logan and Lincoln, southeastern Mayes and western Hitchcock Counties.

The western or alfalfa region embraces all the territory west of this irregular line. Here the soil changes to a sandy loam or drifting sands. Passing from east to west there is a steady decrease in the rainfall and the flora changes to conform to the dryer conditions. The altitude increases westward as the rainfall decreases. From less than 1,000 feet in southeastern Nebraska it rises to more than 5,000 feet in the Pine Ridge district in the northwestern part of the state. The average rainfall varies from 30 to 32 inches in the east to 12 to 15 inches in some sections of the west end.

White Clover Region

Myron H. Swenk (American Bee Journal, May 1922) divides the white clover region into two distinct sub-regions, the northern prairie region and the southern prairie region, with the Platte River as the dividing line. The east end of these regions has different conditions, owing to the influence of the Missouri River, which borders the state on the entire eastern side.

Alfalfa Region

In similar manner Swenk divides the western or alfalfa region into sub-regions, northern plains region and southern plains region, with the intermediate sand-hills region. He also adds a smaller region in the northwest, since the Pine Ridge district offers conditions peculiar to itself.

Honey Plants of the State

In the eastern region, white and sweet clover furnish the principal sources of surplus honey, while in the western region alfalfa and sweet clover furnish the nectar. Sweet clover is the one plant of major importance throughout the entire state where beekeeping is important. Alsike is also important in some portions of the eastern region. Heartsease and false indigo furnish abundant surplus in the eastern region some seasons, and basswood was formerly important in a few localities along the streams. Fruit bloom and dandelion, maples and willows are important sources of early nectar and pollen.

Swenk lists a great variety of minor sources, some of which are the source of occasional surplus in limited areas. Among the more important may be mentioned the following:

Black locust, honey locust, catalpa, box elder, wild plums, wild gooseberry, shoestring (*Amorpha canescens*), horsemint, buckbrush (*Symphoricarpos*), Virginia creeper, poison ivy, vervain, sumac, Indian hemp, aster, goldenrod, wild sunflower, marigold, snakeroot, boneset, gumweed, mustard, partridge pea, buckwheat, clematis, ironweed, Spanish needle, Rocky Mountain bee plant, gaura.

NECTAR, Sugar Concentration in.

The fact that bees often visit the blossoms of one plant freely while neglecting others is readily explained by the difference in the amount of sugar available in the nectar. The sugar content varies from day to day with the same plants and also from place to place depending upon soils or other environmental conditions. In Circular 650 of the U. S. Department of Agriculture, G. H. Vansell discusses at length the factors affecting the usefulness of honeybees in pollination. He has shown that some varieties of fruit are neglected because others blooming nearby have a greater sugar concentration in the nectar. Apples and cherries are very attractive to the bees because of high sugar concentration while pears are low in sugar content and are likely to be overlooked if apples or cherries are within reach. In like manner some varieties of any of these fruits are more attractive than others for the same reason.

The concentration of sugar in the available nectar has an important bearing on the size of the honey crop since less evaporation is necessary to ripen nectar of high sugar content.

Vansell reports results of numerous examinations. The following is copied from the above named bulletin:—

Data on the average sugar concentrations of several plants, collected during a number of seasons and from many locations, have been accumulated and are presented in the following table, the plants being listed according to the descending order of the sugar concentration of their nectar.

Sugar concentrations of the nectars of some California and Oregon plants, arranged in order of decreasing sugar

Plant	Average Sugar Concentration
	Percent
Alfileria (Erodium cicutarium)	65.0
Black locust (Robinia pseudoacacia)	63.2
Willows (Salix spp.)	60.0
Locoweed (Astragalus pachypus)	59.2
Wild buckwheat (Eriogonum fasciculatum)	58.1
Common vetch (Vicia sativa L., extrafloral nectar)	56.5
Sainfoin (Onobrychis viciaefolia)	55.4
Bachelor's button (Centaurea cyanus)	52.5
Melanops vetch (Vicia melanops Sibth. and Sm., extrafloral nectar)	52.1
Bigleaf maple (Acer macrophyllum)	52.0
Deerweed (Lotus scoparius)	52.0
Yellow sweet clover (Melilotus officinalis)	51.6
Dandelion (Taraxacum palustre var. vulgare)	51.2
Rape (Brassica napus)	50.5
Wild buckwheat (Eriogonum sp., annual pink, from eastern Oregon)	50.5
Salal (Gaultheria shallon)	50.5
Chickweed (Stellaria media)	50.0
Common mustard (Brassica campestris)	50.0
Sweet cherry (Prunus avium)	[1] 50-60
Flax (Linum usitatissimum)	49.5
Cultivated buckwheat (Fagopyrum esculentum)	49.0
Horsechestnut (Aesculus hippocastanum)	49.0
Hoarhound (Marrubium vulgare)	48.3
White sage (Salvia apiana)	48.0
Castor-bean (Ricinus communis L., extrafloral nectar)	47.8
Hungarian vetch (Vicia pannonica, extrafloral nectar)	47.7
Crimson clover (Trifolium incarnatum)	47.7
Cantaloup (Cucumis melo)	46.7
Perennial sunflower (Wyethia angustifolia)	45.8
Jackass clover (Wislizenia refracta)	45.3
Oregon grape (Berberis aquifolium)	45.0
Apple (Malus pumila)	45-55
Black sage (Salvia mellifera)	44.9
Toyon (Photinia arbutifolia)	44.8
Jim Hill mustard (Sisymbrium altissimum)	44.1
Hairy vetch (Vicia villosa)	44.0
Alsike clover (Trifolium hybridum)	43.3
Locoweed (Astragalus hornii)	42.4
Vine maple (Acer circinatum)	42.0
Cotton (Gossypium hirsutum, Acala variety, extrafloral nectar)	41.9
Buckeye (Aesculus californica, wide range, 33.0—50.8 [2])	41.5
Alfalfa (Medicago sativa L., except Imperial Valley data)	41.1
White clover, ordinary and ladino (Trifolium repens)	41.0
Spikeweed (Hemizonia pungens)	39.9
Privet (Ligustrum sp.)	39.2
Tamarisk (Tamarix sp.)	38.8

[1] Two figures indicates a range between varieties or species.

[2] Ruth Beutler (2), in Germany, reported red buckeye with a sugar concentration of 66 percent.

Plant	Average Sugar Concentration
	Percent
Wireweed wild buckwheat (Eriogonum sp., California)	37.8
Yellow star-thistle (Centaurea solstitialis)	37.5
Penstemon (Penstemon sp.)	37.3
Wild buckwheat (Eriogonum sp., yellow perennial)	37.3
Milkweed (Asclepias speciosa)	37.2
Incense-cedar (Libocedrus decurrens, honeydew)	36.3
Aster (Aster spinosus)	36.2
White sweetclover (Melilotus alba, except Imperial Valley data)	35.8
Evergreen blackberry (Rubus laciniatus)	35.6
Fireweed (Epilobium angustifolium), wide variations with varying humidity	35.0
Canada thistle (Cirsium arvense)	35.0
Red clover (Trifolium pratense)	34.3
Linden (Tilia spp.)	33.6
Locoweed (Astragalus watsoni)	33.3
Strawberry clover (Trifolium fragiferum)	33.3
Fenugreek (Trigonella foenum-graecum)	32.9
Creeping sage (Salvia sonomensis)	32.4
Monkeyflower (Mimulus sp.)	32.0
Sunflower (Helianthus annuus)	31.6
Red gum (Eucalyptus camaldulensis)	30.0
Peach and nectarine (Prunus persica and P. persica var. nectarina)	30.0
Catnip (Nepeta cataria)	28.7
Snowberry (Symphoricarpos albus)	28.6
Purple vetch (Vicia atropurpurea)	28.0
Himalaya-berry (Rubus procerus)	27.2
Bluecurls (Trichostema lanceolatum)	27.1
Locoweed (Astragalus trichopodus)	27.0
Tree tobacco (Nicotiana glauca)	25.9
Hungarian vetch (Vicia pannonica, floral nectar)	25.2
Turkeymullein (Eremocarpus setigerus)	25.0
Navel and Valencia oranges (Citrus sinensis)	25.0
Cotoneaster (Cotoneaster harroviana)	24.5
Wild vetch (Vicia sp.)	23.1
Common vetch (Vicia sativa L., floral nectar)	22.6
Melanops vetch (Vicia melanops, floral nectar)	22.6
Rocky Mountain bee plant (Cleome serrulata)	21.9
Cotton (Gossypium hirsutum L., Acala variety, floral nectar)	21.5
Figwort (Scrophularia californica)	19.6
Tulip poplar (Liriodendron tulipifera)	19.6
Flowering currant (Ribes sanguineum)	18.9
Blue gum (Eucalyptus globulus)	17.0
Locoweed (Astragalus asymmetricus)	16.0
Manzanita (Arctostaphylos spp.)	16-50
Avocado (Persea americana)	15.2
Madrona (Arbutus menziesii)	15.0
Sour cherry (Prunus cerasus)	15-40
Eucalyptus (Eucalyptus sp.)	13.4
Yellow cleome (Cleome lutea)	11.0
Apricot (Prunus armeniaca)	5-25
Pear (Pyrus communis)	4-30

These plants are shown to be very different in the concentration of nectar sugar, the extremes in these average values being about 4 and 65 percent. In general, one species or variety appears to have a rich or poor nectar in comparison with another, irrespective of geographical location.

NEGUNDO, see Box Elder.

NEVADA—Honey Sources of.

Although much honey is produced in Nevada, nearly all the product which reaches the eastern markets is from alfalfa and sweet clover, which is produced in irrigated valleys. Much of this state is composed of desert, little used except for grazing herds of cattle and sheep. There are a number of desert plants which furnish nectar freely. Among the best known may be mentioned creosote bush, rabbit brush, prickly pear, catsclaw and mesquite, the last two named found more particularly in the southern part of the state. There are large areas of the state which are desert and much remains to be learned concerning the honey-producing flora. In the cultivated districts box elder, maple and willows, together with dandelion, furnish ample forage for brood rearing. Snowberry, gumweed, clematis and wild plums are available in favored locations.

NEW BRUNSWICK—Honey Sources of.

Conditions in New Brunswick are similar to the New England States, with much of the surplus honey gathered in the fall from goldenrods and asters. White and alsike clover furnish good yields under favorable conditions, with a large number of minor sources contributing to the support of the bees.

In spring, fruit bloom, together with willows, maples and dandelion, furnish abundant nectar and pollen, but unfavorable weather conditions often make it unavailable. Fireweed is important in the burned-over areas. Buckwheat, Canada thistle, mustard, sumac, milkweed and button-bush add something to the total of the crop but are not important as sources of surplus.

NEW HAMPSHIRE—Honey Sources of.

There is very little interest in honey production in New Hampshire. There are few localities where large apiaries could be profitably supported and average yields of honey are generally small. In the mountains of the northern section fireweed and wild raspberry offer some pasture, but the most of the bees are found in the farming districts near the Connecticut and Merrimac Rivers. As a rule but few colonies are kept and these in connection with some other business enterprise.

The honey flora of New Hampshire is very similar to that of other New England States. In spring, willows, maples, fruit blossoms and dandelions keep the bees busy in preparation for the main flow, which comes from the clovers. Goldenrod is also important. Sumac, laurel, blueberries and asters are worthy of mention, also. See Maine and Massachusetts for additional plants.

NEW JERSEY—Honey Sources of.

New Jersey is a state of ample rainfall, humid climate and wide variations of heat and cold. It is a small state with little more than 8000 square miles of surface, with several large cities within its borders and a highly diversified agriculture.

It may be divided into three areas for consideration of its beekeeping possibilities. In the northern portion the altitude rises to 1800 feet in the Kittatinny Mountains near the New York State line. This portion of the state is hilly and rocky with farming largely confined to the valleys and the hills overgrown with forests. In this region the variety of plants insures something for the bees at all times from spring until frost. There is a summer flow from the clovers and a fall flow from buckwheat, aster and goldenrod. In spring there is a variety of bloom from such trees as maple and willow and from dandelion and other plants to insure ample forage of brood rearing. There is much sumac also.

In the central region we find the congested population of many large cities with much unused land on the outskirts overgrown with goldenrod and aster. The farming is largely confined to intensive cultivation of vegetables and offers little support for the bees. E. G. Carr reports, (American Bee Journal Dec. 1923), that in favorable seasons the beekeeper is assured good yields in this area, often as much as 100 pounds of surplus from goldenrod and aster. This region lies rather level and there is much fertile soil. In some neighborhoods heartsease yields well.

In the southern district we find a low sandy plain where alsike clover is the principal source of surplus in the farming districts, with clethra and blueberry yielding surplus in the swamps. From Mr. Carr's Manual of Bee Husbandry we select the following list of plants as valuable for the state as a whole:—

Plants from which nectar is gathered in less than surplus quantities:

Maples, mid-March-early April.
Peach, early April.
Pear, mid-April.
Apple, late April early May.
Willows, late April.
Dandelion, early May.
Wild strawberry, May.
Lupine, May.
Milkweed, silkweed, July.
False indigo, July.
Button bush, July.
Burdock, July-October.
Tree of Heaven, pride of China, July.
Catnip, July.
Motherwort, August.

Raspberry, May.
Huckleberry, blueberry, late May-late June.
Persimmon, mid-June late June.
Vervain, late June early September.
Virginia creeper, late June late July.
Grape, late May early June.
Smartweed, heartsease, late August-mid-September.
Horsemint, August-September.
Boneset, mid-August-September.

Surplus honey plants:

Crimson clover, mid-May.
Locust, May 20-June 1.
Tulip-poplar, May 26-June 10.
Poison ivy, mid-May-mid-June.
Holly, late May-late June.
Mountain laurel, late May late June.
Sheep laurel, late May-late June.
Swedish clover, alsike clover, June 1-July 10.
White clover, early June-mid-July.
Dogbane, Indian hemp, early June-late August.
Basswood, linden, late June-early July.
California privet, mid-July-late July.
Sumac, mid-June-mid-July.
White sweet clover, June-November.
Cranberry, June 15-August 15.
August flower, soap bush, sweet pepper bush, late July-late August.
Rose mallow, swamp mallow, late July early September.
Spanish needle, mid-August into October.
Heartsease, smartwood or blackheart, August-September.
Heath aster. white aster or St. Michaelmas daisy, late August-mid-October.
Bushy goldenrod, late August-mid-October.
Buckwheat, early August.

NEW JERSEY TEA or RED-ROOT (Ceanothus americanus).

New Jersey tea grows from one to three feet high from a dark red root, hence the name "red-root." It is a shrubby plant with flowers in white clusters. The leaves were used for tea during the revolutionary war. It is common in woodlands from Ontario to Manitoba and south to Arkansas, Texas and Florida.

It is not often mentioned as a source of honey, but according to H. B. Parks, is regarded as valuable in northwestern Missouri. (See Mountain lilac.)

New Jersey tea is abundant in some sections of north central Texas, where it is valuable for carrying the bees over the period from early brood rearing until the main flow from cotton. Beekeepers in the neighborhood of Waxahatchie regard this plant as very valuable in that region as a minor source.

NEW MEXICO—Honey Sources of.

New Mexico is a high and dry state. The fourth largest in area, it is among the least in population. Its lower river valleys are about 3500 feet above sea level and the greater part of its area is much higher, ranging from 4000 to 10,000 feet elevation. The summits of its highest mountains rise to an altitude of 13,000 feet. Due to the lack of humidity, the air is thin and warms very quickly. In summer the heat is intense, but when the sun goes down the same conditions favor rapid cooling and the difference between day and night temperatures varies greatly at all seasons.

New Mexico, a state of mountain, plain and plateau, is for the greater part a vast expanse of grazing country largely devoted to pasturing cattle and sheep. In the higher elevations where more moisture occurs, are large forests, while the narrow valleys are devoted to agriculture, watered by irrigation.

Alfalfa is the most important source of surplus honey with sweet clover and cotton also furnishing large quantities. Cantaloups yield some surplus in the Rio Grande Valley in the vicinity of Messilla Park and southward toward El Paso.

While about three-fourths of the commercial honey produced in New Mexico comes from the cultivated crops already mentioned, there are many native plants which yield much nectar. Mesquite is the one source of first importance among the wild plants with catsclaw also of major value.

Mistletoe furnishes the first spring stimulation in the valleys where it blooms in February. Ephedra blooms early and furnishes a small amount of surplus in favorable seasons, where it is sufficiently common. From cottonwood come large quantities of pollen and the willows are also valuable along the streams. Creosote bush, which is commonly known as "greasewood" is valuable for early spring brood rearing, although it rarely yields surplus. Creosote bush is one of the most characteristic and widely spread shrubs of the desert region.

There are four species of baccharis found within the state. These grow in moist places and are often mistaken for small willows. The common name is Guatemote. In the northern part of the state, and in the higher elevations south, dandelions are an important early source of pollen and of some nectar. The fruit trees furnish abundant early pasture, but in the Pecos Valley there is much loss of bees from spray poison which compels many beekeepers to move their bees from the vicinity of the orchards.

In the Pecos Valley, Salt Cedar or Tamarix has become naturalized in large areas and furnishes much surplus honey. Arrow-wood, commonly called "Cachinilla," grows in great abundance along the streams and irrigation ditches and is reported as an important addition to the early pasture. There are several species of rabbit brush (Chrysothamnus) which yield considerable honey in late summer in dry areas throughout the state. Lycium in two or more species occurs in the washes where occasional floodings occur and is of some value to the bees.

Sotol, a near relative of the yuccas, is very common over large areas of the southern part of the state and is reported as occasionally yielding surplus. The yuccas also are common but add little to the sum total of production.

In the towns, umbrella tree or chinaberry, tree of Heaven, Bird of Paradise Tree, tamarix and other ornamentals offer a limited amount of bee pasture.

In the high elevations cleome, salvia, snowberry, beard-tongue, wild geranium and numerous others attract the bees.

Beekeeping in New Mexico is mostly confined to the irrigated valleys and only a few dry locations offer sufficient pasturage to insure successful honey production on a commercial scale. The Rio Grande and Pecos River Valleys furnish most of the honey produced within the state, with limited amounts coming from a few of the smaller valleys.

Only a few beekeepers have tried the mountain locations, but their success indicates that many such may prove satisfactory.

NEW YORK—Honey Sources of.

So varied is the climate of New York that the U. S. Weather Bureau divides the State into four sections for discussion of its peculiar conditions. Likewise the surface of the area ranges from sea level on the coast to altitudes above 5000 feet in the mountains. There are varied soil types and a great diversity of plant growth.

There are localities where beekeeping is highly profitable and where commercial honey production is followed extensively. Likewise there are regions where but little forage is available to the bees and few are to be found. Because of the proximity to the best of markets, conditions are especially favorable for the sale of the crop and New York is widely known as one of the most important honey producing states. Good locations are very unevenly distributed and some study of soils and flora are necessary to find the best opportunities.

Rainfall varies from an average of less than 30 inches annually in the neighborhood of Lake Champlain to about 60 inches at the headwaters of the Ausable River. On the whole, the precipitation is ample to support a varied flora.

Although cold weather prevails over most of the state during the winter, the climate is greatly modified by the Great Lakes and the Atlantic Ocean. The growing season extends over about 200 days between killing frosts on Long Island, but is confined to less than half that period in some parts of the Adirondack Mountains.

New York has a wide diversity of conditions within its borders and several plants which are important in one section are unknown in another. Of those common to the whole state may be mentioned white and alsike clover, sweet clover and buckwheat. Buckwheat is the source of most of the surplus honey in a large area. Alfalfa is mentioned as a source of honey in some sections, although it is not dependable, owing to climatic conditions. In the orchard districts an immense acreage of apple, peach, pear, cherry and plum are available in early spring. Black locust is of local importance. Sumac is reported as a source of surplus in some sections. Blue thistle, goldenrod and asters are valuable for late summer forage.

In limited areas the thyme has become so well established as to yield heavy surplus. Purple loosestrife, locally called rebel-weed, is abundant along the Hudson River. Fireweed is found in the burned-over districts in the northern forest area. Canada thistle is quite generally distributed and yields a good quality of honey. Clethra is common on Long Island and in some other localities. Vetches, hawkweed, and numerous other minor sources, should be included in a complete list of the honey plants of the state.

NEW ZEALAND FLAX (Phormium tenax).

New Zealand flax, or flax-lily, is native to the country from which it gets its name. There it is said to grow up to an elevation of 4,000 feet. It is also found on the Auckland Islands, Chatham Islands and Norfolk Islands, according to Von Mueller. It has been introduced into England and the Orkney Islands. In America it is little known outside of California, where it is often grown as an ornamental. There are three varieties of the plant which vary in height, the smaller variety growing not much above five feet in height, while the largest reaches a height of fifteen feet. The plants yield a large amount of fibre, chiefly used for making rope and paper. Since it will not stand a large amount of frost, its culture in America will be confined to the extreme southern states and California. In Australia it is recommended for waste places about railway lines, along sea beaches, or rocky declivities, where it can be left to grow without attention. Since the fibre has commanded a price of from $50 to $100 per ton, under normal pre-war conditions, its introduction for such lands is to be recommended.

It is recommended as an important honey plant where grown in sufficient quantity. Its culture on an extensive scale would add a large amount of available pasture to the beekeepers within reach.

NORTH CAROLINA—Honey Sources of.

North Carolina is divided into three rather distinct natural districts. Along the coast and gradually rising toward the interior is the Coastal Plain. Across the center of the state extends the Piedmont or foothill section and westward the higher elevations or mountain region. An article by Vernon R. Haber, in the American Bee Journal of June, 1921, shows the life zones represented within the state.

North Carolina has a wide range of altitude and temperature and accordingly is rich in plant life. Some sections are extremely favorable for honey production and there is much unoccupied territory within her borders.

The Coastal Plain

Nearly half the land area of the state is included in the Coastal Plain or tidewater region, which extends inland nearly 200 miles from the sea. The soil is largely silt and sand and the annual rainfall is about 54 inches. There are large areas of swampy lands in this region as well as numerous streams and lakes. Many locations will support apiaries of considerable size and from 100 to 250 colonies in a single yard is not unusual. Yields of 50 pounds per colony without the best attention are not uncommon. Tupelo, gallberry, holly, huckleberry and tulip-tree are among the more important sources which Haber credits to this region. It is probable that the asters, goldenrods and Spanish needles are also important in the lowlands.

The Piedmont Plateau

The annual rainfall of this region is about 52 inches, or slightly less than the Coastal Plain. It is a general farming country and includes about one-third of the area of the state. Haber states that the best locations are along the streams and that 25 or 30 pounds of surplus honey per colony are often secured from aster alone near the close of the season. There are a great variety of nectar sources and seasons of entire failure are rare. However, this region is not as favorable as the tidewater section. Most of the sources of nectar common to this section are rather evenly distributed over the entire region. Persimmon, black-gum, goldenrod, aster, willow, cow-peas, sumac, blackberry, maple and fruit bloom are the principal sources listed by Haber for this portion of North Carolina.

The Mountain Region

The mountain region of North Carolina is famous for its fine climate, although the winter temperatures are much colder than the regions already described. The average rainfall of 55 inches is not much different from other parts of the state. There are some fertile valleys in this region which furnish profitable farming areas, but a considerable portion is covered with forest. There is a great variety of trees present, more than 100 different kinds being known. In this region much of the surplus honey comes from forest trees, including tulip-tree, basswood, sourwood, wild cherry, locust and maples. In addition to the trees, the clovers, buckwheat and asters are abundant. Sourwood yields fine crops of splendid honey and is often mixed with basswood. Since the best timber has already been cut in many places, it is important to use care in selecting a location.

General Sources

Since the best honeyflow in many parts of this state comes very early, it is important that early brood rearing be encouraged to get strong colonies in time to get maximum crops.

From Haber's article, already credited, and from Franklin Sherman's bulletin, issued as No. 1 of Vol. 29 of the North Carolina Department of Agriculture, the following list of plants of importance to the beekeeper is selected:

Sourwood, tulip-tree and clovers are of first importance, taking the state as a whole. The tulip-tree is most widely distributed, while sourwood is most abundant in the Piedmont section. The clovers are common to the entire state, although of greater value in the higher elevations.

Gallberry, black-gum, persimmon, basswood, holly, huckleberry, buckwheat, ironweed, locust, aster, cotton, Spanish needles, cowpeas, fruit bloom, sumac, goldenrod, rattan, wild blackberry and maples follow in the order of their importance.

There are numerous minor sources which might be mentioned as yielding small quantities of nectar or pollen. However, with its abundant flora the minor sources are of far less interest to the beekeeper than is the case of many western localities where the flora is less abundant.

NORTH DAKOTA—Honey Sources of.

The greater portion of North Dakota has a fertile soil, much of which consists of a black loam underlaid with clay. Some of the finest agricultural land to be found anywhere lies in the Red River Valley along the eastern boundary, and this section is in a high state of cultivation.

The elevation ranges from about 900 feet in some parts of the valley to above 3,000 feet in the southwest corner of the state. The state as a whole is deficient in rainfall, the average precipitation for the eastern section being only from 17 to 22 inches annually. The season is short, with an average frost-free period of but little more than four months. High winds prevail over much of the year, and extremes of temperature are common.

Under the conditions above described a short, intensive season is to be expected. Sweet clover is the source of the greater part of the surplus honey for the entire state and the flow does not come until July, thus providing ample time to get strong colonies where the bees are given proper attention. Sweet clover is grown in large acreage in rotation with small grain and is important throughout the state. In Grand Forks County, in 1922, it appeared from assessors' reports that there were 40,000 acres of sweet clover. Heavy honeyflows and large crops are the rule. In neighborhoods where sweet clover is not found, beekeeping is an uncertain business.

In early spring North Dakota bees get pollen and nectar from the willows along the streams. Later dandelions furnish both nectar and pollen in great abundance and greatly assist in building up the colonies. In some localities sow thistle is abundant, but it is only in neighborhoods where poor farming is the rule that sow thistle and Canada thistle are likely to be of much importance as a source of nectar. Snowberry is common on the prairies and is the source of good quality honey. The snowberry acreage has been greatly reduced through cultivation. There is much gumweed in places, which is the source of honey of indifferent quality, which granulates quickly: In the fall there is some honey from wild sunflowers, goldenrods and asters. The cottonwoods, elms, maples and poplars along the streams furnish pollen over much of the state.

Caragana has been planted for hedge and windbreak by many farmers and furnishes some nectar. It is seldom found in sufficient abundance to furnish surplus. There are few of the native prairie flowers still sufficiently common in the eastern section to be of much value, owing to the general cultivation of the soil.

The wolf willow is abundant on the prairies in many parts of the state and is generally regarded as a valuable minor source of nectar. The breaking of the prairies is reducing the areas where this shrub is to be found.

NOVA SCOTIA—Honey Sources of.

Nova Scotia is similar to Maine in its natural features and its flora. There are large areas where white and alsike clover are very abundant and yield nectar freely in suitable seasons. There is also much honey gathered from goldenrod and asters in late summer. In some sections fruit growing is an important industry, and the bees find abundant early forage on the blossoms of the fruit trees. Dandelions, willows and maples furnish additional spring pasture. Blueberries are important, as is the wild radish and kalmia.

Canada thistle, sumac, mustard and buttonbush are found in suitable locations. Where buckwheat is grown on sandy soils it is a valuable source of surplus.

O

OAK (Quercus).

The oaks are frequently reported as sources of nectar. The fact is that it is usually honeydew, rather than nectar, which the bees gather from oaks. Pollen is produced in abundance by these trees and all species may be regarded as valuable for pollen. There are many species of oaks, some of which are common to any part of America where trees grow naturally. There are 24 species recorded from Alabama, 24 from New Mexico and 12 from Connecticut.

Richter gives 4 species—field oak (*Quercus agrifolia*), tan bark (*Q. densiflora*), mountain white oak (*Q. douglassii*) and weeping oak (*Q. lobata*), as sources of honey in California. Scholl gives 6 species as sources of honey in Texas, as follows: Post oak (*Q. minor*), live oak (*Q. virginiana*), red oak (*Q. rubra*), Spanish oak (*Q. palustris*), water oak (*Q. aquatica*), and jack or barren oak (*Q. nigra*).

Pollen-bearing blossoms of red oak.

H. B. Parks, of the Texas Agricultural College, has found that post oak yields some nectar from extrafloral nectaries.

The live oak yields honeydew in large quantity in west Texas, from the balls made by the live oak gall. George Schmidt, of Crystal City, reported an average of 25 pounds per colony from live oak balls, on the Nueces River, in 1917. The honey was dark and heavy, but of good flavor. The flow lasted till frost. There are numerous similar reports, from west Texas, of honeydew from this source, from August till late fall. During the drought of 1917 and 1918 there was little else for the bees in many Texas localities, and the live oak saved the bees from starvation.

Occasionally the bees work on white oaks in Illinois. In this case they are attracted by the presence of a soft scale (*Lycanium cockerellii*).

George D. Shafer has described in American Bee Journal, (page 561, Nov., 1923) *Dicholcaspis eldoradensis,* a gall insect found on valley oak, as the source of honeydew. Shafer found the bees working heavily on trees infested with this gall and concludes that it is the source of the honeydew to which Richter refers. The illustration shows the galls on the twigs of valley oak. Shafer also mentions surplus from this source in the Sacramento valley secured by A. Gambs.

In American Bee Journal, (Jan. 18, 1890), S. L. Watkins of Placerville, California wrote:—

> "In Valley Counties of California, the white oaks furnish unlimited quantities of honeydew; the honey is poor quality, of a slightly bitter taste, but makes excellent winter food for bees."

In his book, "The Honey Ants and the Occident Ants" published in 1882 Henry C. McCook reports the honey ants as feeding on the exudation of the galls on *Quercus undulata* in Colorado from which they

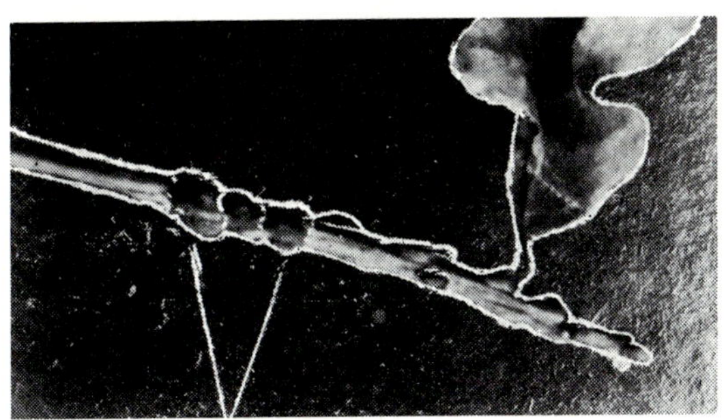

Galls of California valley oak, the source of honeydew.

apparently secure the food which they store in their unique living containers. It is probable that bees likewise gather from this same source.

McCook reports this ant honey as, "very pleasant with a peculiar aromatic flavor, suggestive of bee honey, and quite agreeable."

In American Bee Journal, (March 1939—p. 120), Edgar Abernathy reports the bees getting nectar from post oak in North Carolina.

OGECHE PLUM, see Tupelo.

OHIO—Honey Sources of.

The topography of Ohio is varied, ranging from extensive level plains near the lake shore, through gently rolling elevations to very hilly regions along the Ohio River. It is a large state with an area of 41,040 square miles. There are some rather extended areas of swamp the vicinity of which offer favorable beekeeping locations. While Ohio as a whole is an agricultural state, there are extensive regions where the soils are acid and the legumes succeed only when the fields are limed. There are a few commercial honey producers in the state who have done well, but good locations are unevenly distributed and some knowledge of local soils and flora are necessary to find them. There are large sections of the State but poorly adapted to extensive honey production. Sweet clover is materially improving the bee pasture of many counties. Alsike is largely grown for seed in a few localities in the northwestern part and in this region beekeepers do very well.

The sources of nectar in Ohio are very similar to those of many other States in the great central basin, where white and alsike clover furnish the principal surplus. There is the usual list of plants furnishing nectar and pollen for early spring, such as willows, maples, fruit blossoms, dandelions, etc. The main flow comes from white and alsike clover in June and July. In some sections buckwheat yields surplus. Basswood, which was once an important source of nectar, has been cut until few beekeepers are able to get crops of much value from it. Goldenrod and asters are important in many sections for late fall

See Indiana for longer list of plants.

OKLAHOMA—Honey Sources of.

Oklahoma with an area of about 70,000 square miles, is nearly twice the size of Ohio and varies greatly in climatic conditions from east to west. The eastern part of the state has a rainfall of 40 to 45 inches as an annual average. The winters are short and mild and the region is well adapted to a diversified agriculture. Much of this area is wooded and the altitude varies from about 350 feet above sea level in the extreme southeast to about 1400 feet in the highest mountains.

The western part of Oklahoma is a plains region with gently rolling prairies sloping toward the east. The elevation varies from less than 900 feet in the Red River Valley in the center of the state to more than 4500 feet in northwestern Cimarron county. The rainfall diminishes rapidly toward the west until it drops to about 15 inches annually in the extreme west end. This region is marked by extremes of wet and drouth as well as a greater range of temperature.

In eastern Oklahoma there is a wide variety of plants which offer something for the bees. Beginning with mistletoe which blooms as early as February or March there is little interruption of bloom until November. Elms, maples, willows and box elders furnish early pasture in ample quantity for early spring brood rearing. Add to these the dandelions and the various fruit blossoms and we find ideal conditions for preparing the bees for the later harvest.

Glen V. Mills in articles in the Muskogee Times-Democrat, has listed rather a formidable number of plants furnishing nectar and pollen. (1925). He mentions wild cherry, black locust, honey locust, hawthorn, persimmon, poison ivy, button-bush, false indigo, and sumac among the trees and shrubs. Catnip, horehound, milkweed, horsemint, vervain, gaura, passion flower, boneset, goldenrod, Spanish needle, bitterweed, bluevine, heartsease and crown-beard are among the more important herbaceous wild plants.

Among cultivated plants which are important should be mentioned, alfalfa, alsike and white Dutch clover, sweet clover and cotton. Buckwheat is sometimes mentioned, but it is of doubtful importance under such climatic conditions as prevail in Oklahoma.

OKRA or GUMBO.

Cook's Manual lists okra as a honey plant. It is a well-known garden vegetable, especially popular in the South, but is seldom grown in quantity sufficient to be important to the beekeeper.

OLD MAN, see Rosemary.

OLEASTER, WILD OLIVE, RUSSIAN OLIVE (Elaeagnus angustifolia).

The Russian oleaster, was first introduced into this country by Professor J. L. Budd because of its hardiness and ornamental qualities.

Oleaster or Russian olive.

The young trees branch freely and produce an abundance of white, scurfy foliage, which makes it a most attractive and striking shrub. * * * It is one of the most fragrant of cultivated small trees. The season of blossoming varies somewhat, but with us (Iowa) is about the middle or early part of June.

It is one of the best of our spring honey plants. The bees visit the plant in large numbers when in full bloom from early morning until late evening.—L. H. Pammel, American Bee Journal, November, 1917. See Wolf Willow.

OLIVE (Olea Europaea).

Olives are extensively grown in California and in the Salt River Valley in Arizona but are of little value to the bees. Richter in Bulletin

217 of the California Experiment Station lists olive among the plants yielding an occasional surplus.

In private conversation with the author, T. O. Andrews of Corona, California, reported that he had secured a crop of honey from olives on only one occasion during his long residence in California. The most probable explanation of this unusual case would seem to be that it was honeydew rather than nectar which the bees secured. Richter states that his bees within reach of olive trees have only been observed to collect pollen.

In Flora Apicola de Espana, Barcelona 1904, Miguel Pons Fabregues, lists the olive as doubtful. He states that some beekeepers assert that its blossoms are melliferous, while others say they are not.

There is little evidence to indicate that olives are important to the beekeeper anywhere.

ONION (Allium).

According to Richter, wild onions sometimes yield surplus honey in the vicinity of Sacramento, California. In some parts of that State, where cultivated onions (*Allium cepa*) are grown for seed, they are also important. The onion yields freely and the honey is said to be amber in color with a characteristic onion flavor which disappears after the honey is fully ripened. There are few localities where onions are sufficiently abundant to be important.

Before the original prairie flora was destroyed wild onions were abundant in the middle west and the early settlers found them to be an important source of nectar. The following from the American Bee Journal, page 310, 1878, mentions this early abundance:

> "I write to call attention to a valuable plant for honey, the wild onion. If farmers can tolerate it in their pastures it will be very valuable. It begins blossoming here about July 20, and continues for two or three weeks. I am located six and one-half miles south of the court house, in Chicago, at Englewood, and at present the prairies around, as well as railroad tracks, are covered with its delicate pink-white blossoms, and my bees are gathering it fast. I can smell the onion flavor at the entrances, blown out by the busy wings of the bees ventilating the hives. The onion flavor thus passes off and when the honey is sealed you could not tell it from white clover; though I extracted some of it three years ago, and my better half always insisted that she could taste the onion flavor."—R. J. Colburn.

Raymond Newell who kept bees in connection with the Harris Seed Co., at Coldwater, New York, wrote to the author as follows:

> "Perhaps you would be interested to know that my colony average of onion honey was about ten pounds. I find it is not unlike clover honey and loses its onion flavor when fully ripened."

In an article, "Onion Improvement" (Yearbook U.S.D.A. 1937), H. A. Jones states,

"Most of the pollen is shed between 9 a.m. and 5 p.m. Pollination is effected mainly by insects that go from flower to flower and visit the nectaries at the base of the three inner stamens. * * * Cross pollination is the rule."

ONTARIO—Honey Sources of.

Ontario contains an immense area, much of which is unsettled. Southern Ontario, situated between the states of Michigan and New York and bounded on the south and west by the Great Lakes, is an old settled region where general farming is followed. In this region beekeeping is a highly specialized industry. Many commercial honey producers are located in this area and great quantities of honey are exported to Great Britain.

The principal sources of nectar are alsike and white Dutch clover, buckwheat and basswood. Farther north, wild raspberry and fireweed yield good crops. Goldenrod and aster furnish late nectar in all parts of the province.

Among the minor sources may be mentioned blueweed or viper's bugloss, dandelion, sweet clover, Canada thistle, milkweed, and boneset. For early spring stimulation maples, willows, dandelions and fruit bloom. In the Niagara fruit district there are large orchards of cherries, apples, etc., from which the bees get much nectar under favorable weather conditions. Hawthorn grows in pastures and waste places over a wide area and adds much to the spring pasture.

Blueberry is reported as important in some sections of Northern Ontario.

The Province is well supplied with sources of pollen. Such trees as elm, oak, walnut and birch being widely distributed. Early blooming flowers such as skunk cabbage, bloodroot, and spring beauty are everywhere available, so that Ontario bees seldom lack for plenty of pollen.

In burned over regions of the northern portion of the province large crops from fireweed or willow-herb are often gathered.

OPUNTIA, see Prickly Pear.

ORANGE (Citrus sinensis).

The orange tree is a native of Asia, early introduced by the colonists into Florida. It thrives in a semi-tropical climate and its culture in America is confined to southern Florida, a few small areas along the Gulf Coast, the lower Rio Grande Valley in Texas, Salt River Valley, Arizona, and to California. It is sensitive to frost and the trees are easily killed by freezing.

In California, orange is one of the important sources of honey. The trees bloom in April, when many colonies are too weak to make the most of the crop. The flow is extremely rapid, as at times the nectar is secreted in such abundance that men and horses working in the orchards are saturated with it. The flow lasts about three weeks. In 1919 it continued 23 days in Tulare County. Four hundred colonies in one yard averaged more than 60 pounds per colony from orange. O. F. Darnell, of Porterville, extracted 171 pounds of orange honey from one colony in ten days, after having previously extracted 24 frames which were not weighed. A. W. Gambs, of Anderson, reported that some colonies, made extra strong by drifting bees, stored four full-depth extracting supers of fresh nectar from orange in four days.

In some of the interior valley locations the bees get fair crops from orange about four years in five. Along the coast, the fogs are unfavorable and the crop is uncertain. Good orange locations are much in demand, and many beekeepers move their bees long distances for the flow. With a larger reserve supply of honey, left in the hives for winter to enable the bees to build up early in spring, and with more careful attention to wintering, it would be possible to greatly increase the average returns from orange. This statement is based on the results of a few beekeepers who take special pains to get their bees into condition for the orange flow.

Orange blossoms.

The honey from orange is white in color, heavy in body, of the finest quality, and is much in demand in the markets.

From Florida come reports of an occasional average as high as 75 pounds per colony from orange. Single colonies in Dadant hives have produced as high as 175 pounds of surplus for Doctor Horton.

The pollen yield from orange is said to be light.

ORANGE HAWKWEED, see Hawkweed.

OREGON—Honey Sources of.

We may divide Oregon into three divisions with relation to its beekeeping. West of the Cascade Mountains is a narrow Coastal strip with heavy rainfall and little frost. But a small part of this area is under cultivation, the greatest part being covered with forest. Next we consider the fertile river valleys where farming is extensively carried on, either under irrigation or watered by natural rainfall. East of the Cascades is a wide and fertile region which is semiarid. Much of it remains uncultivated for lack of moisture.

Commercial beekeeping is not extensively followed in Oregon. A few locations in the valleys offer satisfactory returns, as do some fireweed locations where the annual rainfall is 50 inches or more. In American Bee Journal, for September, 1924, appears an article by H. A. Scullen giving information concerning Oregon honey producing possibilities.

Probably more surplus honey is secured from alfalfa and sweet clover than from any other plants in Oregon. Fireweed is important under favorable conditions, and white clover yields in some sections on the west coast. Alsike is also of local importance in the same region.

Of the native plants, vine maple is one of the best, although blooming too early for the bees to make the most of its heavy flow. Willow, chickweed, salmonberry, dandelion, orchard fruit trees, grapes, currants, gooseberries, Juneberry, barberry, huckleberries, laurel, dogbane, lobelia, catnip, snowberry, arrow-wood, Spanish needle and everlasting are all reported as valuable by Oregon beemen.

Chittam, salal, gum-weed, madrona, mustard and wolfberry should also be included in any list of Oregon honey plants. Cascara is also valuable near the coast. The flora is similar to that of Washington and of northern California. See lists under those states for more detailed information.

OREGON GRAPE (Berberis nervosa).

Oregon grape is common in the woodlands near the Coast from Marin County, California, northward. It is abundant in the forest regions of Washington. It is a shrub with prickly alternate leaves and yellow flowers in racemes. Oregon grape is reported as a source of honey of minor importance in the Northwest. Like the agarites or triple-leaved barberry of Texas, it blooms in early spring when it is of special value for stimulative purposes. Another species of barberry (*Berberis Aquifolium*) is common to this region, and most beekeepers do not differentiate between the two species. (See Barberry.)

OREGON MYRTLE, see California Laurel.

OREOCARYA.

A desert plant growing on high, loose, sandy soils. Blooms in May, with a small bur following the bloom. It ranges from Wyoming to Arizona and west Texas. It yields nectar freely and an average of forty pounds per colony has been reported from this source. The honey is light amber, of inferior flavor, according to D. W. Spangler, of Longmont, Colorado, but it is extremely valuable for building up colonies in spring. Extended cultivation is rapidly reducing the area where the plant is to be found.

ORGANY, see Marjoram.
OTHAKE, see Polypteris.

OX-EYE DAISY (Chrysanthemum leucanthemum). Field Daisy, White Weed.

The ox-eye daisy, white weed or field daisy, is a common weed in meadows, pastures and waste places in the Eastern States and Canadian provinces. It has large white flowers with bright yellow centers. It is native to Europe and Asia and has become naturalized over a wide area in eastern North America.

The author can find little to indicate that it is of value to the beekeeper except the following quotation from Moses Quinby (Mysteries of Beekeeping, 1865):

> "Ox-eye daisy (*Leucanthemum vulgare*), a beautiful flower in pasture and meadow, and worth but little in either, also contains some honey. The flower is compound, and each little floret secretes so minute a quantity that the task of obtaining a load is very tedious. It is only visited when the more copious honey-yielding flowers are scarce."

P

PAGODA-TREE (Sophora japonica).

The Japanese pagoda-tree or Chinese scholar-tree is a native of China and Korea which has been generally planted in the northeastern states as an ornamental. It is quite hardy and the showy clusters of pea-shaped flowers which appear in late summer are very attractive to the bees. It is not sufficiently common in this country to be important as a source of honey but Edward Kellner wrote from Czechoslovakia to say that it is highly regarded as a source of nectar in Central Europe. He reported that one small grove on a sandy hill yields nectar sparingly while another on deep black soil yields freely and hives on scales within reach of it show gains as long as the blooming period lasts.

The pagoda tree resembles the black locust but lacks thorns. The blooming time is August while the locust blooms earlier. Like the locust the tree is a legume.

PAINT BRUSH, see Hawkweed.

PALM.

Royal palms, (*Roystonea*) are abundant in the West Indies where they are characteristic trees. In bulletin 15, Porto Rico Experiment Station, Dr. E. F. Phillips states that the Royal palm or Palma real, (*R. borinquena*) is present in all parts of the Island. I quote as follows:

> "When in bloom, if accessible to the bees, the number of bees at work on it and the noise they make might lead the visiting beekeeper to think that a swarm had issued. This species * * is an important honey source and in some localities is doubtless sufficiently abundant to support large apiaries."
>
> "The author saw some comb honey at Rio Pedras which, according to informants was from royal palm. This honey was a light amber color and of excellent flavor. The trees bloom at any season of the year." In a private letter to the author Doctor Phillips adds:—
>
> "The royal palm of Hawaii which is not the same species as the Palma real of Porto Rico, yields nectar in such abundance that when in bloom a northern beekeeper would think a swarm had lighted on the tree. I saw a number of these trees in bloom while in Hawaii and it was the same in every case, and the bees were certainly getting nectar."

Concerning Royal Palm, Ordetx, (Plantas Melliferas de Cuba) writes:—

Cocoanut grove in southern Florida.

> "They produce cream white pollen that bees collect eagerly for their brood; and they secrete nectar copiously.
>
> "The color of the honey is golden yellow; it has a characteristic flavor and aroma, that makes it delicious to the taste.
>
> "The royal palm, because of its particular abundance in all parts of the Island is our most valuable species after the white campanula. (Campanilla).

The Cocoanut Palm, (*Cocos nucifera*) is the most important of the palm trees since the nuts furnish large quantities of food, much used in tropical countries. This species is common to southern Florida and the West Indies. Although reported as an important source of nectar in some places, it is probably not equal to the foregoing species. Phillips states in the bulletin previously quoted, that it is not classed as one of the best honey plants. He mentions one apiary at Mayaguez where it is important as a source of nectar. Like the other it blooms at irregular periods throughout the year.

The date palm yields an abundance of coarse pollen.

> "On my visit to the Dakhla Oasis in mid-September last year, I noticed that the surplus honey removed from the hives had an oak color with a slight red tinge and with a medium consistence. Its aroma and taste were distinctly suggestive of the date fruit, and consequently it did not fetch a good market price.
>
> "There are no hornets in the oases of Egypt; and at the time this honey surplus was being gathered, the dates were not yet ripe, so that there is no question of this surplus, which was abundant, not being sucked by the bees from the damaged fruits. Again, there were no other blooming trees at the time that particular honey was being collected, and the bees, I am told, used to frequent the palm trees in large numbers."—M. Abdel-Azim, Bee Kingdom August 1930.

PALMETTO (Sabal).

The palmettos are the most conspicuous feature of the flora of the south half of Florida. To the man accustomed to dense forests, the open, park-like growth of the palmettos hardly seems like woodland. The illustration gives a good idea of the typical Florida landscape.

This group of plants is not important in America, outside of the State of Florida. A small area in lower Texas, about the mouth of the Rio Grande River, is covered by a species of palmetto closely resembling the cabbage palmetto, but it is thought to be a different species. An occasional tree is also found along the seacoast as far north as Charleston, S. C. They are to be found also as street trees in various southern cities along the Gulf Coast and in south Texas. The small saw palmetto (*Serenoa serrulata*) also extends its range into Georgia and the Carolinas, in open pine woodlands.

In Florida both forms are sufficiently abundant to furnish nectar in quantity worthy the attention of the commercial beekeeper. However, in too many localities there is little else available, so that the season between flows is too long to make beekeeping worth while. To take advantage of the palmetto flows and at the same time get good crops through the rest of the year, the late O. O. Poppleton practiced migratory beekeeping. His apiaries were moved several times during the year, so as to be near different sources in the period of bloom. The great drawback to beekeeping in Florida is the lack of a sufficient variety of honey plants in one location to support the bees profitably throughout the year.

Cabbage palmetto.

The cabbage palmetto (*Sabal palmetto*) gets its name from the cabbage-like formation in the bud at the top of the growing trunk. The tree grows 25 to 35 feet in height and has large fan-shaped leaves several feet long. It grows along the Atlantic Coast to the north line of Florida, but in the interior is not found in abundance more than about two-thirds of the way.

The tree blooms during July and August, the latter date applying to northern parts of the State. The blossoms are very delicate and have been likened by Professor Baldwin to a giant ostrich plume. According to his statement, the flowerlets are sensitive to weather conditions. Too much moisture blights them, while the opposite extreme blasts the delicate bloom. As a consequence, it does not yield abundantly more than about one year in three, although at times it yields very profusely.

Saw palmetto.

> "On the St. Lucie River, Mr. Hill extracted, barreled and shipped 3,500 pounds of palmetto honey from 55 colonies in two weeks."—Page 489, American Bee Journal, 1899.

While palmetto honey is regarded as of very high quality, the honey from the cabbage tree is rather thin and requires some care in getting it properly ripened, as the following quotations will show:

"Cabbage palmetto honey, sealed or unsealed, will foam as though fermentation was in progress; that taken from the combs unsealed will ferment enough to deprive it of all the honey flavor, but the sealed only foams. Thin and acid and amber in color, it will flow bubbling from the cells behind the knife, and it is not a rare thing to see gas bubbles under the cappings of the sealed cells. Whether the colonies are strong or weak, it is always the same, when the bees work on the cabbage trees, as the common palm tree of Florida is called. The name comes from the fact that the bud in the head at the top is eaten in lieu of cabbage.

"The saw palmetto is decidedly different in the nectar it yields. Saw palmetto honey, even unsealed, may be called a good honey, and it is, too. When ripened it is a honey that makes a name for itself when enough care is taken by the producer to have it unmixed with other nectars.

"I write from personal experience on the east coast of Florida."—L. K. Smith, Gleanings, page 39, 1909.

The palmettos are a conspicuous feature of the Florida landscape.

The saw palmetto (*Serenoa serrulata*) is a low growing little palm, found on dry soils in the Gulf Coast region. In the southern portion of its range, in peninsular Florida, it attains the proportions of a small tree. There it sometimes reaches a height of 20 feet, with erect or inclined trunk. Further north the stem is almost invariably underground. Large areas of pine lands are covered with it.

The blooming time is April and May. O. O. Poppleton wrote concerning his calendar of the year:

"April—Saw palmetto flow commences early in the month and continues until last of May. Our apiary work these two months is extracting, building

Bloom of the cabbage palmetto.

up all colonies and replacing poor queens."—Beekeepers' Review, page 11, 1893.

Concerning the honey flow from saw palmetto we quote E. G. Baldwin as follows:

> "The honey from saw palmetto is lemon-yellow in color, thick and waxy and of pronounced but delicious flavor. It is not quite so transparent as pure orange honey, but seldom candies, and makes a choice table article. Mr. O. O. Poppleton pronounces it the best honey in Florida, with the possible exception of tupelo. It is liked by almost everyone at first taste; is a trifle milder, even, than orange."—Gleanings, page 177, 1911.

Forest fires frequently destroy many square miles of the saw palmetto, thus removing this source of nectar for one year. However, according to Baldwin, the burned-over portions usually produce the most honey the following year.

Concerning the flow from palmetto, E. B. Rood, of Bradentown, writes as follows:

> "We have been having the heaviest honey flow from palmetto for ten years. One colony on scales brought in 50 pounds in four days, and 80 pounds in ten days. I expect 20,000 to 30,000 pounds. I have extracted 13,000 pounds now and am just starting on another round."—Gleanings, page 703, 1908.

The dwarf palmetto or bluestem (*Sabal glabra*) has a horizontal underground stem and is found in low grounds from South Carolina to Florida and Louisiana. The scrub palmetto, often mentioned as an important source of nectar, is *Sabal megacarpo,* commonly found in peninsular Florida. This species also has a creeping stem and grows freely in the pine lands and inland sand dunes.

Palmetto is reported as the best source of surplus honey in central Louisiana but there is disagreement among the beemen as to its color. This is apparently due to the fact that it blooms with peppervine and the honey is usually mixed with that.

Dr. Everett Oertel studied the flow from *sabal minor* over a five year period, 1931 to 1935. This is reported in detail in Beekeepers Item May 1937. He reports a net yield of from 65 to 200 pounds per colony. Average daily gains of colony on scales during flow ranged from one pound to 16.6 with 21.5 the highest gain recorded in one day by a single colony.

Oertel records the honey as white to light amber.

PALO VERDE (Cercidium torreyana.) Green Bark Acacia.

The Palo Verde or green bark acacia is a tree of striking and unusual appearance common to a limited area in southern Arizona and adjacent California and southward. The bright green color of the bark of both trunk and branches gives the appearance of luxuriant vegetation when the tree is entirely leafless. The leaves appear in early spring, but fall again very soon. The range of this tree is restricted to desert regions where nature makes careful provision for the conservation of moisture. Since much moisture is lost from transpiration through the leaves, many desert plants shed them quickly after the spring rains have passed. The bright green of this tree offers a striking contrast to the dreary and dead appearance of most desert shrubbery throughout most of the year.

The palo verde attains a height of twenty or more feet, and with its spreading and bushy top provides welcome shade to the traveler in a

The Palo Verde tree.

region where shade is scarce. The wood is soft and of little commercial use except for fuel.

As a source of honey, palo verde is far more important than seems to be generally recognized. H. Wedgworth of Florence, Arizona, in private conversation informed the writer that he sometimes gets a carload of honey from this source. He describes the honey as of light yellow color, equal to valley alfalfa in both body and flavor. The flavor of palo verde honey is distinct, somewhat similar to the taste of the bark of the tree. Beekeepers report that the yield is much more dependable on lower lands than on the dry hills, due to the greater quantity of available moisture. On low lands it is regarded as quite dependable, with a honeyflow from trees on higher lands only in seasons of more than average rainfall.

C. floridum is a tree similar to the above found along the Rio Grande in Texas and Mexico. For related trees see Jerusalem thorn and retama.

PARADISE FLOWER, see Acacia, also Catsclaw.
PARKINSONIA, see Jerusalem Thorn.
PARSLEY, see Cogswellia.

PARSNIP (Pastinaca sativa).

The cultivated parsnip is a valuable honey plant. In the seed belt of California, where it is grown largely for seed, it is important as a source of surplus. Writing in Gleanings (November, 1919) E. R. Root mentions having seen hives five and six stories high, jammed full of honey from parsnip and celery. He states that the honey is not of the best, and that honey from parsnip is inferior to that gathered from celery.

Parsnip.

The wild parsnip, introduced from Europe, has spread over a wide area from the Atlantic to the Pacific Coast. It is common along railroads and highways everywhere. The small yellow flowers, which are borne in clusters like an open umbrella, are attractive to a large variety of insects.

In the Bee World for October 1931 is an article "Honey from Wild Parsnip" by Dr. F. Thompson from which the following is quoted:—

> For some years there has been approximately twenty acres of fallow ground behind my house largely filled with wild parsnip. * * For four years hardly any bees had been seen on it. This year, however, during the last half of July when weather permitted the bees were smothering over it and obviously it was the main flow.
>
> The honey is quite light in color, of moderate density and firm but distinctive flavor.

Wm. S. Barclay of Beaver, Pennsylvania, in American Bee Journal, Page 142, July 30, 1891, wrote as follows:—

> There I found vast numbers of bees busily engaged upon the bloom of the wild parsnip with which the river bank was literally lined. * * Except upon yellow poplar bloom or buckwheat I have never seen bees in such numbers. * * * I failed to see any pollen upon their thighs but found each little worker loading up his store of honey to carry home. It is not often an important source of honey, although the plant is sometimes very plentiful.

PARTRIDGE PEA (Cassia Chamaechrista).

Partridge pea, known also as sensitive pea, is an important source of honey in Georgia and Florida. The flowers are of an attractive yellow color. This plant is common along sandy roadsides in the Middle West, and may at times be found for miles at a stretch. It is seldom reported as an important source of surplus honey, except in the South-

Partridge pea.

eastern States. While the bees visit it freely when in bloom in the Northern States, the honey stored from this source is seldom noticeable.

The plant is peculiar in that in addition to the flowers, nectar is secreted by extra floral nectaries at the base of the leaf stalk or petiole. The bloom lasts for several weeks in midsummer. As the bloom comes for the most part after the close of the clover harvest, its chief value in the North is to keep the bees occupied till the later flowers furnish a fall flow. The quality stored from this source is reported as poor in the North.

In contrast, the following report indicates that it is valuable in Georgia and Florida:

> "Bees store from one to three supers, during its flow, of light honey. The flow begins in June and lasts until October." Wilder in American Bee Journal, page 369, 1912.

E. G. Baldwin describes honey stored from partridge pea in Florida as "very dark red and strong in flavor." (Gleanings, July, 1917).

In a private letter to the American Bee Journal, George W. Pillman of Centaur Station, Missouri, not far from St. Louis, wrote as follows:—(August 1925).

> "About the time I had given up all hope that there would be any kind of crop, partridge pea blossomed and saved the day. * * *
> I have found partridge pea to be a reliable source of surplus honey, though the surplus from this source may not average one season with another more than 25 to 35 pounds per colony. It is more certain than white clover in my location. Though buckbrush (symphoricarpos) contributes something to this partridge pea surplus, I am satisfied that the greater part of the surplus honey produced at this time of year comes mainly from partridge pea. The honey is almost water white."
>
> From the above it will be seen that there is a difference of opinion regarding the value of this plant. It is possible that more of Mr. Pillman's honey came from buckbrush than he realized.

Dr. Everett Oertel reports that the partridge pea which yields surplus in North Central Florida is identified as *Chamaechrista fasciculata*. Plants from which the bees secured pollen and some nectar near Shreveport, Louisiana, proved to be *C. mississippiensis*, while the common variety in the vicinity of Baton Rouge that yields only pollen is *C. robusta*.

The different species which appear to often be confused may explain conflicting reports as to the value of this plant.

R. E. Foster reports that partridge pea is one of the best of honey plants in parts of Florida yielding a honey that is light amber.

PASQUE FLOWER, see Prairie Crocus.

PASSION FLOWER (Passiflora).

There are numerous species of *Passiflora* native to tropical America which are widely cultivated as curiosities or for their showy flowers. There are about a dozen species found within the United States, but although they secrete nectar freely, they are of only incidental interest to the beekeeper, since they are seldom sufficiently abundant to be important.

In the South the Maypops (*P. incarnata*) is a common climbing vine from Virginia and Kentucky southward.

P. caerulea, which is native to the South American forests, has been cultivated for more than two centuries. It is said that when early explorers found the festoons of these flowers in the forest they saw in them an emblem of the passion of the Christ.

There are numerous references to this plant in old books. Francis Huber, in his "New Observations," writes as follows:

> "We regret extremely to announce that some honest humble-bees of our acquaintance have taken to drinking, and to such extent that they are daily found reeling and tumbling about the door of their houses of call—the blossoms of the passion-flower, which flow over with the intoxicating beverage; and there, not content with drinking like decent bees, they plunge their great hairy heads into the beautiful goblet that nature has formed in such plants, thrusting each other aside, or climbing over each other's shoulders till the flowers bend beneath their weight. After a time they become so stupid that it is in vain to pull them by their skirts and advise them to go home instead of wasting their time in tippling."

Kirby and Spence, in their Introduction to Entomology, Seventh Edition, London, 1856, refer to the visits of the hive bee to the passion-flower for nectar.

From the above notes it will be seen that these plants have long been recognized as attractive to the bees.

PAULOWNIA.

Paulownia is a tree, native to Japan, which is somewhat similar in appearance to the catalpa. *Paulownia tomentosa* has escaped from cultivation in New York and New Jersey and established itself in the wild state. It is now found southward to Florida and Texas. It is related to the figwort, which is a famous honey plant.

The leaves are very large, the wood is soft, coarse-grained and of poor quality. The tree grows rapidly and in ten years reaches a height of 30 to 40 feet. It is grown for shade and ornament.

T. B. Parker, writing in the Beekeepers' Exchange (Nov., 1879,) praises the tree as a source of nectar as follows:

> "The bees literally swarm on it, but I do not know what the quality of the honey is. * * * * It is a profuse bloomer, and so dense are the umbels of flowers that they form a shade almost equal to a tree having leaves. It blossoms before the leaves appear. * * *

"It is very fragrant while in bloom. The flowers are a light purple color, about two inches long, and bell-shaped and in clusters. As a shade tree it is very ornamental. I believe it is one of our greatest honey-producing trees."

PEACH (Prunus persica).

The cultivated peach is an important source of nectar over a wide scope of country. In localities where large peach orchards are found, it is important for building up colonies in spring. Like most of the tree fruits, the blooming period comes early, before the bees are strong enough to profit to the fullest extent from the abundance of nectar.

PEA, GARDEN (Pisum sativum).

The only report of honey from garden peas to come to this author is from Ernest A. Fortin, a bee inspector of Rougemont, Quebec. He reports the honey as crystal white and the most transparent he ever saw but having a strong taste of peas. He says: "In this locality many extensive fields of peas are grown for the canneries and no doubt much pea honey is gathered but we seldom notice it since it is generally mixed with other nectars."

PEANUT (Arachis hypogaea).

Lovell in his "Honey Plants of North America" lists the peanut as the source of large quantities of nectar in Florida. He describes the honey as a trifle darker than white clover, as thick as orange honey with a characteristic flavor which is very mild.

The author has not been able to verify this report from any beekeeper living in regions where the peanut is commercially cultivated. One man who has spent a lifetime growing peanuts reports that he has never observed the bees to work them. (Texas.) Another says that as far as he has seen bees get neither pollen nor nectar from peanuts. (Arkansas.) Likewise a Georgia beekeeper says that peanuts are not a source of honey worth talking about.

C. C. Champion of Argyle, Georgia, reports that he can recall bees working peanut blossoms only on one occasion, (1945) when they stored a shallow super of honey from this source.

W. E. Buckner who had charge of Tanquary apiaries at Lena, South Carolina, in 1945, reported that in fifteen years spent where peanuts are grown extensively he has never seen anything resembling a surplus of honey or pollen from peanuts.

PEAR (Pyrus).

The pear is of old world origin but widely distributed in America. In the mid-west it usually blooms ahead of apples and often attracts the bees when little else is available. At such times the bees visit the flowers in large numbers and appear to be getting abundant nectar. Recent investigations have shown that pear nectar is usually low in sugar content and for this reason the blossoms are often neglected if other flowers are within reach.

Scullen and Vansell, (Bul. 412 Oregon Ex. Sta.) report that with 23,000 acres of commercial pear orchards in Oregon they find no reports of pear honey. They found the sugar concentration to vary from four per cent to as high as 25 per cent depending upon variety and weather conditions. They conclude that bees visit the blossoms mostly for pollen. Since most varieties are self sterile honeybees are necessary to insure pollination.

PEA-TREE, see Caragana.

PENNSYLVANIA—Honey sources of.

Pennsylvania has an area of more than 45,000 square miles of greatly diversified surface configuration. The western part of the State is hilly or mountainous with the tillable lands mostly in small and scattered areas. There are rugged mountains and deep and beautiful valleys which combine to make an attractive countryside, but rather poor farming country. The rainfall is ample, but the weather changes are rather severe with high summer temperatures and frequent spells when the thermometer registers below zero in winter. The hills are covered with second growth timber and much coal is mined along the streams.

The central portion of the state contains much territory which is mountainous, with deep, fertile and well cultivated valleys. In the southern part of this region is a rolling agricultural section which is fertile and productive, and in a high state of cultivation. The eastern part of Pennsylvania is mountainous in the north with some rugged forest areas which are wild and unsettled. Southeastern Pennsylvania is a highly cultivated agricultural region with well kept and prosperous farms. Here the winters are mild with zero temperatures but rarely reached. The rainfall for the entire state averages from about 40 to 46 inches annually.

Beekeeping is not an important industry in the State, although there are some neighborhoods where commercial honey production is profitable.

Pennsylvania is in what is generally known as the white clover region. White and alsike clover are of first importance. Buckwheat, in some sections, is also a main source of surplus. In spring, the willows and maples, followed by dandelion and fruit bloom, furnish both nectar and pollen for early brood rearing. In the mountains, azalea and laurel furnish some honey. Black locust, tulip-poplar and sumac are valuable in certain localities. Basswood, goldenrod and asters should also be included as of local importance.

Because of the acidity of the soil there are large areas in Pennsylvania where the clovers do not thrive and little surplus is gathered until autumn, when Spanish needle, heartsease and other fall flowers bloom.

PENNYROYAL (Mentha Pulegium).

The pennyroyal of Europe has been introduced into numerous localities. Jepson mentions it in California as found in Sonoma and Marin Counties and on the islands of the lower San Joaquin. It is a

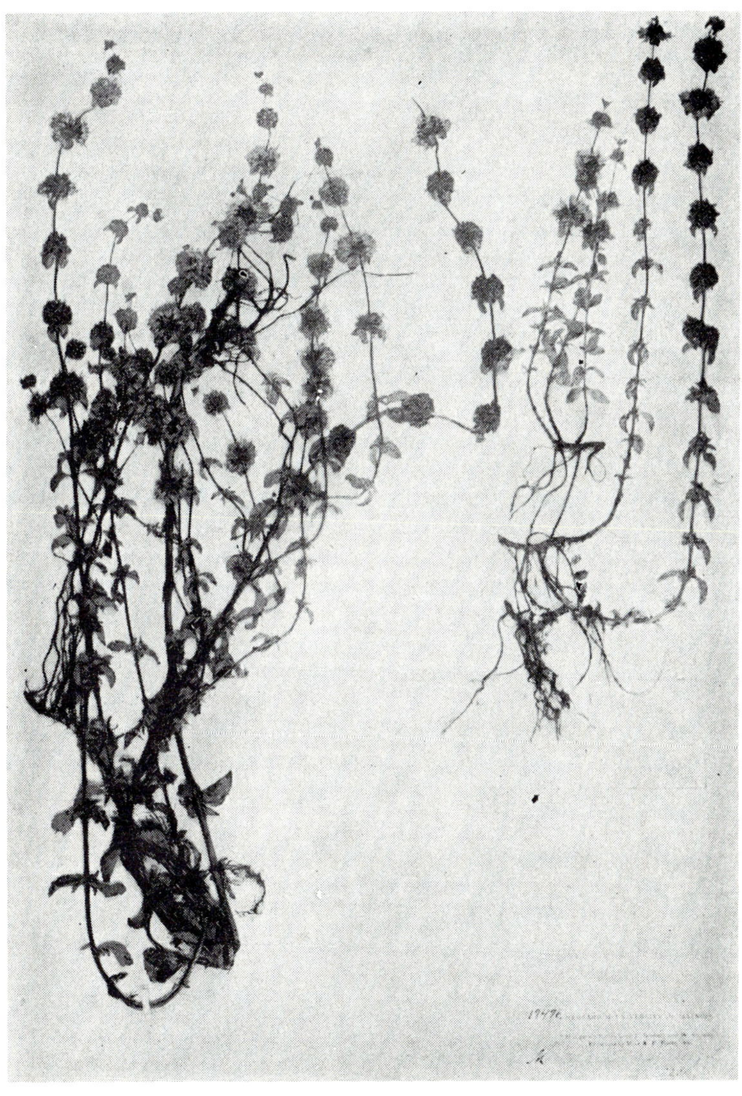

Pennyroyal is an introduced plant.

valuable source of honey, where abundant. Rayment states (Money in Bees in Australasia) that it succeeds in Australia. We quote as follows:

"A perennial scent herb yielding ethereal oil. Medicinally it is a powerful stimulant; it is also an insecticide. This herb furnishes a crop of honey in those districts where it abounds. It thrives in moist climates such as Warragul in Victoria. It has practically overrun the North Island of New Zealand, and has materially altered the flavor of the Dominion butter. The honey is pale in color, but of thin body. Like the nectar gathered from horehound, it does not appeal to all palates."

This plant should not be confused with the wild pennyroyal of Florida, which is an entirely different plant. (See Wild Pennyroyal.)

The following concerning pennyroyal in New Zealand is copied from New Zealand Smallholder, Jan. 16, 1935:—

"Pennyroyal generally starts yielding nectar about the time the flow from white clover begins to lessen. As the pennyroyal honey is objectionable in flavor to the ordinary palate, and has a strong, disagreeable aroma, it is advisable to make an effort to keep this class of honey separate from the good white clover honey. Pennyroyal honey is very dark in color when liquid, although when granulated is much lighter than one might expect."

PENTSTEMON or BEARD TONGUE.

The beard-tongue (*Pentstemon laevigatus*) has been brought to the attention of the beekeepers under the name of "Wonder honey plant" by J. J. Wilder in the Dixie Beekeeper, (Nov. 1923). Although the plant is found sparingly from Pennsylvania to Florida and west to Missouri and Arkansas, it has attracted attention as a source of nectar principally in Florida. In the high sandy cut over lands of Escambia county, it has assumed so much importance that the beekeepers have scattered the seed in an effort to further its spread. Wilder reports it as especially important in the region about Branford in LaFayette County, where it was the source of an average of 200 pounds per colony of comb honey in 1922.

Wilder states that the bees work it mostly in the morning and that few bees are found on it after the middle of the day. The honey he reports as mild in flavor, of heavy body and light color, slow to granulate.

There are many species of pentstemon found in the higher elevations of the west from Colorado to California. In the higher altitudes in Colorado the bees are reported to work freely on two or more species of this plant. Some honey is probably secured from this source in New Mexico, Utah and Arizona.

Concerning *P. laevigatus* C. W. Wood writes in Gleanings, (Jan. 38) that he had found a low moist field of ten acres or more in Michigan which was a mass of this plant. The field was swarming with bees. Two weeks later when sweet clover was in bloom it was found that most of the bees had deserted the beard tongue.

He states that as a usual thing it attracts the bees from the time it begins to bloom in June until the last flowers fade in July.

One of the most attractive bee plants in our test garden is the scarlet turtlehead or shell flower, (*Pentstemon barbatus coccineus*), which grows in the mountains of Utah and southward to Mexico. It blooms through most of the month of June and is constantly alive with bees. The bees crawl into the deep red flower for the nectar which is apparently secreted in abundance. It is highly ornamental and quite desirable for the garden.

Pentstemon grandiflorus is reported as the source of considerable honey in the Missouri River hills south of Sioux City, Iowa.

PEONY.

The cultivated peonies are introduced from Asia and are commonly grown for ornament. Most varieties are double and produce little pollen. The single varieties, however, produce pollen in abundance, and at times the bees seek them eagerly. The writer has seen as many as six to eight bees gathering pollen on a single blossom. The opening buds also seem to exude a nectar-like substance sought by ants and bees.

PEPPERBUSH or White Alder (Clethra alnifolia).

The sweet pepperbush occurs from Maine along the coast to Florida and west to Louisiana. It is a small shrub with white flowers, which appear in midsummer. In Alabama, Georgia and north Florida, it is common in the coast plain, on swampy banks of streams and in low, wet thickets. It is very fragrant when in bloom and is often used for ornamental planting on lawns and in parks. The honey is thick, white and of fine flavor. In localities where the plant occurs abundantly, in the wild state, it seldom fails to bloom, since it grows in wet places and is unaffected by drought.

The War Food Administration semi-monthly honey report No. 609, October 1, 1943 reported the pepperbush flow in southeastern Connecticut as the best in years. Individual colonies stored 150 to 170 pounds which was said to be an exceptional yield from clethra for this section.

PEPPERIDGE, **see Tupelo.**
PEPPERMINT, **see Mint.**

PEPPER-TREE (Schinus molle).

The pepper-tree, a native of western South America, has been widely planted in California for ornament and shade. Its bright red berries are a substitute for pepper, hence the name. It is a well-known tree on the Pacific coast, as far north as Martinez, on the Suisun Bay. It is grown in parks, on lawns and along the streets of nearly every Southern California city.

While a small amount of bloom may be seen at almost any time of the year, its principal period is in late summer when there is a scarcity of nectar-secreting blossoms. The blossoms are rich in nectar and the bees gather some surplus from this source. The honey is amber in color, strong and rather peppery in flavor.

Blossoms of clethra, or sweet pepper bush.

The flowers are small and of a greenish or yellowish color, in large sprays. The berries ripen in November and December.

In Bees and Honey, January 1934 P. C. Chadwick writes:—

> "The week of November 13 was unprecedented for a nectar-flow from pepper-trees in this city. * * The flow of nectar was equal to any I have

ever seen, even in July which is the month of its usual heavy yield. The temperature of mid-day during this period ranged around ninety degrees. It always yields best in extremely hot weather. * * I have had colonies here in the city store as much as fifty pounds in a season. Few cities in the state have as many pepper-trees as Redlands."

Blossoms and fruits of pepper tree.

PEPPER-VINE, see Ampelopsis, also Cow-itch.
PEPPERWOOD, see California Laurel.

PERILLA.

A drying oil much used in the paint and varnish industry is made from perilla seed most of which is imported from foreign countries, principally Japan. An effort has recently been made to encourage cultivation of perilla in this country.

Although perilla is a native of eastern Asia one species, *Perilla frutescens* has become established in many southern localities through accidental distribution. It is to be found in pastures and waste places in Kentucky, Tennessee, Arkansas, Louisiana, Oklahoma and Texas, as well as Alabama and less commonly northward to District of Columbia.

Doctor Oertel reports that bees get honey from perilla in some Louisiana neighborhoods and it appears probable that should the effort to establish it as a farm crop be successful, it may prove of advantage to the beekeeper.

PERSIAN CLOVER (Trifolium resupinatum or glomeratum).

Persian clover is an annual which will not endure severe cold. In the United States its range is confined to the extreme south. It was introduced to this country in 1906 and has become naturalized in Louisiana and to some extent in Mississippi.

It is particularly adapted to low moist soils but is not satisfactory on soils of low fertility. It blooms in April and May and appears very attractive to the bees. Reports of its value vary with different observers. Some report that the bees work it but lightly for nectar and pollen while others are enthusiastic in saying that it is one of the most promising honey plants.

The blossoms are lavender pink in color and the plant is reported as reseeding freely.

PERSICARIA, see Heartsease.

PERSIMMON (Diospyros).

The persimmon, or possom-wood (*Diospyros virginiana*), grows from southern New England south and west to Missouri, Arkansas, Florida and along the gulf to Texas. It is a tree of medium size, reaching a height of 50 feet and rarely exceeding 12 inches in diameter. The fruit is composed of a rich and palatable pulp and a few large seeds. When green it is very astringent and very disagreeable. The flowers appear in May in the southern part of its range, and later northward. Where abundant, persimmon is a valuable source of nectar. The Mexican persimmon (*Diospyros Texana*), called also date plum, is a shrub or tree 10 to 30 feet in height. It thrives best in canyons and ravines, and is common over much of south Texas. It is frequently reported as an important source of early nectar in many parts of Texas.

Franklin Sherman, Jr., lists the persimmon as sixth in importance of the honey sources of North Carolina. He states that it is irregular in yield and lasts but a short time, but does well while it lasts.

A sample of persimmon honey received from J. C. Elliott of Columbus, Kansas, is of very fine quality. It is amber in color with delightful flavor. He states that it is usually mixed with blackberry and adds that persimmon yields but little in eastern Texas while it is one of the best sources of honey in part of Kansas. He writes (1933):—

> "Persimmon is one of my main flows. The great trouble with it is that it comes so early that the bees are not ready for it. This year it started blooming May 20, and lasted two weeks. Some colonies gathered as much as seventy-five pounds surplus. The bees worked from before sunrise in the morning until after sunset in the evening. I have known the blooms to appear as early as May 10."

PHACELIA (Phacelia).

Jepson described thirteen species of phacelia as native to California. Of these Richter lists four as of value to the beekeeper. Hill

Persimmon blossoms.

vervenia (*Phacelia distans*) he lists as common in the plains and foothills, yielding both nectar and pollen. The caterpillar phacelia (*Phacelia hispida*) is listed as common in the chaparral belt, yielding a water-white honey of fine flavor that candies soon after extracting.

This is common in Ventura County, and M. H. Mendleson reported that he extracted a carload of honey from this source before the blooming of the sages. *Phacelia ramosissima* is reported as a fair honey plant, but not equal to the others. The valley vervenia, or fiddle neck (*Phacelia tanacetifolia*) is listed by Jepson ("Flora of Western Middle California") as found in the Sacramento Valley and southward to southern California. It blooms in April and furnishes bee pasture in about six weeks from seed, and the bloom lasts about six weeks.

Blossoms of *Phacelia tanacetifolia.*

"The nectar flows all day. The honey is amber in color, sometimes light green and of a mild aromatic flavor. Cows fed on it show a marked increase in flow of milk, but will not eat it alone at first."—Harry E. Horne, page 342 of above book.

This species was introduced into Germany and had quite a boom there in the early nineties. Much attention was given it in the German bee magazines and it was endorsed as valuable both for forage and for bee pasturage.

Thos. Wm. Cowan writes as follows, in American Bee Journal, in regard to the growing of phacelia in Europe:

"The one grown in Europe, *Phacelia tanacetifolia,* is literally covered with bees from morning till night. The species was introduced into Europe from California in 1832, and is called tanacetifolia (tansy-leaved) from the resemblance of its leaves to those of tansy. It is an annual with bluish pink flowers, racemes spike-formed, elongated, corymbose; height of plant two feet. It is grown in Europe as a bee plant for its nectar, and is the only one which produces an appreciable quantity of it."—November 20, 1902, page 751.

An evidence that European beemen have not lost interest in *Phacelia tanacetifolia* is found in the appearance of a book of 64 pages devoted to this plant in 1935. It was published by the Beekeepers Society at Prague and contained 18 illustrations.

There have been recent reports also of experiments in Germany with this plant as a source of forage for livestock. It is used in either the green state or as ensilage but apparently does not make satisfactory hay. In nutritive value it is reported as ranking between red clover and crimson clover and to be more palatable than alfalfa.

Scholl lists *Phacelia congesta* and *Phacelia glabra* as yielding sparingly in Texas.

A beekeeper from Indiana reports that *Phacelia purshii* grows freely in the wheat fields in his locality and that the bees work freely on the blue flowers. He states that in places it grows so abundantly that the wheat takes second place to the phacelia. On the 27th of May, 1919, strong colonies already had full depth extracting supers almost filled. The honey from this source is of good quality.

Herbert Sanborn writes that on May 4, 1923, at Nashville, Tenn., the bees were working on *P. purshii* almost exclusively. He stated that the fact that humming birds were also visiting the flowers furnished proof that it was rich in nectar. He stated further that the year previous, when locust was cut off it was the main source about the Middle of April. Although it lasts but two or three weeks he ranks it high in the list of important nectar sources for middle Tennessee.

L. A. Schott in a letter to the author at about the same time stated that it covers large areas in the sandy soils of Scott County, Missouri, and that the bees seemed to be getting much honey and some gray colored pollen from it.

Phacelia globata is common in south Texas in winter but is not of much apparent value except to provide early stimulation.

The author has found *Phacelia laxa* to be common in some places in the lower Rio Grande Valley, blooming in February. There it is reported as the source of some honey, where it grows along with *Stachys drummondi* locally called "pinkmint." Since the two plants are found together the credit for such yield as it may provide is usually credited to the pinkmint.

Hurd reported that it is locally important as an early source of nectar in the Hebronville and Falfurrias neighborhoods. He described the honey as light amber and of good flavor.

This Phacelia is commonly known as Little Baby Blue Eyes.

PHYSOSTEGIA.

The False Dragon Head, (*Physostegia virgininia*), is one of those minor honey plants which while attractive to the bees are nowhere sufficiently abundant to be important to the beekeeper. This native

False dragonhead while very attractive to the bees is of minor importance.

plant common to wet places widely distributed in the Northeastern States is of greater interest as a garden plant than as a source of honey.

PHYLLODOCE.

In the Cascade and Olympic Mountains and northward are found two species of *Phyllodoce*. They are low growing, evergreen shrubs, commonly known as heather, red fill-o-do-see, red false heather, etc. *P. empetriformis,* the red flowered variety is very showy and occurs in abundance at high altitudes. The other, *P. glanduliflora,* has a yellow blossom. Both yield nectar according to western beekeepers who call the product "heather honey." The author has been unable to secure information which would separate the honey from these and from *Cassiope,* all of which is known as "heather." The honey samples received have been of amber color and good quality. (See heather.)

PICKEREL-WEED or WAMPEE (Pondeteria cordata).

The pickerel-weed, in the South often called wampee, is a widelv-distributed plant found on the borders of streams and ponds and in swamps from Nova Scotia to Ontario and Minnesota and south to Florida and Texas. In Texas it is often called tule. It grows in great profusion in swampy regions and the spikes of violet blue flowers persist through a long season.

The bees work the flowers freely over a long period, but it is the author's impression that it is chiefly in search of pollen. However, there are numerous reports of nectar from this source. The following reports are characteristic:

> "Bees work it heavily, but no data as to honeyflow is available."—H. B. Parks, Texas.
>
> "Bees work it freely for both nectar and pollen. Blooms all summer."—E. G. LeStourgeon, Texas.
>
> "Sample of wampee went to you by mail. Lots of it here; said to carry nectar until the middle of November."—F. M. Baldwin, Georgia.

PHORMIUM, see New Zealand Flax.
PIGEON BERRY, see Buckthorn.
PIGEON-BERRY, see Serviceberry.
PIGEON CHERRY, see Wild Cherry.
PIGEON PLUM, see Sea Grape.

PIN CLOVER or FILAREE (Erodium cicutarium).

The pin clover, or alfilaria, or filaree, is widely distributed in the Old World, and in this country has been naturalized from Europe. It is especially well known on the Pacific Coast from British Columbia to southern California. There it is said to be one of the most valuable wild pasture plants.

It is also called pin grass and heron's bill. It has a long period of bloom, beginning in February or March in California, and in some places continuing through the summer. It produces an abundance of pollen and considerable honey of good quality. In Gray's botany it is listed as "storksbill," and is mentioned as scarce in New York and Pennsylvania. It is recorded as occurring in Alabama, where it apparently was carried with railroad ballast. June is given as the blooming period in the vicinity of Mobile.

It is also known in several places in Connecticut, where it is said to bloom in May and June. Professor Pammel states that it is abundant in the dry soils in the Salt Lake basin and from Colorado to Texas.

The seeds cling to the wool of sheep and this aids in its wide distribution.

The Figure shows the plant with blossoms and seed pod. It is from the peculiar shape of the latter that it gets the name of "storksbill's" and "heron's bill."

Vansell has found the sugar concentration in the nectar from pin clover to be higher than in any other source examined. The plant is

Filaree or pin clover.

of special value as a source of stimulation although surplus has been reported from filaree in California. There are several other species in that state some of which are of value to the bees.

PINE (Pinus).

Occasionally honeydew is reported from the pine trees. The following references are typical of those to be found in current literature:

> "We are having a real flow of water-white honeydew from the pine here in Polk County. It has been on now for two weeks. Bees in general are in bad condition."—Luther Presswood, Reliance, Tenn., Jan. 18, 1907. American Bee Journal, page 98, 1907.

Large quantities of honey are often secured from pine woods in certain parts of Germany. The honey is nearly black in color, still it finds many admirers, and must, therefore, be of better quality than the honeydew gathered here at times. The Emmendingen Beekeepers' Society furnishes all the honey for the Grand Duke's table (in Karlsruhe), and it is specified that the honey must be this black honey of the pine woods."—Bztg. for Schlesw. —Holstein. American Bee Journal, page 616, 1906.

Similar honeydew is harvested in Switzerland and appears to be also much prized by the consumers.

In an article in American Bee Journal in November, 1929, C. E. Burnside discussed at length the occurrence of Melezitose in Honey. He states that such honeydew is often secured from the scrub pine, *Pinus virginiana* in Maryland and Virginia and probably Pennsylvania and South Carolina also.

Honey from this source is reported as granulating very quickly and when removed from the combs as thin, of poor flavor and amber color. A special demand has developed for medical purposes and the limited supply is likely to find a favorable market.

In 1938, A. J. Zastrow of Spout Spring, Virginia, secured about 5000 pounds of honeydew honey from pine, which appeared to be unsuited for table use.

PINKMINT (Stachys drummondii).

Pinkmint is the name given by beemen of the Lower Rio Grande Valley to a common hedge nettle which is of considerable importance in that region and to some extent in portions of East Texas.

It is an annual common to moist ground which blooms freely in seasons of more than usual rainfall. It begins blooming at about the Christmas season and at times continues until late spring. In seasons of unusual abundance it is the source of a good crop of winter honey. In one such year when conditions were especially favorable beekeepers in the San Benito neighborhood secured as high as one hundred pounds per colony averages from this source.

The rose red flowers are borne in terminal racemes on stems twelve to eighteen inches in height. The minute stinging hairs at times give rather painful stinging sensations which give rise to the name "wound-wort."

The honey is reported by local beemen as very light in color and of heavy body with pronounced minty flavor. (See Hedge Nettle).

PINK-VINE, see Coral Vine.

PLUM (Prunus).

Plums, both wild and cultivated, are important sources of nectar over most of the United States. There are eight or more species which are native to North America and which are generally known as wild plums. The blooming period comes early and with other tree fruits they are of great value as the source of nectar and pollen at a season when the colonies most need stimulation. The nectar yield is sufficiently abundant for the bees to store surplus where colonies are strong and weather conditions favorable. In Sacramento County, California, surplus is sometimes reported from prunes. There are probably

Plum blossoms.

few places where bees are kept in America where plums of some variety are not present.

According to C. F. Kinman, Farmer's Bulletin 1372, USDA,

> "The presence of bees in the plum orchard at blossoming time has been demonstrated to be almost an economic necessity. * * Poor crops or perhaps failures may be expected of self-sterile varieties where no bees are present, and even with self-fertile varieties the presence of bees has caused a decided increase in the crop."

POINSETTIA (Euphorbia pulcherrima).

The poinsettia is a well-known ornamental plant. It is a shrub, native to tropical America, grown to a considerable extent for ornamental purposes in California. The blossoms are small and inconspicuous, but there is a striking cluster of brilliant red leaves, surrounding the flowers, which give an impression of flowers.

The nectaries are very conspicuous cups at the side of the blossoms. Nectar is secreted in abundance, and if the plant was sufficiently common it would be an important source. The nectar gathers in large drops.

POISON IVY, see Laurel, also Sumac.
POISON OAK, see Sumac.

POISONOUS HONEY.

Much has been written in regard to poisonous honey. Well authenticated cases of serious poisoning from honey are rare, so rare, indeed, that many persons doubt whether such cases occur. There are persons with a peculiar susceptibility to honey from any source. To such, honey may seem to be poisonous, which can be eaten by others without any ill effect. Prof. A. J. Cook, writing in the American Bee Journal (October 12, 1905, page 711), mentioned having on several occasions received samples of so-called poisonous honey, which he ate without inconvenience.

An extended discussion of the poisonous effect of nectar from buckeye in California written by George H. Vansell, appeared in the American Bee Journal, December, 1925. Mr. Vansell proved experimentally that honey from this source was poisonous to the bees. (See buckeye.)

In December, 1880, issue of American Bee Journal, page 552, a case is mentioned where a native of New Zealand died from eating honey gathered from the wharangi bush (*Melicope ternata*), which is said to be one of the two poisonous plants to be found in New Zealand. The symptoms are reported similar to strychnine poisoning. A letter from New Zealand published in Gleanings, page 435, 1908, reads as follows:

> "Some one, usually a native, gets poisoned every year about here through eating bush honey, usually not capped, and puka-puka usually gets the blame. Mr. Hopkins had a look at a case of Moaris honey poisoning last year, and I think puka-puka got the blame; but the fact is significant, nevertheless, that it is the Maoris only, or principally they, that get poisoned, and in that case the honey eaten is never capped."—Stephen Anthony, Wastete, New Zealand.

The mountain laurel (*Kalmia latifolia*) is the plant most frequently reported as yielding poisonous honey. This shrub is common to the mountain regions of the Eastern States, and it would seem that cases of poisoning would be reported much more frequently if there was good reason to suspect the honey from this source. It is a well-known fact that disagreeable odors disappear from honey that is well ripened. In this connection a writer in American Bee Journal, page 664, 1884, suggests that there is no evidence of poisoning from well ripened honey. He further states that uncapped honey from the yellow jasmine is actually poisonous and has produced death, but that after it is capped there is no honey more wholesome. It should be noted that special emphasis was placed on the fact that the cases of poisoning in New Zealand were from eating uncapped honey.

Writing in Gleanings January, 1875, Dr. J. Grammer who had been a surgeon in the Confederate army writes as follows concerning honey from Mountain Laurel:—

> "Wherever the Mountain Laurel grows the bees are very fond of it. * * It is dangerous for any one unable to detect the taste to eat the honey. It has a highly poisonous effect, being an extremely distressing narcotic, varying in its effects in proportion to the quantity eaten. During the war I had many opportunities of witnessing its effects and on one occasion, personal experience gave me the right to say that I know something about it. * * Some time after eating a queerish sensation of tingling all over, indistinct vision, caused by dilation of the pupils, with an empty dizzy feeling about the head and a horrible nausea which would not relieve itself by vomiting.
>
> "The first case or two that I saw were entirely overpowered by it, and their appearance was exactly as if they were dead drunk.—* * the enervation of all the voluntary muscles was completely destroyed. The usual remedies for narcotics partially restored them in a few hours, but the effects did not completely wear off for two or three days, and I was assured that fatal consequences have been known to follow a too free indulgence."

The lime-tree (*Tilia petiolaris*) is popularly credited as poisoning insects which gather its nectar and numerous references to this effect have appeared in the British Bee Journal (July 13, 1922). The following quotation from "The Trees of Great Britain and Ireland," by Elwes and Henry, is of interest:

> "*Tilia petiolaris* is a beautiful weeping tree which has not been nearly so generally planted as it deserves to be. There are good examples in the botanic gardens at Kew, Cambridge and Glasnevin. I noticed many dead bees under it in 1911. In 1908 the bodies of innumerable bees, poisoned by the flowers of a tree of *T. petiolaris* at Tortworth had so manured the ground under its outer branches that a very green ring of turf was visible in the autumn following and was noticed by the Earl of Dulcie to be even more conspicuous in 1909."

Pammel states (Manual of Poisonous Plants) that honey obtained from *Euphorbia marginata,* the well-known snow-on-the-mountain, is poisonous and unfit for use. He also states that the Indians of Brazil use honey gathered by wasps from flowers of *Serjonia lethalis* for poisoning their arrows, and also as a fish poison. It contains a narcotic poison which causes death.

The following resume of the subject is copied from Pammel:

> "Prof. Lyman F. Kebler, who has made a somewhat extended investigation with poisonous honey has given an excellent bibliography with reference to the early literature on the subject. It has been known for centuries that the honey collected from *Ericaceae* acts as a narcotic irritant, producing giddiness, vomiting and purging. Poisonous honey was described by Xenophon. He gives a fairly accurate description of how the soldiers of his army acted that ate honey that was poisoned. He states that they lost their senses, vomited and were affected with purging, and those who

had eaten but little were intoxicated, but when they had eaten much they were like mad men. Strabo and Pliny spoke of poisonous honey, the latter writer, an early naturalist noted for his accurate observations, records poisonous honey, which he called "aegolethron" (goat's death), which bees collected at Heraclea. He gives a description of the honey, which is said to have had a peculiar smell and produced sneezing. It is generally supposed that this honey came from a species of *Rhododendron,* the *R. pontica.* This and allied species are the chief source of poisonous honey in Asia and Asia Minor, but it may be said in this connection that honey collected from the heather in Scotland is not poisonous.

"Barton, an early American botanist, reported poisonous honey in New Jersey as early as 1794."

In October 1885, the Beekeepers' Magazine reprinted an article from Country Gentleman which credited the death of persons near Branchville, South Carolina, to the eating of honey gathered from Yellow Jessamine (*Gelsemium sempervirens*).

"A boy eleven years old was the first of the family to eat some of the honey. In an hour afterward the child became giddy and staggered as he walked, and could not see. He was affected with general lassitude and slight nausea. In two hours he was seized with convulsions and died. The honey was given to a negro woman who gave it to her children. In one hour two of these children died, after an attack of dizziness, blindness and nausea. Several other persons who ate the honey were made sick in similar manner but vomited, and thus escaped fatal consequences."

The article above quoted states that the scent of the flowers will produce nausea, headache, dizziness of vision and trembling in some persons.

Writing in Gleanings in Bee Culture, (Dec. 1935), C. E. Burnside gives poisoning by nectar or pollen of southern leatherwood (*Cyrilla racemiflora*), as the cause of purple brood disease of bees.

POISONWOOD (Metopium metopium).

The poisonwood is a tree found in southern Florida, the West Indies and Central America. It is known by a variety of names, bum-wood, hog-gum, coral sumac or doctor-gum. Its sap is dangerously poisonous, and some persons are affected by a near approach to the tree. The tree reaches a height of more than forty feet and in south Florida it is the source of large quantities of surplus honey of good quality. The honey is usually blended with that of other sources blooming during the same period.

The tree has a thin bark which splits into large scales as it grows older. These are of a reddish brown color, brighter on the inside. It is sometimes confused with manchineel and is often classified as a *Rhus.*

POLECAT-TREE, **see Buckthorn.**

POLLEN.

Pollen represents the male element of reproduction in seed plants. The flowering plants are normally reproduced by seed. Fertilization is brought about by the intervention of the grains of pollen which are borne by the stamens or male organs. In the corn plant the stamens are produced in the tassels where pollen is developed in great abundance. The pistils, or female organs, are represented by the silks attached to the ear. There is a separate silk attached to each kernel, and each must be separately fertilized. In many plants both stamens and pistils occur in the same blossom, as is the case with apples, pears, etc. In some cases the stamens or male organs are borne on a separate plant or tree, as is the case with the persimmon. Cucumbers have the separate organs in separate flowers. In order to insure fertilization it is necessary that the pollen grains be carried from the stamens to the pistils. In the grasses and similar plants this is usually accomplished by the wind, while in most fruits, insects are the principal agents.

Pollen grains are very minute and are produced in small sacs in the tip of the stamen, commonly called anther. When ripe, the breaking of the sac sets the grains free in great abundance. A single grain is sufficient to fertilize an ovule and produce a seed; yet, because of the distance between the stamens and pistils, large quantities of the pollen or flower dust are necessary to insure pollination. Ragweed and corn produce pollen in such abundance that one brushing by the plants is dusted so freely as to appear to be covered with flour.

Pollen is of special importance to the beekeeper, since it serves as food for young bees. It is the sole source of nitrogenous food for the growing larvæ, and is stored in large quantity in the open cells of the brood combs. Without an abundance of pollen available for food for the developing brood, the colony cannot prosper. Pollen is second only in importance to surplus honey to the beekeeper, and an abundant supply of plants which yield pollen during the brood-rearing season is very desirable.

POLLINATION.

Pollination is the fertilization of the blossom of a plant, resulting in the development of seed. As ordinarily used among the beekeepers, pollination refers to the transfer of the pollen grains from the stamens of one plant to the pistils of another. Since many plants are self-sterile, it is necessary that pollen from other plants and sometimes from other varieties be brought to them to insure fruitage. As already stated under "Pollen," the honeybee is the principal agent in the pollination of many of the edible fruits.

So important is the bee regarded by horticulturists that cucumber growers contract for colonies of bees to be placed in their greenhouses, cherry growers often lease apiaries to be moved to their orchards at the beginning of the blooming period, and apple orchardists contract with beekeepers to furnish sufficient bees to insure fertilization of their orchards.

Darwin was among the first to realize the importance of cross pollination. He showed that continued self-fertilization resulted in inferior fruit, while cross fertilization increased the vigor of the offspring. Beach has shown that many varieties of grapes are self-sterile, and at the California experiment station it was shown that "bees are a necessary aid in pollination with the French and Imperial varieties of prunes." It was proven that practically no fruit was produced when all pollen-carrying insects were kept from visiting the blossoms. (Bulletin 291, California Experiment Station).

It has been shown by numerous experiments that many varieties of fruits, especially of apples and pears, are likely to be unfruitful if deprived of the services of the honeybee as pollen carriers.

POLYPTERIS.

Polypteris callosa, described by Britton and Brown as *Othake callosum,* or rayless othake, is common on dry soil from Missouri to Texas and New Mexico, blooming from June till October. Mr. Preston Blair, a beekeeper from Hercules, Missouri, sends specimens to the author with the following information concerning its value to the bees:

> "We had a severe drouth here, but it did not hurt this plant. It will grow where the soil is not over two inches deep. It grows up to 20 inches or more in height and blooms in early September, lasting till frost. The bees gather pollen from it that is very near white; the bloom is pink. The honey is white and has a little peach flavor. It was the richest honey plant this fall (1918) I ever saw. The bees would fly all day long and come in heavily loaded."

POME-DE-TERRE, see Climbing Boneset.

POND LILY (Nymphaea advena).

The pond lily, also called beaver-root and spatter-dock, is found in ponds and streams from Nova Scotia to the Rocky Mountains and south to the Gulf of Mexico. It has a long blooming period and is reported as an important source of pollen in some localities.

POPLAR, see Tulip-Poplar, also Aspen.

POPPY (Papaver).

The garden poppy (*Papaver somniferum*) is a native of Asia, but is widely cultivated in many countries, including our own. It is the source of opium and morphine, drugs widely used in the practice of medicine. The bright-colored flowers are very attractive to the bees, which seek them in large numbers and fairly revel in the abundant pollen masses. The pollen secured from poppy blossoms is very dark. There are numerous reports to the effect that the bees show evidence of a narcotic when working on this plant.

The California poppy (*Eschscholtzia californica*) is a common and widely diffused plant in California. It is a gorgeous plant of variable habit, especially abundant in spring, but in some parts of the State may be found in flower at almost any season. Richter lists it as the source of some honey and of large amounts of orange-colored pollen.

The California poppy is a gorgeous flower.

The prickly poppy (*Argemone*), often called poppy thistle, is widely distributed, especially in the southwest. It is the source of large quantities of pollen.

The Arizona Poppy, (*Kallstroemia grandifolia*) is a common flower in the desert regions in seasons of sufficient rainfall. It is freely visited by the bees but is probably of little importance.

POPPY THISTLE, see Poppy.

PORTULACA.

Portulaca is commonly grown as a garden flower but is too little cultivated to attract attention as a bee plant in this country. The following note from Bee Kingdom, Cairo, Egypt, September 1930, gives an idea of its importance abroad:—

"Our tests on Portulaca show that it is eagerly sought by the bees for both honey and pollen. It seems to yield abundantly under varied conditions. * * *"

It gives a sainfoin-like delicious honey.

The above article lists four species of Portulaca, mostly from Tropical regions. Of these the Sun Plant of Brazil (*Portulaca grandiflora*) is the one commonly cultivated in American gardens.

POSSUM HAW, see Holly.

POSSUM-WOOD, see Persimmon.

PRAIRIE CLOVER (Petalostemon).

There are several species of prairie clover native to the western prairies from Indiana west to the Rocky Mountains and south to

Prairie clover in bloom.

Texas. There are frequent reports to the effect that they are valuable honey plants, though no longer sufficiently plentiful to be important except in a few localities.

In Botanical Gazette, April 1890, Charles Robertson lists the honeybee as gathering both nectar and pollen from *Petalostemon violaceum*.

In his list of Honey Producing Plants of Nebraska, Chas. E.

Bessey lists *P. candidum* and *P. purpurea* as the source of both honey and pollen in that state. (Bul. Agr. Ex. Sta. Vol. 7.)

PRAIRIE CROCUS (Anemone patens). Pasque Flower.

The prairie crocus or pasque flower is common on the prairies from Wisconsin to Texas and of western Canada from Manitoba to Alberta. The pale blue flowers appear very early in spring. It is especially valuable as a source of early pollen on the Canadian prairies, where it is the first bloom to supply the bees with material

Prairie crocus are among the first spring flowers.

for early brood rearing. It is extremely abundant where the native prairie is still unbroken, but disappears as the soil is turned for cultivation.

PRAIRIE PARSLEY (Polytaenia Nuttallii).

From H. B. Parks of San Antonio, Texas, the author received a report of honey from prairie parsley in the season of 1941. He described the honey as of fair flavor, with a very peculiar shade of amber color with reddish reflections. He reports further that it yields quite heavily. From his letter the following quotation is taken:

"Going eastward we ran into this plant about one hundred miles east of San Antonio. All through the river bottom country and into east Texas, we found this plant to be very abundant. In talking with beekeepers all of them agreed that it had contributed to their big crop this year but they had never known it to be a honey plant before."

PRICKLY ASH (Xanthoxylum Clava-Herculis). TOOTHACHE-TREE.

The prickly ash or toothache-tree is found from North Carolina to Florida and west to Texas. It is a small tree with the bark armed with short, warty thorns, while the branches have longer ones.

In east Texas it is frequently mentioned as a source of honey. The honey is reported as pungent in taste, sharp and peppery and light in color. The blossoms yield freely, and, where sufficiently abundant, may be expected to yield surplus. At Palestine, Texas, surplus is reported from prickly ash. (See also Colima).

PRICKLY COMFREY (Symphytum asperrimum).

There is a plant native to the Caucasus and Persia which produces an enormous quantity of green forage which might well be grown in

Prickly comfrey, a forage plant which has not met with much favor in America.

many parts of America. The prickly comfrey (*Symphtum asperrimum*) will, perhaps, yield as large a quantity of forage as any plant known. It roots deeply and continues to grow vigorously almost throughout the year in regions of mild climate. It is hardy, however, and survives cold winters without difficulty. Once established, neither cold nor drought affects it seriously.

From three to six crops per year can be harvested and an established field will last for fifteen years or more without replanting. A yield of 10 to 40 tons of green feed can be expected annually. According to a bulletin of the U. S. Department of Agriculture, its greatest value is as a soiling crop for dairy cows. The leaves are so watery and gummy that it is difficult to cure the crop for hay and when used for ensilage there is danger that it will heat and acquire a disagreeable odor.

The plant is not a legume and consequently requires heavy manuring to maintain a heavy yield.

It is said to thrive on sandy or clay soils, but to succeed best on moist or even boggy land. It is not equal to either alfalfa or sweet clover on lands where either of these plants will grow. In Australia it is recommended for naturalization along the banks of streams and the margins of swamps. Here is a hint for American beemen.

It is difficult to grow from seed and is usually propagated by root cuttings. They are planted in rows from 1½ to 2 feet apart in the row.

While the plant is attractive to the bees, there is no available information as to the quality or quantity of honey which may be secured from it. It has been tried by several experiment stations and grown to some extent on a few American farms, but as yet not much enthusiasm has been shown for it except for soiling purposes, as above stated. Some foreign publications recommend it for hay or ensilage, but the writer has not been able to find anyone in this country who has been satisfied with results when used in this way.

PRICKLY PEAR or INDIAN FIG (Opuntia).

Plants of the cactus family are widely scattered in the arid regions from Dakota to Washington and south to Texas and California. Of the prickly pears (*Opuntia*) there are about 150 species, mostly found in the warmer sections of North America and southward. A few are to be found in sandy soils further east, ranging from Ontario and Massachusetts south to Florida. They are also sometimes grown as ornamentals. The blossom of the prickly pear is of pale yellow color and very attractive. It is reported as a source of nectar in both Texas and California. *Opuntia engelmanni* is reported by Scholl in "Honey Plants of Texas," as "of much importance to the beekeeper, especially during a season of partial drought. Both an abundance of pollen and honey was obtained, the honey being light amber in color, of heavy body but 'stringy,' so much so that it fairly draws out into strings when very thick. The flavor is very rank."

In some parts of the Southwest it is valued more for pollen than for nectar. It blooms for four to six weeks. E. G. LeStourgeon reports that in the vicinity of San Antonio, Texas, it yields surplus honey about one year in four, but that the flow is usually short, seldom more than four or five days. A peculiarity of this honey is that of granulating in large crystals in clear liquid. It is often spoken of as "buttermilk honey," because of this peculiarity. LeStourgeon secured an average yield of 87 pounds per colony of cactus honey in Atascosa County. Such yields from this source are rare. Cactus grows in great abundance over large areas of Texas and is probably of more importance to beekeepers in that State than in any other.

Blossom of the cactus or prickly pear.

Opuntia leptocaulis, known in some localities as *Tassijilla,* is widely distributed over Texas and adjacent Mexico. Its small flowers appear from April to July and are visited freely by the bees for both nectar and pollen. Because of the long blooming period the nectar is so fully blended with that from other sources that little information can be obtained concerning its value.

Alex R. Brown in the Australasian Beekeeper, September 15, 1928, writes as follows concerning prickly pear in Queensland:

> "During a season of average rainfall, and given sultry weather while in bloom, the pear will give pollen and nectar enough to force the bees along well. While if the season is wet, and followed by sultry days, during the blooming period, strong colonies will gather between 30 and 40 pounds besides increasing their brood nest considerably. I got three crops of prickly pear honey in nine years. I have spent nine years of the best part of my life in the pear belt of the west, hoping and hoping that next year would be the fulfillment of my dreams.
>
> "Prickly pear has broken more hopes, hearts and homes than any other plant I know in this beautiful state of sunshine and flowers—Queensland."

PRICKLY POPPY, **see Poppy.**

A prickly pear plant in bloom.

PRINCE EDWARD ISLAND—Honey Sources of.

Alsike and white clover are the principal honey plants of the Island, with *Tilia europea* also important at Charlottetown.—F. W. L. Sladen.

PRINCE'S PLUME (Stanleya).

There are nearly a dozen species of stanleya common to the dry western plains from Montana, Oregon and Nevada to California and New Mexico. They are leafy perennials with flowers in many-flowered racemes. They belong to the mustard family which furnishes many good honey plants. A specimen was received in early June from C. E. Andrews of Fallon, Nevada, who stated that it seems to yield quite heavily in a desert region which is short of sources of nectar for spring brood rearing.

PRIVET (Ligustrum).

There are several species of Ligustrum cultivated for ornament most commonly as hedge plants. The common privet of Europe has become naturalized in many localities in the eastern states.

L. japonicum, is reported from Houston, Texas, as much visited by bees in search of nectar. The privets, like most ornamentals, are only of incidental interest in most localities.

Privet honey is dark in color, fairly thick and distinctly bitter in taste according to F. N. Howes (Plants and Beekeeping). He states that the bitterness is liable to spoil good honey with which it may be mixed.

PROPOLIS.

Propolis is a resinous gum gathered by the bees from buds of trees, and used to close crevices about the hive. It becomes brittle and hard in cold weather, but when warm is sticky and very tenacious. The beekeeper finds it quite a problem to keep his hands free from propolis when manipulating the hives in mid-summer. Some races of bees deposit propolis much more freely than others, the Caucasians being especially inclined to gather a great surplus of the cement and place it in lumps about the hive. If, perchance, a mouse, beetle or other object which the bees are unable to remove enters the hive, they frequently coat it freely with propolis, and there it remains.

Cracks seem obnoxious to the bees, and the small spaces between the frames are quite likely to be filled with the cement. In the production of comb honey, the tendency of the bees to fill every crevice with it leads to much labor on the part of the beekeeper in cleaning his finished sections for market. This is especially true of honey left on the hive till late in the season, as propolis is gathered in quantity in late fall in anticipation of cold weather.

Cottonwood is thought to be one of the chief sources of propolis wherever that well-known tree is found.

Bees gather much material from a great variety of sources. In addition to the fresh wax gathered from plants and trees, they are often attracted by fresh varnish or other substances from which they can get sticky material to serve as glue.

PROTEA.

Protea mellifera is an ornamental shrub 6 to 8 feet high, which is popular in Southern California. Beekeepers from that region report it as very attractive to the bees and as the source of considerable nectar.

Langstroth and Dadant, in "The Hive and Honeybee" speak of Protea as follows:—

"In the vicinity of the Cape of Good Hope, there is a blossom, the *Protea mellifera* which probably surpasses all others in the abundance of its nectar. Indeed it is so abundant that the natives gather it by dipping it from the flowers with spoons."

This particular Protea is one of about 100 species native to that region. Some are trees of moderate size, while others hug the ground. Most of them are shrubs of moderate size. In his book, "Plant Hunting," Ernest Wilson, famous explorer comments:

"Overflowing with honey are the pink and white heads of the *Protea mellifera,* known to the Boers as Honeypots. The honey is collected and made into a kind of sugar, the blossoming season being a great occasion for picnics.

"In my judgment the handsomest inflorescense in the world is that of *Protea cynaroides,* seen on its native heath."

PRUNE, see Plum.

PSORALEA.

There are a few references to the psoraleas as attractive to the bees.

Concerning *Psoralea Onobrychis,* Chas. Robertson writes in the Botanical Gazette, April, 1890:—

"The plants grow in large patches and bear many racemes of blue flowers, which are very attractive to bees. Greenish lines on the banner form path-finders. The wings and keel are depressed together, and return so as to cover the stamens. The stigma is raised considerably above the anthers and so strikes the bee in advance of them. The calyx tube is about two mm. deep so that small bees which know how to force their way into the flower can reach the nectar. The flowers are sought by many insects, especially bees of the genus Megachile."

J. Skovbo reports that in Oregon the bees work on *Psoralea lanceolata* all summer but that he cannot see that they get much from it.

In Arizona *Psoralea lanceolata* is locally called "Lemon Weed" and reported as a valuable source of nectar. This species is found from Alberta and Saskatchewan south to Texas and Arizona.

George E. Capwell of Cottonwood Falls, Kansas, reports honey from Silver Leaf Psoralea, (*Psoralea argophylla*) which is native to prairies from Wisconsin and Alberta south to New Mexico.

Psoralea esculenta is the well known Indian Bread Root which provided an important source of food for the red race but is not regarded as of much value as a honey plant.

The best reports of honey from psoralea come from Oklahoma where considerable surplus honey is reported from them on the prairies.

PUCKERBUSH, see Max Myrtle.

PUMPKIN (Cucurbita pepo).

The pumpkin is a well-known gourd-like fruit, usually of deep yellow color that is widely cultivated as a food for stock and for pie. There are numerous varieties of various sizes and colors. The blossoms are large and showy and very attractive to the bees. The plants yield an abundance of pollen as well as nectar. The honey is amber and not of high quality, and granulates readily.

Where pumpkins are grown on a large scale for stock feed or for canning factories, they are a valuable source of bee pasture.

> "Where the proper atmospheric conditions prevail it has few superiors, producing a fine straw colored nectar of excellent flavor, and very heavy body, weighing 12 pounds per gallon.
>
> Our large apiary in Cuba was surrounded on two sides by a corn field of eight or ten acres in which pumpkin seeds had been 'stuck' in every third or fourth hill, producing vines which completely covered the ground. During the month of February we took five tierces (6,000 pounds) of honey from these blossoms alone."—A. J. King, American Bee Journal, Page 551, 1886.

PURPLE-FLOWERED MINT (Mesosphaerum spicatum).

M. spicatum is known as purple-flowered mint in Florida. According to Small it is found also in Alabama. It has a long blooming period and yields some nectar from March to December. Frank Stirling regards it as very important, since it often tides the bees over periods when little else is to be had. He states that the honey is of good quality and light amber color. It is found in practically all sections of Florida, but does best on soils having a lime formation, as is evidenced by the way it thrives along the roads built of shell and lime rock.

The swamp basil (*M. rugosum*) is abundant throughout the pine region of Alabama and is found from North Carolina to Florida and Texas. It has white flowers which appear in July and August. It is to be found on the margins of ponds in the pine barrens.

E. G. Baldwin states that purple flowered mint is often confused with wild pennyroyal by Florida beekeepers. The mint blooms further north than does the pennyroyal, although both are found in the southern area. The pennyroyal blooms in winter while the mint blooms in summer and fall. Honey from the former is nearly white, while that from the purple flowered mint is amber. (See Wild Pennyroyal).

PURPLE LOOSESTRIFE (Lythrum).

There are two species of lythrum frequently reported as honey plants. The European species, *Lythrum Salicaria,* the spiked loosestrife, has become naturalized in wet places from Nova Scotia to Delaware. According to H. D. House, State Botanist of New York, it is common in and around the inlets and backwaters of the Hudson River all up and down that stream. It is also common up the Mohawk Valley at Oneida Lake and along the railroads westward across the State to Lake Erie. It is a tall plant of vigorous growth, but as it confines itself largely to wet places, it is not likely to be any more of a nuisance than the usual coarse weeds growing in such situations. The bees work this plant freely and occasional reports of honey from this source are received from New York State. The honey is very dark and of strong flavor, having a slight tobacco-like taste as it gets older. It is commonly called rebel-weed.

Massachusetts beekeepers question the reports of poor quality from

Purple loosestrife grows luxuriantly along the Hudson River in New York.

New York and say they regard it as of light color and good flavor.

William Sumnick with bees in Orange and Ulster Counties, New York reports moving one yard into a 100 acre bog of purple loosestrife each season. He says the bloom lasts five weeks but that yield of honey is disappointing. An average of about a super per colony is all that is secured at best. The honey is strong in taste and in the comb looks green. Wax drawn from this plant is golden yellow. He regards it as the best fall stimulant and that its presence insures good wintering.

Rex Ballard of Augusta, Michigan reported a substantial portion of his crop of 16,000 pounds from 120 colonies in 1944 had come from purple loosestrife.

Lythrum alatum, a native species, is found from Ontario and Minnesota south to Georgia, Florida and Texas. A report from Oklahoma is to the effect that it blooms with sweet clover, lasting till frost. The bees work upon it from daylight till dark in either wet or dry weather. Similar reports come from Texas. It is found on low lands in the southwest.

It is also reported by beekeepers from New Orleans as attractive to bees in Louisiana. *L. ovalifolium* is also reported from Louisiana.

PURPLE THISTLE, see Blue Thistle.

Q

QUEBEC—Honey Sources of.

In Quebec alsike and white clover furnish the principal surplus honey. In the southern part of the province buckwheat, basswood and sweet clover are also important. Fireweed, blueberry, goldenrod and aster are the source of surplus honey over much of the province and willows and maples are valuable for spring brood rearing—F. W. L. Sladen.

QUEEN ANNE'S LACE, see Carrot.
QUERCUS, see Oak.

QUINCE (Cydonia oblonga).

The quince is a small crooked fruit tree which comes from Asia and is cultivated to some extent for its fruit which is used in preserves.

Like other orchard fruits the flowers are freely visited by honeybees for both nectar and pollen but since the trees are usually few in numbers it is not important.

The Japan quince (*C. japonica*) is much used as an ornamental. It attracts the bees in large numbers and in the region of the Gulf Coast is valued for winter brood rearing.

QUININE TREE, see Hop-Tree.

R

RABBIT BRUSH (Chrysothamnus).

Richter lists rabbit brush (*C. nauseosus*) as yielding nectar in the vicinity of Independence, California, from September till November. He states that the bees work vigorously on it, but that the honey is dark, of poor flavor and disagreeable odor, and that when the bees are evaporating it, it can be smelled all over the place.

There are forty-two different varieties of chrysothamnus in the southwestern desert from New Mexico to California and north to Utah and Colorado, Nevada and Nebraska. They are described as coarse plants, usually shrubby and growing to a height of as much as six feet. They have small heads of yellow flowers. The Navaho Indians use the heads of the various species to dye wool yellow.—(Flora of New Mexico.)

From the dry plains from eastern Washington to Montana and Colorado *Chrysothamus lanceolatus* is a valuable source of nectar. A Utah beekeeper writes as follows:

> "Rabbit brush abounds here on the desert heather, vales and hillsides, growing apparently with little or no moisture, and blooming profusely from the middle of September to the end of October. It has the fragrance of honey and is now fed upon largely by the bees. It stands much frost and blooms on.
>
> "It grows from 2 to 3 feet high in massive clumps with many branches, all covered with terminal bloom. The honey and wax made from it are of a light straw color. It gives a month or more of forage, let the season be wet or dry, when there seems nothing else to feed on."—J. E. Johnson (The Beekeepers' Journal, Dec., 1871).

RADISH or JOINTED CHARLOCK (Raphanus Raphanistrum).

The wild radish or jointed charlock is a troublesome weed in the fields of eastern Canada and United States from New England to Pennsylvania. Like many other introduced plants, it has been widely scattered with grain seeds. It has been introduced into northwestern Iowa with oats, where it is spreading in fields and waste places.

Sladen lists it as one of the important sources of nectar in Nova Scotia.

The cultivated radish (*Raphanus sativus*), common in every garden, is the source of some nectar and pollen in spring. It has run wild in old fields and gardens and in poorly cultivated land both east and west. It is mentioned as especially common in towns and cities in the vicinity of San Francisco Bay, California. The author found it very common in young orange groves in Arizona where it was blooming in late February and the bees were busy on the blossoms. In a few neighborhoods where garden radishes are raised for seed it may offer enough pasture to be important but in general it is a minor source.

RAGGED LADY, see Red Gaura.
RAGGED SAILOR, see Centaurea.

RAGWEED (Ambrosia trifida).

The great ragweed is often called horseweed. This is a very common roadside weed, growing to a height of 10 or 12 feet. It is common in Quebec and Ontario, west to Manitoba. In the United States it occurs from New England west to Colorado and south to the Gulf. It is also found in Cuba and Mexico. It is especially common in the rich lands of the Mississippi Valley from Minnesota to Texas.

The ragweed does not produce nectar, but furnishes large quantities of pollen in late summer and fall.

RAGWORT, see **Butterweed.**

RAPE (Brassica napus).

In Europe it is highly regarded as a honey plant, as the following letter from Baron Von Berlepsch, which appeared in American Bee Journal in April, 1874, will show:

"During the years between 1841 and 1858, that I was a practical agriculturist, I cultivated rape, to a large extent, and can in consequence thereof, and from knowledge otherwise gained testify most assuredly, that in all Germany there is no plant yielding more honey than rape. I know of instances, occurring in my own experience, where a very populous colony of bees, during time when rape was in blossom, gained a weight of 20 pounds and over in one day.

"On the 10th of May, 1846, there was near me a 65-acre field in blossom. The weather was excellent, and my strongest colony, which I placed on a platform scale, gained that day more than 21 pounds in weight. I know of only one other plant that can be compared with rape as a honey-yielding plant, and that is esparcet (*Sainfoin*).

Bloom of the tall ragweed.

A Wisconsin beekeeper, writing in American Bee Journal the same year stated that rape in Wisconsin is scarcely second to linden. He described the honey as of golden color and good flavor. He further cited the fact that rape blooms about the middle of August, when there is little else as a great advantage in cultivating it for its honey.

The few references to it in our literature indicate that it may be of more value to the beekeeper than is generally recognized.

From New Zealand Beekeeper, October 20, 1941, the following is quoted:—

"A strong colony is capable of storing up to two supers of surplus honey from rape if the weather is reasonably settled.

"The honey is of the finest quality, water white in color, with a mild flavor and fine grain. Very little variation in quality has been noticed in

South Canterbury, but samples taken from rape honies gathered from heavier land farther north indicate that the quality of the honey is subject to soil conditions, as variations in color are apparent. A characteristic of this honey is the remarkable rapidity with which granulation occurs. * * The honey granulates about seven days after extraction and if subjected to even slight agitation will crystallize in four days. Frames of honey not removed from the hives will granulate within a month of being stored."

RAPHANUS, see Radish.

RASPBERRY (Rubus). THIMBLEBERRY.

The wild raspberry (*Rubus occidentalis*) is a very common plant in the woods of the Northeastern States. Mohr gives its natural range as New England to Quebec and Ontario, Minnesota, Nebraska, Colorado and Oregon, south to Ohio and West Virginia, and along the mountains to Georgia.

There are several species of wild raspberries, and probably all are good honey producers. Raspberry honey is produced extensively in northern Michigan, where the plant is abundant on cut-over lands. It blooms following the tree fruits and is usually ahead of the white clover. In localities where it is plentiful it is a most valuable honey plant and phenomenal yields have sometimes been reported from this source. A good raspberry location is very desirable. Beekeepers who chance to be near large plantations of raspberries cultivated for market are equally fortunate. The honey is white and of a superior quality.

Some surplus honey is reported from cultivated raspberry in Oregon where the crop is extensively grown.

RATTAN VINE (Berchemia scandens).

Rattan vine, or supple-Jack, is a common climbing vine in low thickets throughout the Southern States from Virginia, Kentucky and Missouri, south to Florida and eastern Texas. The flowers are greenish yellow, appearing in late spring. The plant is slender and of vigorous growth, frequently climbing high trees.

In east Texas it is one of the important sources of surplus honey. There the plant is abundant along the streams, and is reported as yielding honey for a long period. The author has found rattan vine highly regarded in many sections of the South, especially in Alabama.

Scholl states that the honey is dark amber in color and used mostly for manufacturing purposes.

RATTLE BOX or RATTLE WEED, see Loco Weed.
RATTLEBUSH, see Indigo-weed.

RATTLESNAKE-ROOT (Nabalus altissimus).

Rattlesnake root is listed by the late A. J. Cook as "which swarms with bees all the day long," but the author can find so few references to it that it is probably of little importance.

RED BAY (Persea Borbonia).

Red bay, or Florida mahogany, is also known as tisswood, sweet bay and laurel-tree. This tree should not be confused with the magnolia, which is also called sweet bay. (See Magnolia). The alligator pear (*Persea persea*) is a near relative, which has been introduced from Central America and extensively cultivated for its fruit. It has run wild in some parts of Florida.

Red bay is reported as yielding surplus honey in considerable quantity, but of poor quality along the Gulf Coast of Texas. The honey is said to be very dark, suitable for baking purposes. Beekeepers report that the plant is quite dependable, but of limited range.

In Florida, Wilder reports sweet bay (Dixie Beekeeper, Oct. 1922) as a poor source of honey. He probably refers to *P. Borbonia.* At the same time he speaks of red bay which grows to about the height of a man, as the source of abundant nectar. The author assumes that he here refers to shore bay (*P. littoralis*), or to scrub bay (*P. humilis*), which are shrubs or small trees common to Florida. These may be confused with the swamp red bay (*P. pubescens*), which is very similar to the red bay, except that it is smaller. This species is found in swamps and wet places near the coast from Virginia to Florida and Texas.

Concerning honey from this source Wilder is quoted as follows:

> "The blooms are small, very delicate and easy to blight, yet no plant secretes more nectar for the time that it is in bloom. * * * Bees gather more honey from this plant for the short time it yields than any other plant known. The nectar is very thick as it strings from the bloom, and before the apiarist is aware of it every available place in the hives is chock full of this bay honey, capped over as fast as it is stored. The honey is dark and rather strong in flavor and can only be sold as No. 2 grade. For the most part it is left with the bees and furnishes a great amount of our winter stores throughout the low, flat country along the Coastal Plains.
>
> "Red bay honey never granulates and is always very thick."

RED-BERRY, see Buckthorn.
RED BOX-TREE, see Eucalyptus.

RED-BUD (Cercis canadensis).

The red-bud, or Judas-tree is a common shrub or small tree in the Southeastern States. It is found occasionally from western Pennsylvania to southern Michigan, southern Iowa and Nebraska south to western Florida and west to Texas. It grows along streams and in the woodlands where the soil is moist and rich. In Alabama it blooms late in February, and in southern Iowa in April. The rose-pink blossoms appearing in early spring, before the leaves are out, make the tree very conspicuous during the blooming period. Where the tree is abundant it furnishes a liberal pasture for the bees for early spring brood rearing. Blooming so early, it is rarely the source of surplus. In the northern part of its range, it often blooms with fruit trees and dandelion, so that it is not as important as farther south.

There is another species in south Texas and Mexico which blooms in March, the Texan red-bud (*Cercis reniformis*), and one, the west-

The blossoms of red-bud appear before the leaves.

ern red-bud (*Cercis occidentalis*), which occurs in the mountains of California and occasionally in Utah.

The red-bud is also known as salad-tree, or June-bud.

Concerning the California form, Coleman writes (Western Honeybee, July, 1921), as follows:

> "Along streams in Mendocino County and the region about Mt. Shasta southward to San Diego County, growing singly and in shrubby clumps, interspersed with California buckeye, ceanothus, manzanita and other chaparral. Flowering March to May, according to location. Honey is light and of good flavor. Where abundant it furnishes an important source of honey for early brood rearing."

RED CEDAR (Juniperus virginiana).

Red cedar is a well known evergreen forest tree which is found on dry soil from eastern Canada to South Dakota and south to Florida and Texas. It reaches a height of 80 to 100 feet and a diameter of four or five feet, under favorable conditions. It is more commonly seen as a small tree and is much prized as an ornamental, and in the middle west for windbreaks.

The wood is valued for fence posts because of its long life; it is also used for household chests and for pencil wood.

Honeydew is sometimes gathered from red cedar where there are extensive plantings or forests. In the vicinity of Starkville, Miss., in February, 1920, there was a heavy flow of honeydew from this source.

According to W. A. Stephen of the Central Experiment Farm, bees were working freely on the cedars in the vicinity of Ottawa, Canada, the first week in June, 1942. Honeydew was excreted by an insect identified by the Division of Entomology as *Lecanium fletcheri*, Cockerell.

The tree is also known as Savin and Juniper.

RED CLOVER (Trifolium pratense).

There have been so many conflicting statements regarding the question as to whether or not the honeybee is able to secure honey from red clover (*Trifolium pratense*) that it has seemed worth while to investigate the subject with some care. There have been so many reports of honey from this source, that it is desirable to learn whether the honey did come from red clover,

Red clover blossoms.

or whether the beekeepers have been mistaken, and some explanation of the confusion is necessary. There is no question but that the plant secretes nectar in abundance, but since the corolla tubes are much longer than the tongues of the bees, they are unable to reach it under ordinary conditions. It is a well-known fact that plants behave very differently under different climatic conditions, so an effort has been made to secure evidence from as many localities as possible, and from a great variety of conditions.

In Iowa the writer has sometimes found bees working freely on red clover in extremely dry seasons. At such times the bees were apparently getting some nectar, although it could not be detected in the hive. However, one year, Mr. C. H. True, of Edgewood, Iowa, had on exhibition at the State Fair, a generous quantity of honey which he thought was secured from red clover. It was slightly tinted with red, and had a flavor different from white or alsike clover honey. The explanation often given is that in dry seasons the florets are somewhat dwarfed, and because of the shorter tube the bee is able to reach the honey. Dr. L. H. Pammel, botanist at the Iowa College of Agriculture, has made a special study of bees and red clover under Iowa conditions. After having many measurements made, he has reached the conclusion that the effect on the length of the corolla tube, as a result of drought, is so slight that the bee would not be able to reach the nectar from this cause. He goes on record as follows:

> "I have for several years closely observed honeybees and red clover, and from these observances I am still inclined to the opinion, earlier expressed, that honeybees do not get nectar from the flowers of the red clover, notwithstanding the opinion of many beekeepers in Iowa."—Third report Iowa State Bee Inspector.

At the 1917 Convention of the Illinois Beekeepers, Mr. Frank Bishop, of Virden, reported that one season he secured an average of 100 pounds per colony from red clover. According to his statement, there was no other bloom within reach at that time. He further stated that he visited the red clover fields, investigated the matter carefully, and was fully satisfied that red clover was the source of the honey.

Mr. Wm. McEvoy, of Woodburn, Ontario, wrote in Gleanings in Bee Culture, page 468, 1907, as follows:

> "In September, 1905, I extracted over 3,000 pounds of pure red clover honey, after giving the bees plenty to winter on. This honey was a light amber color, and good in flavor, and sold for the same price as honey gathered from white clover. My bees being Italians, worked well on the second crop of red clover, which was not injured by the midge in my locality, in 1905, on account of the first crop being cut early."

Adrian Getaz, of Knoxville, Tenn., makes the following contribution to the subject in Gleanings, page 660, 1909:

> "In regard to bees gathering nectar from red clover, several opinions have been advanced. Generally, it is supposed, that, owing to dryer weather,

the second crop has blossoms with shorter corollas, and that the bees can reach the nectar on that account. Another theory is that the nectar is more abundant, and fills up the corollas better, and thus comes within reach of the bees. A German apiarist a few years ago undertook to settle the matter, and spent a part of the summer lying down in the clover fields to see how it was. He reported that very few insects take the nectar through the corollas; but some kinds cut a hole near the bottom and help themselves through it. The hole once made, a number of insects, including bees, take advantage of it; and if the bees do not work on the first crop, it is because there are few hole-boring insects present."

Here follows a brief report with nothing to indicate whether the bees were seeking nectar or pollen:

"Last year was very dry and there was scarcely any white clover in blossom here; but the bees went fairly wild on the red clover, and it was the first crop, too."—J. F. Brady, Deerfield, Minn. Gleanings, page 149, 1911.

That the subject is not new will be found by examining the files of the bee magazines of many years ago. Apparently, it has been a controverted subject since beekeeping has been followed seriously in America. In the first volume of the American Bee Journal, page 228, 1861, we find the following:

"I noticed in August and the beginning of September, while the bees were gathering honey from buckwheat, that they obtained pollen of a brownish color from some source. On investigating the matter, I found that they collected it from red clover. This somewhat surprised me, as I had never seen them gathering honey from the red clover to such an extent, particularly while other forage was plenty. * * * I have also noticed that the bees visited only those heads that were imperfect, the tubes being shorter in consequence."

The principal interest attached to the above is the statement that the bees visited only the imperfect blossoms. On page 9 of the same volume is a statement somewhat similar, reported in one of the German journals, of Italian bees getting honey from red clover in 1858. It is said that the season was very dry and the blossoms somewhat smaller as a result.

In 1899, page 15, American Bee Journal, we find another report of bees working on it in dry weather:

"My bees work more or less on it almost every year during hot and dry weather, but it does not produce as fine honey as white clover; when candied it is coarser grained, and has a water-soaked appearance. I wish that my bees would let it alone, for we have plenty of white clover when the red is in bloom."—Fred Bechle, Poweshiek County, Iowa.

Again, on page 27 of the same issue, Theo. Rehorst, of Fond du Lac County, Wisconsin, reports:

"The mammoth red clover produces good honey and all our honeybees can reach the nectar, although the corolla is far longer and deeper than the common red clover. I never saw any honey from common red clover; only thin, red stuff, thin as water."

In 1903, E. E. Hasty, of Ohio, wrote, in the American Bee Journal, that while he admitted that bees worked on red clover at times, he was extremely doubtful about their ability to get much honey from it. The same doubt has been expressed by numerous observers from time to time, the usual explanation being that the bees are gathering pollen, rather than nectar.

On page 49 of the 1903 volume of the American Bee Journal is reported an interesting case of honeydew from red clover. Since it is the first case of the kind found in all the literature consulted it is quoted quite fully:

> "For about ten days my bees have been bringing in honey from the second crop of red clover. Now this is nothing remarkable, for I have seen them doing so for more than twenty years past; but recently, passing through a field of red clover in bloom, I stopped to watch them, and, to my surprise, found them working, not on the blossoms, but on the leaves. This, I confess, I had never seen before. On closer examination I found the clover leaves covered with small plant lice, and the under leaves covered with honeydew, very similar to that frequently found on the leaves of the hickory, oak and other trees, though the honey is not so dark-colored as from leaves of trees."

On page 839 of the American Bee Journal for 1906 is found a rather convincing discussion of the subject of honey from red clover. It was at a convention of the National Association, and several men of wide reputation took part in the discussion, and testified to the fact that they had secured surplus from red clover. Hutchinson stated that he had secured 500 pounds from red clover at a time when there was nothing else in bloom, and that it was a light amber or dark white color. Messrs. Townsend, Stone, Davenport and others agreed that they had secured red clover honey, Townsend reporting as much as 2,000 pounds stored in two weeks' time.

The subject is discussed at length in Bulletin No. 46 of the New Zealand Department of Agriculture, by Isaac Hopkins, whose experience in this connection is interesting. We quote him in part:

> "In my early days of beekeeping it was a moot point whether Italian bees worked on red clover or not. At this time I had a unique opportunity of testing the matter thoroughly, an opportunity which would rarely occur; therefore, I feel myself on safe ground when dealing with Italian bees and red clover.
>
> "For five years (1882-87), I was located on the late J. C. Firth's estate at Matamata, where I started large bee farms. My bees, which were chiefly Italian, were near to thousands of acres of red clover. * * * Now and again we saw a few here and there gathering pollen from the blossoms, and sometimes a good deal of pollen from red clover was brought in when, no doubt, it was scarce elsewhere.
>
> "In order to make a thorough test, I shifted, on one occasion, a number of strong two-story colonies to the center of a 700-acre paddock of red clover. The first crop had been cut for hay, and the second crop flowers were just opening. There was no ordinary bee forage anywhere near. After the fourth day, I examined the hives and found from the odor that came

from them on removing the covers that some nectar had been gathered from the surrounding clover. I also observed that some clover pollen had been stored.

"There were two seasons out of the five when my bees worked more freely on the red clover than in others. In these seasons it was noticeable that myriads of small-sized moths flitted about the clover, while they were rarely seen at other times. I was much interested, and in casting about for a reason, I became satisfied, after very many tests, that the red clover was secreting at times much more nectar than usual, and it may have been that it reached a higher level in the tubes on these occasions, and so came within reach of the tongues of the bees. Be that as it may, some red clover nectar was gathered from second crop flowers in these seasons."

In the annual report of Illinois Beekeepers' Association for the year 1925, A. C. Burrill gives an account of a harvest of honeydew from red clover in South Idaho in 1916. He states:—

"Crystallizing honey-dew was on almost every blossom before the middle of August, and so much honey-dew was stored in the combs of leading apiaries (as E. F. Atwater, Meriden Idaho), in that vicinity, a wine red nectar, making red tinted honey, that few questions were asked as to other sources of sweets, when thousands of bees were working the clover fields.

Bulletin 391, Colorado Experiment Station, by R. G. Richmond, states that honeybees are a major factor in the pollination of red clover and that red clover is a convenient and prolific source of pollen for bees in Colorado. It states further that the first cutting sets a good crop of seed when conditions are favorable for the honeybees.

Extension Bulletin 253 Ohio State University states that red clover blossoms are 82% pollinated by honeybees.

The Zofka Clover

Dr. Joseph Zofka of Prague spent about twenty years in an effort to develop a red clover with a short corolla tube which would permit the honeybees to reach the nectar. Reports from Europe indicate a considerable measure of success and that his clover yields more hay as well as more seed.

In the spring of 1937 a plot of one tenth of an acre was planted in the American Bee Journal test garden from seed secured direct from Doctor Zofka. Germination was satisfactory and a good stand was the result. The plants began blooming in about six weeks. Soon after the clover began blooming, Dr. J. N. Martin of Iowa State College made measurements to ascertain how much variation there might be in the flower tubes. He found some as short as five millimeters with others as much as eight millimeters in depth. The average was about six millimeters whereas ordinary red clover is from nine to eleven.

An examination of the flowers disclosed that the nectar was about two millimeters in depth in the tubes of the Zofka which was about double the amount present in common red clover.

Over a period of five years the Zofka clover attracted the bees freely at times while they deserted it at others. It yielded seed freely but proved to be unequal to our winters and we found great difficulty in maintaining a stand.

It appeared to be very palatable as the rabbits sought it so eagerly that we found difficulty in protecting it from them. No other clover in the plots was equal to it in attraction to the rabbits.

RED DAISY, see Hawkweed.

RED GUM, see Eucalyptus.

RED-HAW, see Hawthorne.

RED MAIDS (Calandrinia caulescens).

Red Maids is an annual native to the western United States. It is one of the common early spring wild flowers of the Arizona deserts and is reported also as common in cultivated fields west of the Cascade mountains in Oregon. Scullen and Vansell report that it yields considerable yellow-amber nectar in the afternoons only.

REDROOT (Gyrotheca tinctoria).

The red-root here described should not be confused with New Jersey tea, also called red-root. (See New Jersey Tea). *Gyrotheca tinctoria* is an herb with colored rootstock common in swampy lands near the coast from Massachusetts to Florida. The flowers are yellow and loosely woolly in terminal corymbs and appear from June to September.

From southern Florida it is reported as the source of considerable honey, yielding for a long period. Bees are said to work on the plant all day long. The honey is described as of unpleasant taste and of decidedly poor quality. There are reports of palmetto honey spoiled by a mixture with honey from this source.

RESEDA ODORATA, see Mignonette.

RETAMA (Parkinsonia aculeata).

Retama is a small tree common throughout southern and western Texas. It has slender branches, bearing the yellow petaled flowers in axillary racemes. It is frequently mentioned as a source of nectar by Texas beekeepers. Scholl states that the bees work on it more or less all summer. Like many Texas shrubs, it has a habit of blooming at irregular periods, from spring till September.

The retama is sometimes known as horse bean and is found also in the Colorado River Valley in Arizona and California. While there are frequent references to it as attractive to the bees, the author has nowhere found it regarded as important as a source of surplus honey.

RHODE ISLAND—Honey Sources of.

The sources of honey in Rhode Island grouped in the order of their appearance are willows, maples and other less numerous trees, which furnish bees with the early supply of pollen and honey, so useful and so needful in building up the bee population preparatory to the harvest in which the beekeeper shares.

Next come the fruit blossoms, plum, peach, cherry, pear, apple, raspberries, huckleberries and blueberries which, when the spring is favorable, yield good crops of honey. In some places, dandelions are an important addition to the fruit bloom, though not always opening at the same time. In several parts of the State there are large areas of locust. This blooms the latter part of May, and when conditions favor, yields for about eight days a heavy water-white honey. The clovers usually follow this, but are of consequence only under favorable conditions of rainfall, save in a few sections where soil conditions afford abundant moisture.

Blossoms and leaves of Retama.

In many sections sumacs furnish the next crop, and where they are sufficiently abundant the beekeeper may rightly look for a good crop of a very fair honey.

In some of the more swampy and less settled sections, button bush, clethra (sweet pepper bush) and clematis yield a white and highly-flavored honey, that from clematis being of the very highest quality. But the yield from these plants is irregular, in some years being almost absent.

In some of the villages and cities the European lindens are numerous and yield heavily. The bloom ceases toward the end of the clover flow, though the time of flowering of different trees in the same neighborhood varies greatly. Native linden, basswood, is now found only in a few places. The season closes with the goldenrods and asters, which yield a rich, aromatic honey which, though not acceptable to many persons, commands a fancy price from others. The crop from these two sources is not always to be depended upon, being more affected by the weather than some of the others.—Arthur C. Miller, Bulletin, State Board of Agriculture.

RHODODENDRON, see Azalea.
RHUS, see Sumac.
RICHARDIA, see Mexican Clover.
ROBINIA, see Locust.

ROCKBRUSH (Eysenhardtia amorphoides).

Rockbrush is a small shrub common to southern and western Texas, and extending into Mexico. It blooms after heavy rains, several times during the year, and yields honey in surplus quantity. It is reported frequently throughout the region south and west of San Antonio to the Rio Grande.

Colubrina Texensis, an entirely different shrub, which is common from the Colorado River to the Rio Grande and west to New Mexico, is also known by the name of rockbrush. It is reported as yielding both pollen and nectar, but not as a source of surplus honey.

ROCKY MOUNTAIN BEE PLANT (Cleome integrifolia).

The Rocky Mountain bee plant, also known as stinking clover, is principally confined in its distribution to the plains region west of the

Cleome, or Rocky Mountain bee plant.

Missouri River. It is also reported from north Pacific Coast States. While it is a dry land plant, it is occasionally reported from Illinois, Iowa and Minnesota. In Iowa on the Missouri River bluffs, it is plentiful in some localities. This plant is reported as especially valuable in Colorado, where it is said to produce considerable quantities of honey.

It is an annual with large, showy, pink or purple flowers. At one time there was much interest in this plant on the part of eastern beekeepers who tried to introduce it by sowing seed. At the Michigan Agricultural College a small field was planted to ascertain whether it could be grown profitably for honey alone. It is acrid and pungent and said to be distasteful to animals, which seldom eat it. If the plant had any value for any purpose besides honey production, an effort to extend the area of the distribution might succeed, but the introduction of plants that are essentially weeds in their nature seldom meets with favor.

According to Frank Rauchfuss, the cleome is erratic in its yield. If there is a wet spring the seeds germinate early. When this is followed by good rains in June the plants are vigorous and spreading in their growth and each will have many blossoms. One year he extracted an average of 116 pounds per colony from a ten-days' flow. The honey is white in color, with a greenish tinge. It has a rather sickening flavor when fresh, but improves with age. When pure it is a first quality honey. It is rare that a good crop is secured from his source.

The plant thrives best on sandy and gravelly soils. See also cleomella and spider plant.

In American Bee Journal, April 1939, N. Pankiu wrote that this plant yields nectar very heavily in cool and cloudy weather, especially when the clovers do not secrete in Manitoba. In his home apiary the bees work cleome from about five A.M. to about ten A.M. when the clovers start to yield. The bees did not work cleome continually but only when the sweet clover was not yielding but this yard gave a higher return than outyards with no cleome in reach. Brood rearing continued much later also.

ROMAN CANDLE, see Yucca.
ROSA DE MONTANA, see Coral Vine.

ROSE (Rosa).

There are many species of roses of wide distribution. They yield pollen abundantly and are frequently valuable, in localities where they bloom when pollen is scarce. There are numerous reports of rose honey, but in most cases the observer has probably mistaken the object of the bees when working on the flowers.

"Some time ago quite a discussion was brought about by the assertion of Gaston Bonnier, that one never saw bees upon roses, no matter how colored or how fragrant. Dr. Miller replied that he had often seen them upon the crimson ramblers and that they even tore the buds open.

"The magazine 'L'Abeille de l'Aube,' in its August number quotes the different assertions which were made upon the subject since then in Europe.

"Mr. Bonnier came back with the assertion that the bees were only hunting pollen, for, according to him, there is no nectar in roses.

"Joan Ruppin, of Fountenay-Aux-Roses, saw his bees take pollen on the roses, but never any nectar.

"A. Martinot saw the bees often on the crimson ramblers and on other similar roses, never on the double flowers.

"Mr. Pitrat believes they find both nectar and pollen on the simple flowers.

"Louis Rosseil, Consul of Belgium in Athens, says that in the Island of Eubia, the bees work upon fields of roses and produce a white honey much esteemed."—American Bee Journal, October, 1912.

Prof. Hermann Muller in his book, "Fertilization of flowers" states that "Rosa rubiginosa produces an obvious secretion of honey."

Beekeepers in the Rocky Mountains of Colorado and on the Pacific Coast report that their bees get honey from roses.

Frank Stirling states that the swamp rose (*R. palustris*) is found in low, damp lands of Florida, blooming very early in spring, and is an important source of pollen for early brood rearing.

Charles E. Bessey, in his "Honey Producing Plants of Nebraska, lists both wild and cultivated roses as the source of both honey and pollen.

The Cherokee rose (*Rosa laevigata*) is found from the Coast Region of the South Atlantic States westward through Georgia, Alabama, Mississippi and Louisiana to eastern Texas. It occurs also in California. It was in Alabama that the author first heard it mentioned as a source of nectar. Later reports from California indicated that bees sometimes find nectar from this source, though most of the roses yield only pollen. H. B. Parks states that he has observed the bees getting honey from this source, and M. B. Talley, of Victoria, Texas, states that he has observed the bees gathering nectar from it in Mississippi and also at Victoria.

Everett George of Independence, Oregon wrote to the author that once in about five or six years he got a honeyflow from wild roses in that part of Oregon. He stated that while the bees visit roses every year it is only on rare occasions that there is a honeyflow from that source, sometimes as much as seven years passing between such flows. The flows come in June in seasons following a May which is cold with heavy rainfall. In 1933 some of his colonies stored as much as a hundred pounds from wild roses.

He described the honey as having the aroma of wild roses and being of a dark red color and good quality. (American Bee Journal, page 370, July 1936.)

ROSEMARY (Rosmarinus officinalis). Old Man.

Rosemary is native to the countries around the Mediterranean Sea, extending to Switzerland. It is a hardy evergreen shrub and well known garden plant with aromatic leaves, used in seasoning. In our grandmothers' gardens it was often known as "old man." The small light blue flowers are much sought by the bees. So famous is it as a bee plant that nearly every description one reads of it makes mention of its attraction to these insects. Oil of rosemary distilled from its foliage is used in certain perfumes and other preparations.

In this country it is not extensively grown, though often found in gardens. In southern California it is recommended for hedges, especially for dry places near the coast.

Branches of rosemary are often packed away with wearing apparel, because it is reputed to keep the moths away.

In the American Bee Journal of June 24, 1897, page 390 the following quotation from Frank Benton appears:

> "Rosemary of the mint family is an immense yielder of exceedingly fine honey. The honey of Narbonne, so famous in the Parisian markets, is said to come from rosemary. During two winters which I passed in Tunis, North Africa, the hills were blue with the fragrant rosemary, and though the bees were black as night and sometimes cross, they revelled in this royal flower and weighted their hives with its delicious nectar."

ROSIN WEED, see Cup Plant, also Gum Weed.
ROYAL PALM, see Palm.
RUBUS, see Raspberry.
RUDBECKIA, see Coneflower.
RUSSIAN OLIVE, see Oleaster.

RUSSIAN THISTLE (Salsola pestifera).

Russian thistle is a widely distributed annual weed which came originally from western Asia. It is especially common in the dry plains region. It belongs to a group of plants in which one would hardly look for important bee pasture.

Mr. C. E. Ghehiere of Williams, Montana, reports that on three occasions his bees have gathered honey from this plant. In each case a killing frost in September was followed by a period of fine weather when the bees worked on Russian thistle. The first was in 1918. Again in 1923 the bees found pasture on this plant when they harvested a surplus of about fifty pounds per colony. The third time was in 1939 when an unusually long period of mild weather followed a September freeze. Some honey was again secured.

Honey from Russian thistle is unusual but an occasional report of honey from unexpected sources indicates that given the right conditions something may be secured from a great number of plants not commonly known to yield nectar.

S

SABAL, see Palmetto.

SAFFLOWER (Carthamnus tinctorius).

Concerning safflower grown at the California Experiment Station at Davis, George H. Vansell writes in a letter to the author:

"A plot of it grown here at Davis by the agronomy division attracts a greater number of bees in comparison to any other plant available at the present time. It produces an abundance of both nectar and pollen. An individual floret in the compound head produces so much nectar that it fills up the tube and runs out onto the bases of the petals. At times there are as many as eight bees per square yard estimated to be visiting this plot, which simply hums with activity. The nectar is quite rich in sugar, exceeding by 10 to 15% samples taken from neighboring alfalfa fields.

Safflower is an annual native to Asia grown for red dye and also for drying oil used in the paint trade. It is grown in considerable acreage in a few localities, notably New Mexico.

In the American Bee Journal test garden at Atlantic, Iowa, in the season of 1941 the growth was satisfactory, seed yield abundant and the plant proved attractive to the bees. Where grown commercially it is a promising source of bee pasture.

SAFFRON (Crocus sativus).

The dye-saffron is native to southeastern Europe and Asia Minor. The costly dye substance comes from the stigmata of this crocus which is produced in large quantity in Spain and other mild climates.

It has been noted as a bee plant since ancient times and in regions where the plant is grown commercially the beekeeper must profit.

SAGE (Salvia).

When sage is mentioned, we of the east are likely to think of the common garden sage (*Salvia officinalis*), which for at least three centuries has been cultivated for its aromatic leaves. Of this there are several varieties, some with broad and some with narrow leaves. The garden sages are good honey plants, but seldom sufficiently abundant to amount to much as honey producers. The honey from the garden sage is said to be nice and white like that from catnip or motherwort.

Concerning salvia officinalis E. O. Peterson writes in New Zealand Smallholder, Jan. 16, 1935:—

"The honey yielded in October to December, is very pale amber, rather heavy in body, and has the peculiarity when gathered in the pure state, it is said, of remaining in liquid form for a lengthy period. Sage occurs as a garden plant all over New Zealand, and in many places as a garden escaped plant is fairly plentiful."

The name sage is derived from its supposed power to make people wise by strengthening the memory, for which it was used in ancient medicine.

There are upwards of five hundred species of sages, widely distributed in the temperate and warmer regions of both hemispheres. Probably most of the species yield honey, although but few are known to be important. Rayment mentions the wild sage (*Salvia verbenacea*) as introduced into Australia from Europe, but now yielding honey during the dry months of the year. (Money in Bees in Australasia.) There are more than two hundred species known to occur in Mexico and Central America and it is very probable that when beekeeping is developed on a commercial scale in those countries, the sages will be found to be very important honey plants.

It is probable that one or more species of sage occur in nearly every State, but they increase in abundance westward. It is quite likely that sage honey in small amounts is mixed with honey from other sources, and so not detected, in many many localities outside of California. The fact remains, nevertheless, that sage, as an important source of surplus, is not reported outside of that State.

The quality of sage honey is of the best, being water-white in color, of a heavy body and delightful flavor. Since it does not granulate, it is much sought for by bottlers in the east, who blend it with clover or alfalfa. There are many who regard sage honey as the finest in the market. In this connection A. I. Root, in an early edition of his A B C of Bee Culture, wrote:

> "I well remember the first taste I had of the mountain sage honey. Mr. Langstroth was visiting me at the time, and his exclamations were much like my own, only that he declared that it was almost identical in flavor with the famed honey of Hymettus, of which he had received a sample some years ago. Well, this honey of Hymettus, which has been celebrated both in prose and poetry for ages past, was gathered from the mountain thyme, and the botany tells us that thyme and sage are closely related."

Although there are several species of sage which yield honey in California, the quality does not differ materially, as far as can be ascertained from printed reports. It is all described as "water-white, unexcelled flavor, of heavy body and does not granulate."

Prof. A. J. Cook wrote to the American Bee Journal (June 21, 1906) concerning the sage as follows:

> "Chief among the honey-bearing mints are the incomparable sages of California. These are not excelled even by the clovers or linden. The honey is white, delicate of flavor, and must ever rank among the best in appearance and quality. Not only this, but the quantity is often phenomenal. This comes from the fact that flowers are borne in long racemes of compact heads, and as the separated flowerets do not bloom all at once, but in succession, the plants are in bloom for weeks. The sages, then, are marvelous honey producers, first, because of the generous secretions of each floweret, and second, because of the immense number of these flowerets and the long period of bloom."

White sage is uncertain in its yield.

At another time Mr. Cook wrote that the honey from all sages is so much alike that it would be indistinguishable.—American Bee Journal, August 3, 1905.

Richter, in his Honey Plants of California, speaks of the white sage (*Salvia apiana*) Jep., as "very common on the dry plains toward the foothills, and ascending these to about 3,000 feet."

Writing in Gleanings in Bee Culture, P. C. Chadwick describes a journey which he made in the San Bernardino Mountains with the intent to find out the highest elevation at which bloom could be found in sufficient quantities to support bees. Up to an elevation of 7,000 feet he found white sage in abundance, and all alive with bees. (Western Honey Bee, September, 1914.) Richter gives its range as common from Santa Barbara County southward, blooming from April to July. "As abundant as the black sage, but not as good a yielder, nor has the honey as fine a flavor."

Black sage (*Salvia mellifera*), Greene, also known as ball sage, or button sage, is generally credited as being the principal source of sage honey, most of the honey which goes to market under the name of white sage being produced from this plant. Quite probably it is the best honey plant on the Pacific Coast. Richter says of it: "As a general rule, every fifth year an excellent crop is obtained, and every third or fourth year a total failure is experienced, the flow being dependent upon winter rains, with warm spring quite free from cold winds and fog. When in bloom a certain amount of warm weather is required before it produces nectar."

The range of black sage is given as "Mt. Diablo, Los Trampas Ridge, near Hayward, San Mateo County, Glenwood and Brieta, southward to southern California. April-May." (Jepson). "Coast ranges and ascending to 5,000 feet in the San Bernardino Mountains. March to June. San Diego County, February to May."

Mr. J. E. Pleasants, of Orange, California, writing in American Bee Journal (June, 1914) describes the peculiarities of the sages as follows:

> "The black sage is king of them all. When climatic conditions are favorable I think black sage can be relied upon to produce more 'gilt edge' than any other plant in the West, and for body and flavor it is hard to excel. It blooms for weeks. The blossom is small and inconspicuous, but what a flow of nectar it can yield.
>
> "The white sage is a much prettier plant. Its soft, grey leaves and tall blossom spikes make it quite showy, while its pleasing aromatic odor breathes the very essence of wild perfumes. But this queenly plant is much more inconstant than its plainer sister. Some years it produces a good harvest, others very light.
>
> "The silver or purple sage, which has silvery leaves and brilliant light purple blossoms, is usually a good producer, but is much restricted as to locality."

The purple sage (*Salvia leucophylla*), also called white-leaved sage, or silver sage, is reported as a good yielder, although not as abundant as either of the foregoing species. The Richter catalog gives the range as occasional in the foothills of the Santa Monica and San Fernando Mountains, April to July, and from San Luis Obispo to San Diego Counties and not extending inland beyond the coast ranges.

Black sage is the finest honey plant on the Pacific Coast.

Salvia amabilis, loving sage, is reported from Santa Barbara, March-June, but probably not important.

Salvia carduacea, thistle or annual sage. "Inner coast range valleys and throughout the San Joaquin Valley, southern California, June." (Jepson). "A well-known honey plant." (Richter).

Salvia columbariae, annual sage. "Throughout the coast ranges, Sierra Nevada and southern California, on hill and mountain slopes." (Jepson).

The annual sage is known as "chia" and blooms in April and May. It is reported as an important source of surplus wherever the plant grows in abundance. Like the other sages, the honey is light in color and of fine quality.

Salvia sonomensis, Greene, creeping sage. "Montana species at middle altitudes. Ramona Mountains west of Calistoga, Mt. Shasta, Calaveras and Mariposa Counties, San Diego County. May." (Jepson). "Also June, Sierra foothills from Sierra to Inyo Counties, main source of honey in many districts." (Richter).

Concerning the regularity of yield from sage, Mr. P. C. Chadwick wrote in Gleanings in Bee Culture (January 1, 1911), as follows:

> "South of the Tehachapi Mountains lies practically the entire sage of our State, notwithstanding eastern people and many of our westerners term every form of small growth on the vast slopes of the Rocky Mountains 'sage brush.' There is no denying that the button (or black) sage is, of all honey plants, our chief surplus producer. Neither does it average a crop more often than every other year, regardless of rainfall, for it seems necessary, from its semi-arid nature, to be dried out or rested before it comes back to its prime yielding condition. I have seen it return some surplus for three consecutive seasons; but the middle season was not what could be considered a crop, even after a sufficient rainfall."

Again he writes to the same journal to the effect that the sage ranges soon give place to other crops (Dec. 15, 1911):

> "If I should predict that thirty years hence the sage ranges of California would be almost a thing of the past there would doubtless be criticism of my views; but I firmly believe that we shall face such a condition, for emigration to this part of California is increasing rapidly. Hillsides are yielding to the plow, where twenty years ago it would have been thought almost impossible."

Some writers give two hundred pounds per colony as a fair average in a good sage year, so that with even one good year in three it comes well up with the yield of many plants more constant in their production.

Scullen and Vansell, (Bul. 412 Oregon Ex.'Sta.) report desert sage, (*S. carnosa*) as found in limited areas on sandy and gravelly soils in Oregon and yielding only in moist years.

Among the ornamental species commonly grown in gardens, *Salvia superba* is especially attractive to the bees. *Salvia silvestris* and *Salvia pratensis* are likewise swarming with bees during their period of bloom.

SAHUARO, see Giant Cactus.

SAINFOIN or ESPARCET (Onobrychis sativa).

Sainfoin is grown to a limited extent in Canada, and although given numerous trials in various parts of the United States, has not, till the

Purple sage.

present, succeeded in establishing itself successfully. It is a splendid forage crop, somewhat similar to alfalfa, and is an important source of nectar in Italy and other parts of Europe.

Several years ago it was given an enthusiastic endorsement as a honey plant for America by John Fixter, at that time apiarist at the Central Experiment Farms at Ottawa, Canada. A test plot at the Farms proved so attractive to the bees that it attracted much attention on the part of eastern beekeepers. He reported that it yielded nectar in the morning and that the bees began work upon it fully an hour before they did on alsike or white clover. A further advantage was stated to be that the first bloom came between fruit bloom and white clover and the later bloom at a time when there was a dearth from other sources. On plots side by side he reported that there would be something like a hundred bees on the sainfoin to ten on the white clover.

Flower and leaf of sainfoin.

The following article by C. P. Dadant, reprinted from the American Bee Journal, page 790, 1904, will give some information about the value of the plant in European countries:

> "Sainfoin, otherwise called esparcet, is widely cultivated in Europe, especially in France. Its name, "sainfoin," is French, and literally means "healthy hay"—sain-foin. It is a perennial, gives a splendid hay crop, and in some sections of the European continent it is a first-class honey-producer.
>
> "The small province of France, formerly called "Gatinais," is the leading producer of sainfoin honey. According to the best authorities, the honey of Gatinais has the reputation of being of the whitest color and sweetest taste, and is said to be in no way surpassed by white clover honey.
>
> "Gaston Bonnier, the eminent professor who was President of the International congress of beekeepers at Paris in 1900, says in his book, the 'Cours Complet d'Apiculture,' that sainfoin honey is one of the best appreciated grades. He ranks it next only to the honey of the Alpine hills of eastern France and Switzerland.
>
> "From immemorial times the honey crops of Gatinais have been considered as leading in the amount of production, and this was all credited to the sainfoin, which is grown there in immense quantities, somewhat as alfalfa is grown in the irrigated plains of the West."

G. F. Pearcey of British Columbia reported a field of sainfoin of about forty acres in extent that has stood for forty to fifty years. (American Bee Journal, Aug. 1935, p. 385.) He reported that it does not winterkill as easily as alfalfa and that it had withstood temperatures as low as forty degrees below zero. He found it a valuable plant with the bees swarming on the blossoms during its period of bloom.

In the American Bee Journal test garden at Atlantic, Iowa, several strains of sainfoin have been under test for several years. The bees visit the bloom constantly from the time the first flowers open until the last ones fade. It blooms between dandelion and white sweet clover, overlapping to some extent the white Dutch flowering time. With us it does well but is short lived. Since it yields only one good cutting of hay it is not likely to prove popular with farmers who are accustomed to cutting two or three of alfalfa.

Flowers of salal.

For the beekeeper it would prove a valuable acquisition if it should come into popular favor among farmers.

SALAD-TREE, see Red-Bud.

SALAL (Gaultheria shallon).

Salal is an evergreen shrub, one to five feet high, common from central California north to Washington. It grows abundantly in the redwood forests commonly covering the ground. The flowers occur in racemes three to six inches long and are pink or pinkish white.

It blooms with salmonberry, wild blackberry and chittam, so the honey is seldom secured unmixed.

SALIX, see Willow.

SALMONBERRY (Rubus spectabilis).

Salmonberry is common to the margins of the woods and along streams in the vicinity of the Pacific Ocean from Northern California north to Alaska. It is frequently reported as a source of honey in the northwest, especially in Oregon, Washington and British Columbia. Scullen and Vansell report that in Oregon it is an important minor source of nectar but that it blooms too early for surplus. (Bul. 412 Oregon Ex. Sta.)

SALT CEDAR (Tamarix gallica).

Sanborn and Scholl's Bulletin lists the salt cedar as a source of honey in Texas. "Common European shrub which seems to have escaped in many places in Texas." (Coulter). "On roadsides, in thickets and waste places in warmer parts of southern United States, naturalized from southern Europe." (Small.) The author has reports of this plant being common in the vicinity of Corpus Christi, Texas. Reported also as a honey plant in places in California.

It is quite generally cultivated in the southwest under the common name of "Tamarisk." In Arizona and New Mexico it has often escaped from cultivation and has become established in large areas in the river bottoms. W. R. Atkinson of Artesia, New Mexico, reports that it is very abundant in Pecos River valley within reach of his apiaries and that it yields much honey through a long blooming period from May until midsummer. It is an abundant source of pollen according to his report and sometimes an average of 100 pounds of surplus honey per colony is harvested in his valley yards, most of which comes from this source. He reports the honey of a rather low grade, dark amber in color and with a minty flavor somewhat similar to horehound.

SAND CHERRY (Prunus pumila).

The sand cherry is a low spreading shrub common to sandy grounds of the northern states and Canada. It blooms in early spring and is of some value to the bees for spring brood rearing. It is common to the sand hills of northern Nebraska and furnishes early nectar and pollen in this region where spring forage is scant.

SANDVINE, see Bluevine.
SAN MIGUELITO, see Coral Vine.
SAPINDUS, see Wild China.

SASKATCHEWAN—Honey Sources of.

The Province of Saskatchewan offers very good opportunities for honey production in its more favored sections. In the areas devoted exclusively to wheat growing there are many places where but little bee pasture is available, but in the bush country and in the neighborhoods where sweet clover and alfalfa are generally grown good crops are harvested. The province covers an immense area and much of the northern portion is unsuited to beekeeping. This is especially true in the coniferous region and in the swamps. In the central region, which is covered with mixed bush of poplar, tamarac, willow, etc., there are many good locations. Fireweed is perhaps of most importance among the native flora, with wild raspberry also yielding nectar in important quantity. The fragrant giant hyssop and several other members of the mint family, saskatoon, wolfberry and several other plants are valuable in this section. On the borders of the woodlands alsike and white clover are becoming common and yield much nectar.

Cottonwood and poplar are the source of a plentiful supply of pollen over much of the province. On the prairies, the pasque flower is the source of the first pollen available in spring. Dandelions are rapidly extending their range in the more thickly settled regions. The Siberian pea-tree or caragana is very generally planted by settlers for windbreaks and yields both pollen and nectar abundantly.

As the system of farming tends more toward live stock and less toward small grain, the acreage of alfalfa and sweet clover is rapidly increasing and the bee pasture is correspondingly improved. Large yields are gathered from both plants. Conditions are in many respects similar to those found in Manitoba and a reference to that province will give some additional information.

SASKATOON, see Serviceberry.

SASSAFRAS (Sassafras officinale).

Sassafras is an exceedingly well-known tree in the Southeastern States. It is known occasionally as far north as Ontario and Massachusetts southward to southeastern Iowa, eastern Kansas and south to the Brazos Valley in Texas. It is a conspicuous feature of the old fields of the southern plantations. It has spicy aromatic bark and mucilaginous buds and leaves which many people enjoy chewing. Oil of sassafras is distilled from the roots.

The flowers are small, yellow and inconspicuous, but are attractive to the bees. The writer has found occasional reports of bees working on sassafras in the Southern States. At Buffalo, Tex., it is reported as blooming about two weeks. H. B. Parks reports that it yields well in Missouri. The blooming period is early and short, hence its principal value is for spring brood rearing. It was blooming with willow on March 20, 1918, in Leon County, Texas, when the author visited that section, and the bees were apparently getting considerable nectar from sassafras at that time.

SATUREIA, see Wild Pennyroyal.

SAVORY (Satureia).

The summer savory (*S. hortensis*) is found in waste places from New Brunswick west to Nevada and south to Kentucky.

Summer savory is an introduced plant which has been naturalized from Europe. It is native to the regions around the Mediterranean Sea, but is cultivated as far north as Norway. It is used as a condiment and is also the source of an aromatic oil.

Richter lists winter savory (*S. montana*) as a source of nectar in California. This species is also native to arid lands near the Mediterranean Sea. While used as a condiment it is said not to equal the former species for that purpose.

Thyme is sometimes erroneously called savory. (See Thyme).

SAURURUS, see Lizard's Tail.
SAW PALMETTO, see Palmetto.

SCILLA.

The *Scilla siberica,* or Siberian squill, is an early blooming old world flower commonly naturalized in the grass of lawns, parks, etc., for ornamental purposes. It has small blue flowers which appear as soon as the ground is free from frost in the spring. The bees seek them eagerly and, coming at a time when little is to be had, they keep the bees busy on the few sunny days of early spring. Flower lovers will find this plant a desirable addition to their garden list. Aside from planting the bulbs no care is necessary. It will thrive in a stiff bluegrass sod, where few other plants would grow. It is only necessary to avoid cutting the grass till the plants have matured.

Summer savory.

SCREWBEAN (Prosopis pubescens) Tornillo.

The screwbean, also known as screwpod, tornillo and, in Nevada, as mescrew, is a small tree 15 to 20 feet in height found from western

Texas, through New Mexico, Arizona to southeastern California and adjacent Mexico and north to southern Utah and Nevada. It is a close relative of the mesquite and is frequently called by that name. The distinguishing feature of this tree is the twisted seed pod which gives rise to the name. Since these pods hang on the trees for most of the year the tree is readily recognized by this means.

It is an important source of nectar in the southwest and yields large quantities of amber honey. Some beekeepers report it as more dependable than mesquite, yielding something every year. Since it is less common, the total amount of honey secured from screwbean is not sufficient to attract attention in the markets. It is sold as mesquite honey and is not recognized as distinct. See Mesquite.

Pods of screwbean.

SCROPHULARIA, see Figwort.

SCRUB PALMETTO, see Palmetto.

SEA FIG, see Ice Plant.

SEA GRAPE (Coccolobis uvifera).

The sea grape also known as seaside plum, is a characteristic tree of the tropical coasts of America. It is found on both the east and west coasts of southern Florida, in the West Indies, the Bahama Islands, Bermuda and in South America south to Brazil.

A related species, the pigeon plum (*Coccolobis laurifolia*) is also found in southern Florida, the West Indies and in northern South America.

The sea grape probably gets its name from its clusters of grape-like berries. It has short stalked, leathery leaves which are often wider than long. The flowers appear almost throughout the year, so that the tree or bush is almost continuously in bloom. It is evergreen and grows in a bushy, spreading form. The fruit is used to some extent for jelly.

It is reported as the source of amber honey, but is probably of minor importance.

In Cuba the sea grape, there known as "Uvero" is regarded as among the most important sources of honey. Ordetx in "Plantas Meliferas de Cuba" is quoted as follows:—

> "Its flowers are small, of greenish color, arranged in terminal panicles that secrete nectar copiously from morning until well past noon; in excep-

tional cases, bees gather great quantities of nectar until five o'clock in the afternoon. It flowers from April to May but later it has another flowering period. * *

"In some years the uvero may have as many as three flowering periods. As it flowers for a long period of time it is possible that the average quantity of nectar may be greater than that of the mangrove.

"The honey of uvero is of amber color and of good quality."

SEA-LAVENDER (Limonium)

In salt marshes along the Atlantic Coast from Labrador to Texas we find large areas of sea-lavender also known as marsh rosemary or statice limonium. It is also sometimes called ink-root or canker-root. There are two species, *L. brasiliense* with white flowers and *L. carolinanum* with purple ones.

Related species commonly cultivated in gardens are often listed in seed catalogs as statice.

In a private letter to the author under date Dec. 6, 1938, Fred W. Schwoebel of Pomona, New Jersey reported placing bees on a barge

Sea-Lavender.

about two and three quarter miles from the mainland near Northfield. He had noticed the bees working sea-lavender several miles from shore and wished to determine its value for honey. He writes as follows:

> "On August 19 I placed the first of these hives. * * Two days later I found the honey just dripping from the combs. I examined them at approximately one week intervals during the next month. On some days there was practically no honeyflow and on others it was coming in strong. I removed the hives on the 19th of September and on that date new honey had again started coming in and the combs were dripping. The first hive had nearly filled a shallow super beside packing the broodnest with honey. The second hive had possibly ten pounds above the excluder. The honey is of excellent quality, better than the very good grade of honey that we get in my section of New Jersey, being light in color and mild but very pleasing flavor."

The root of sea-lavender has long been used in medicine.

SEA-MYRTLE, see Baccharis.

SELF-HEAL (Prunella vulgaris).

The self-heal, also known as heal-all and carpenter-weed is a common weed of wide distribution introduced from Europe. It is not regarded as troublesome although found in waste places and old fields from New Foundland to Florida and westward to Alaska and California. The purple flowers appear in summer and are freely visited by the bees.

There is an occasional mention of this plant as a source of honey in the bee magazines of years past. Of minor importance.

SENISA (Leucophyllum texanum).

Senisa is a low much branched shrub with silvery foliage and showy violet purple flowers which is common to southern Texas. It is also sometimes known as barometer-bush, wild lilac and ash plant.

It blooms rather sparingly for short periods during the spring months and yields principally pollen at that season. In late autumn, usually October it blooms for several weeks and often yields as much as a super per colony of surplus. Where the senisa is abundant it is regarded as a dependable source of winter stores. The honey is said to be amber without any particularly characteristic flavor.

SENNA (Cassia).

There are several species of the genus *cassia* which yield some nectar. They occur principally in the South. A specimen of *C. tora* has been received from Louisiana with the statement that the bees were working on it in September. There are numerous references to senna as a source of honey from that state. The name senna is applied to *C. marilandica* in many southern localities. (See also Partridge Pea).

SENSITIVE PEA, see Partridge Pea.

SERRADELLA (Ornithopus sativus).

Serradella is a leguminous forage plant of much value which is but little grown in this country. It is grown as an annual in cold countries but may in mild climates stand for several years. A warm moist climate is most favorable to its growth but it is said to stand drouth well and to suceed well on light soils.

Senna in bloom.

Serradella is reported to yield honey freely under favorable environmental conditions and to be an important honey plant where largely grown. The yellowish-red blossoms appear in June and last till August yielding abundantly, according to the American Bee Journal of July 1, 1897. The honey is light yellow in color and of good quality.

According to L. W. Kephart, Agronomist of the United States Department of Agriculture, the plant has not responded favorably to trials in this country. Plants grown in Michigan proved attractive to the bees but results did not encourage the use of serradella as a forage crop in preference to those already in use.

SERVICEBERRY or JUNEBERRY (Amelanchier).

There are about nine species of serviceberries to be found in North America. They are small-sized trees or shrubs, not of much importance to the beekeeper, but are occasionally mentioned as sources of nectar. W. J. Baerg, in Bulletin 170 of the Arkansas Experiment Station, lists *A. canadensis* as the source of a limited amount of nectar in March in that state. This species is known also as Juneberry or shadbush in the northeastern states. In parts of Pennsylvania it is known as Indian cherry, in Newfoundland as wild Indian pear, and in the southeast as currant-tree. Some species are known also as pigeon-berry and sarvice. H. B. Parks credits the above species as the source of both honey and pollen in Missouri.

The serviceberries have a wide range, being found from Newfoundland south to Florida and west to the mountains of Mexico. The western serviceberry is found from Alaska to southern California and eastward through British Columbia to Manitoba and Michigan. The fruits of most species are edible, having an agreeable flavor. The small size prevents their general use.

A. alnifolia is widely distributed in woodlands of the north through Manitoba, Saskatchewan and Alberta and is commonly known as "saskatoon." It is generally recognized as of some value to the bees, but probably not as a source of surplus honey. It is a very common shrub in the bush lands and, because of its abundance, is more important than is the case with the Juneberry in most localities farther south.

SHADBUSH, see Serviceberry.
SHEEP LAUREL, see Laurel.
SHITTIM-WOOD, see Gum Elastic.
SHOESTRING VINE, see Bluevine.
SHOOFLY, see Indigo-weed.
SHRUBBY TREFOIL, see Hop-Tree.
SIBERIAN SQUILL, see Scilla.
SILKWEED, see Milkweed.
SILPHIUM, see Cup Plant.
SILVERBERRY, see wolf willow.
SIMPSON'S HONEY PLANT, see Figwort.
SKUNKBUSH, see Sumac.

SKUNK CABBAGE (Symplocarpus foetidus).

The skunk cabbage is found in wet places from Nova Scotia to North Carolina and west to Iowa. It receives its name from the strong odor which it gives off. Skunk cabbage is one of the very first plants to bloom as frost is leaving the ground in spring, and its principal value to the beekeeper comes from this early appearance. The late G. M. Doolittle wrote (Gleanings, 1909, page 200) that he had seen the bees gathering pollen from this plant when the temperature stood at 42 degrees. He stated further that he valued it more highly than any other pollen-yielding plant or tree, and that there was nothing with which he was familiar so eagerly sought by the bees, nor any source of pollen which so greatly stimulated brood rearing.

SKUNKWEED, see Marsh Fleabane.
SMARTWEED, see Heartsease.
SNEEZEWEED, see Bitterweed.

SNOWBERRY (Symphoricarpos racemosus). WAXBERRY.

The snowberry is a low branching shrub with conspicuous white berries which hang on through most of the winter. It is of wide distribution in the Northern States from New England and Pennsylvania to the North Pacific Coast. It is also common in the hill country of much of California, in the White Mountains of New Mexico and in parts of Colorado and other mountain States. The blossoms are small and inconspicuous, but they are attractive to the bees, nevertheless. The snowberry is a valuable plant in many localities in the West, but is of special importance in Washington and British Columbia. There it is reported as blooming during the last half of June and well into July and as furnishing an important secondary flow. According to H. A. Scullen it is very important in northern Idaho and in Stevens County, Washington, the bees work upon it in preference to white clover.

In a letter from Leslie H. Walling of Cedar City, Utah, he stated that in 1941 to prevent swarming, he reduced an apiary of 100 colonies in a location where alfalfa was the principal source of surplus, each to five frames of brood and bees, in May. Two weeks later he noted a heavy flow of nectar which was traced to snowberry in bloom two miles back in the mountains. In spite of the weakened condition of the colonies he harvested 75, sixty pound cans of snowberry honey. The honey was extra light in color and of excellent flavor. He reports that usually his honey from snowberry is spoiled by mixture with honeydew harvested from quaking aspen which is abundant in his snowberry locations.

Snowberry honey is slow to granulate. In a later letter Mr. Walling reported that a sample of honey from this source gathered in June 1941 was still liquid in March, 1943.

SNOW BRUSH (Ceanothus velutinus).

The snow brush is one of the mountain lilacs common to the coast ranges of California, found also in mountain regions from South Dakota to British Columbia. It is a large shrub eight or ten feet in height, noted for its pleasant odor. This is one of a group of plants which yields nectar freely. (See Mountain Lilac and Buckthorn for description of related species). This species is reported as the source of white honey of good flavor and also as the source of pollen in spring or early summer.

In western Washington this species is known as fragrant laurel. There it occupies areas of sandy and gravelly soil and is an important source of nectar in early summer.

SNOW-ON-THE-MOUNTAIN (Euphorbia marginata). SPURGE.

Snow-on-the-Mountain is a showy plant easily recognized by the white margined leaves. It is native from Minnesota to Montana and south to New Mexico and Texas. It is also cultivated for the peculiar foliage. It is probably nowhere important as a honey plant, and is of special interest from the fact that its nectar is so often reported as poisonous. (See Poisonous Honey.)

In the Beekeepers' Item, September, 1921, Miss A. M. Hasslbauer writes as follows concerning this plant in Texas:

> "During the height of the drought beekeepers reported that they were having a flow from some unknown source. The honey was very dark amber and had a peculiar but not unpleasant taste. Investigation showed that the

source of this honey was *Euphorbia marginati.* This plant is well known, as it occurs almost all over the state. It goes under various names of milkweed, ice plant, snow-on-the-mountain. It is easily known by its milky sap, its smooth green leaves and its cluster of small white flowers surrounded by white-edged leaves, which gave to it its peculiar name. Reports of surplus from this plant have been received from several parts of the state, including eastern, central and southwestern Texas. In the Beaumont district it is said to give from ten to fifteen pounds of surplus during certain years, and a closely related species of this plant which grows to a height of only two or three inches, has been reported as giving as high as twenty pounds of surplus southwest of San Antonio."

This plant occurs over much of Texas. Since it blooms there during the dry part of the summer even a small yield is valuable. According to H. B. Parks the honey is strong-flavored and of a red color. He ridicules the notion that the honey may be poisonous. The new honey in the combs appears to be dark red.

SNOWVINE, **see Ampelopsis.**
SOAPBERRY, **see Wild China.**

SOAPBUSH (Guaiacum angustifolium).

The soapbush is native to the dry lands of south Texas and northern Mexico. It blooms after rains in both spring and fall. At Crystal City, Texas, it is reported as the source of the first honey in spring, coming in March. Beekeepers report that with very little moisture

Soapbush (*Guaiacum augustifolium*).

soapbush can be depended upon for honey and that it yields nectar every time it blooms. The honey is of fine quality, mild flavor and light color. The flow lasts from ten to fifteen days. The soapbush has very hard wood, crooked and knotty branches and in places become a small tree known as lignum-vitae.

SOAP PLANT or SOAP ROOT (Chlorogalum pomeridianum).

The soap root is a plant of the lily family found from Southern California north to Oregon. The stems grow from two to ten feet tall with a spreading panicle. The bulb is three or four inches in length and up to two inches through. The blossoms open in late afternoon and remain open through the night. Scullen and Vansell, (Bul. 412 Oregon Ex. Sta.) report that soap root is of considerable value where common. The source of both nectar and pollen.

SOLIDAGO, see Goldenrod.
SORE-EYE, see Yellow-Top.

SORGHUM.

The sorghums are widely planted as forage crops. They yield pollen in great abundance and at times honey dew is gathered from this source, also.

SORREL (Rumex acetosella). Sheep Sorrel.

Sheep sorrel is a common weed naturalized from Europe. Quinby, in his "Mysteries of Beekeeping," edition of 1865, mentions it as an important source of pollen, stating that the bees gather from it only in the morning.

SORREL-TREE, see Sourwood.

SOTOL (Dasylirion).

Sotol is closely related to the yuccas which are characteristic plants of the dry lands of the southwest. There are about three species of sotol common to western Texas and New Mexico and westward. They may readily be distinguished from the yuccas by the curved thorns which occur along the margins of the leaves. The trunk is usually five or six inches in diameter and two to four feet in height. The flower stalks are tall and bear large numbers of small white flowers.

The numerous leaves form a thick crown about the stem. These are often removed and the head used as food for the cattle or other livestock. They are also roasted and used for food for the natives of the region. A strong alcoholic drink called "sotol" is manufactured from the same source.

Sotol is abundant in only a limited range and in the region near Messilla Park, New Mexico, is regarded as of some importance to the beekeeper. J. W. Powell of that place reports that one year he averaged a super of honey per colony from it but that it is not a dependable source.

SOUR CLOVER (Melilotus indica).

Sour clover, sometimes called Indian yellow clover or King Island melilot is a low growing annual with yellow flowers somewhat similar to sweet clover. It is common in the south being grown as a green manure crop in the citrus orchards of California and Arizona and in the cane fields of Louisiana, and other southern localities.

The bees work the flowers freely and it is commonly reported as a honey plant, although of far less importance than the related sweet clover.

SOUR-GUM, see Tupelo.
SOUR-TOP, see Blueberry.

SOURWOOD (Oxydendrum arboreum).

The sourwood tree reaches a height of thirty to forty feet and a diameter of twelve inches, on high lands, but seldom exceeds twenty feet in height on the low lands. It is a common tree from West Virginia to

Sourwood is the source of the best honey in the southeast.

north Georgia and west to Arkansas. The white flowers grow in racemes and appear in July. It is sometimes called sorrel tree, because of the acidity of its leaves. The wood is soft and is not of much value except for light fuel.

Sourwood honey ranks high, both in quality and in quantity of yield. Many people regard it as the finest honey produced in America. It is light in color, of heavy body, fine flavor and slow to granulate. The bloom lasts from two to three weeks and comes in mid-summer, when bees have ample time to build up to maximum strength for the flow. Although sourwood honey is produced in limited areas in several States, North Carolina and eastern Tennessee probably contain the finest forests and remarkable yields are sometimes reported in this region. The crop from this source seldom fails and it is regarded as one of the most dependable sources of nectar. A lady beekeeper, writing from North Carolina, stated in the American Bee Journal that she had never known it to fail. An average of as high as 75 pounds per colony from sourwood has been reported and the local demand usually takes it all at prices above the open market.

SOUTH CAROLINA—Honey Sources of.

South Carolina has an area of 30,989 square miles, more than one-third of which is under cultivation. There are large areas of timberland covering the hills in the higher elevations and of swamp near the coast. There is much good beekeeping territory in the vicinity of the swamps where tupelo yields abundantly. In the low country the gallberry, holly, huckleberry, tulip-tree and asters also yield surplus.

In the highlands of the Piedmont region, fruit bloom is important in spring. Persimmon, sourwood, sumac and asters are also valuable. Cotton is the most important field crop of the state as a whole and on some soils yields honey. Sweet clover is coming into use in some neighborhoods and is an important addition to the flora.

Beekeeping is not extensively followed in the state, although because of the mild climate bees do very well. The long season and mild climate favor brood rearing through much of the year, swarming is general and natural increase abundant. Commercial honey production can be expected to increase greatly in the future.

North Carolina and Georgia are so very similar to South Carolina in general that it hardly seems necessary to repeat in detail the description given in connection with those states. The reader is referred to them for further details.

SOUTH DAKOTA—Honey Sources of.

Situated in the northern plains region South Dakota has a climate of extreme heat and cold, wet and drouth, yet it is rapidly becoming one of the best honey producing states, due to the acreage of sweet clover.

For the most part the state is composed of comparatively level prairie with rich soil. The general average elevation east of the Missouri River, is about 1500 feet above sea level. There is a rise westward to about 3200 feet at the eastern border of the Black Hills. This mountainous area rises 6000 to 7200 feet.

The state as a whole is deficient in rainfall, the general average of the west half being slightly more than 18 inches annually with about four inches more in the eastern section. Most of the precipitation, however, comes from April to September, so that good crops are produced in normal seasons.

There are marked extremes of temperature. Days when the thermometer registers above 100 degrees Fahr., above zero are not uncommon in summer and below zero temperatures in winter are frequent. The lowest recorded is 57 below zero at Camp Crook, in the western end of the state.

The good beekeeping locations in the eastern half of the state are mostly confined to neighborhoods where sweet clover is extensively grown. While alfalfa yields some nectar it is not equal to sweet clover in this region. Large yields of sweet clover are the rule where sufficient pasturage is available.

In the Black Hills section, farming is confined to the rich valleys, but there is a much greater variety of nectar plants available to the bees than elsewhere. Because of the higher elevation there is more rainfall and a greater variety of native vegetation. The hills are covered with trees, mostly pines with a rich undergrowth of shrubs and plants.

There is practically a continuous honeyflow in this area from the time growth starts in spring until frost kills vegetation in fall, beginning with the pasque flower and early willows. Dandelion is abundant and blooms until first white clover blossoms appear. Both white Dutch and alsike clover are common, as well as alfalfa and sweet clover. There are numerous native sources such as vervain, catnip, Virginia creeper, wild cucumber, box elder, bearberry, and gumweed, as well as heartsease and goldenrod. Some of these plants are seldom visited by the bees, due to the abundant sweet clover bloom. Both white and yellow sweet clover are present. The Black Hills region is probably the best beekeeping area in South Dakota and is not often excelled elsewhere.

In other sections of the state there is a much smaller variety of plants on which the bees can build up, and in neighborhoods where sweet clover and alfalfa are not grown there is little for the bees.

SOUTHERN BUCKTHORN, **see Coma.**

SOUTHERN CRAB APPLE, **see Crab Apple.**

SOW THISTLE (Sonchus).

The sow thistles are weeds which are widely distributed from eastern Canada to Florida and from British Columbia to California. They are reported as valuable in east Texas, and Richter lists two species as yielding nectar in California.

It is in the grain fields of the northwest that sow thistle is most abundant. There are areas in southern Manitoba and North Dakota

Sow thistle is a troublesome weed on the northern plains.

where the fields are yellow with the blossoms as far as the eye can see during the month of August. With the bright yellow blossoms, like big dandelions, spreading above the heads of the wheat it is a pretty though discouraging sight to the farmer. Sow thistle, together with Canada thistle, has so fully occupied the land in many places that it is difficult to harvest a crop of grain. The writer has seen the seeds floating away from the threshing machine and carried on the wind in such quantity that the air seemed filled with them and the sky overcast as though light clouds were present. With even a slight breeze the seeds are carried for miles. A careless farmer thus spreads the pest over the land for miles around, and so common has the plant become in this region that in some sections the value of the land has been greatly reduced because of it. When the plants are in bloom the fields often give the impression, at a little distance, of being occupied with sow thistle only.

Although there are three species of sow thistle present in this region, it is the perennial sow thistle (*Sonchus arvensis*) which is most difficult to eradicate from cultivated areas. Summer fallow, with frequent cultivation is about the only method. Once established it is likely to remain permanently, since there are always some plants even with the most thorough methods of cultivation. The plant spreads by means of a perennial creeping rootstock as well as by seed. It is common from Nova Scotia to Saskatchewan, though in only a comparatively small portion of its range has it been permitted to run riot in cultivated fields to such an extent.

Manitoba beekeepers visited by the writer nearly all credited sow thistle as an important source of amber honey in midsummer. Some North Dakota beekeepers, on the other hand, doubted its value and stated that they seldom found the bees working on it. Since its blooming period is similar to sweet clover, it is not surprising that the bees do not work it during the bloom of the latter plant. Even though the secretion of nectar is not as abundant in sow thistle as in some other plants, where it occurs as abundantly as in portions of the northwestern prairies, it is still important. Aside from the prairie provinces of Canada the writer has been unable to find any place where it is considered as an important source of surplus. Press reports credited Consul J. I. Brittain, Winnipeg, with the statement that sow thistle was the principal plant from which Manitoba's 1,800,000 pounds of honey produced in 1922 was harvested.

The honey is of decided flavor and would be regarded as of inferior quality in markets accustomed to the light-colored product from alfalfa or the clovers.

SOYBEAN (Glycine soja).

Of late the soy or soja bean is attracting wide attention as a desirable forage plant, and in some localities large acreage is being planted. The plant has long been cultivated in the warm portions of Asia for the beans. There the beans are used as the main ingredient of the condiment or sauce known as soy. Although cultivated in China and Japan from remote antiquity, it is probable that it originally grew wild in the East India Islands, south Japan and on the coast of southern China. Of late it has been widely disseminated throughout the temperate regions. In addition to its use as a forage crop it is valuable for the oil in which the beans are very rich. Some authorities state that some varieties contain as much as 20 per cent of oil.

Soybeans were first brought to America by the Perry expedition from Japan and the seeds were distributed by the Patent Office in 1854.

It was not until quite recently, however, that farmers became generally interested in this plant. With the extension of the area planted to this crop it becomes of increasing interest to the beekeeper.

Information concerning its value as a source of nectar is meagre. It is well known that under some conditions it yields nectar, yet under other conditions the bees do not seem to find it attractive. Just what conditions of soil and climate are most favorable are not yet entirely known. Reports from different sections are very conflicting. While the author has not found the bees to work upon the plant in Illinois during the time of his observation, beekeepers from some localities write to say that bees work upon them from morning until night. Another report (from Tennessee) states that the bees work soy beans freely from about 9 A. M. until sunset. A North Carolina beekeeper reports that they work from early morning until about noon, or not later than 1 o'clock P. M. Further observation concerning the behavior of the plant under different conditions will be necessary to reconcile these reports.

The honey from this source is light in color, of peculiar but pleasing flavor and rather thin and light in body. A sample received by the writer from Mr. J. R. Pinkham, of Washington, N. C., granulated rather quickly, but would be graded as a high quality honey in most markets. The flavor is very distinctive and should command a ready market once the trade becomes accustomed to it.

Concerning the yield Mr. Pinkham writes:

> "Notwithstanding that the bees only seem to work on soy beans part of the day and that it does not seem to yield regularly every day, a strong colony of Italians will store from 100 to 250 pounds in thirty or forty days, which about covers the blooming period of the plant. I had one colony which filled 175 sections this year." (1922).

In contrast to the above report Mr. Joe Gass, of Tyner, Tennessee, writes that he had been unable to secure any surplus from soys, although the bees worked them freely and they bloomed at a time when nothing else was to be had. He found them valuable, however, as they kept the bees busy at a time when otherwise they would have been robbing, and stored some very light honey in the brood chamber. Mr. Pinkham writes that soybeans do not seem to yield as heavily on uplands as on the black swamp or Pocosin silt.

Harold Kelly of Forest Glen, Maryland, writes that his bees within reach of about fifty acres of soybeans worked the blossoms freely, gathering more than their immediate needs but not storing surplus. The honey was amber in color, slightly on the dark side but of mild flavor. While not important the beans do supply a light honeyflow in his locality at a time when there has usually been a dearth.

In 1944 there were numerous reports of light yields of honey from soybeans. In most cases the surplus from this source was small but Alfred P. Johnson of Illinois reported in American Bee Journal, (September page 306), that some colonies gained 5 to 7 pounds per day with totals of 30 to 85 pounds, honey of good flavor, body and light color.

In American Bee Journal, August 1934, Cecil J. Lent reported that in Iowa he had observed the bees and soybeans for four years. In three seasons they had worked to some extent on the beans while on the fourth they ignored them completely. For the most part they worked one variety, the Virginia and paid but little attention to the Manchu. For the first few days of bloom 75% of the bees carried pollen. Later a larger percentage carried nectar and for a short time the scale hive showed a gain of about four pounds per day. This gain was short and soon slowed down until there was no gain by the tenth day.

In a letter to the author, J. E. Eckert reported that while living in North Carolina he observed that the bees gathered a surplus from soybean in the coastal section east of Washington. The honey was extra light amber in color, of good flavor and granulated quickly. An average of one to two supers were produced in summer when no other source was available.

In Louisiana Everett Oertel reported that he had observed the bees to work soybeans lightly some years but had seen no honeyflow from this source.

Western bur marigold.

There is little to indicate that the soybean is an important honey plant anywhere although it does at times yield some nectar. The fact that bees may work one variety at times while neglecting others blooming near by, indicates a variation in nectar yield which might be increased by selection.

SPANISH BAYONET, see Yucca.
SPANISH DAGGER, see Yucca.

SPANISH NEEDLE (Bidens).

The Spanish needles, also known as bootjacks, beggar ticks, stick tights and marigolds, are very widely distributed plants, and are of interest to the beekeepers from Nova Scotia to California. Most of the species are weeds growing commonly on low and swampy lands. Not all of them produce honey in appreciable quantity, and possibly some of them are not sought by the bees at all. *Bidens aristosa,* which has an attractive yellow flower is most frequently mentioned as a source of honey. This is particularly valuable on the lowland along the Mississippi and Missouri Rivers. During the seasons of 1915 and 1919 much honey was gathered from it.

Two species are reported among the honey plants of California by Richter, *B. frondosa* and *B. pilosa.* The former is one of the most widely-distributed species and closely resembles the one shown, but has a wider leaf. *Frondosa* is seldom reported as yielding nectar, and it is of doubtful value to the beekeeper.

The western bur-marigold (*B. involucrata*), which occurs from Illinois and Iowa south to Texas and Louisiana is reported as a good honey plant. August is the month of flowering with this species. The Spanish needles are all late bloomers, and where they occur yield nectar, and add something to the fall honey flow.

Along the Delaware River and in the adjacent swamps *B. trichosperma* and *B. laevis* grow abundantly. *B. trichosperma* is reported by beekeepers in that region as the main honey plant, blooming from August 20 to mid-September. It is known locally as coreopsis and also as tickseed-sunflower. *B. laevis* is commonly called showy bur-marigold in that region, and begins to bloom in late August. Growing on wetter land it has a more restricted range, being generally confined to the swamps near the coast from Massachusetts to Georgia. *B. trichosperma* is also found as far inland as Illinois and Kentucky. Beekeepers report that a complete failure of the flow from the two sources in the Delaware region has never been known. The honey is light yellow and has the charactersitic faintly spicy odor of the flowers.

Spanish needle (*Bidens aristosa*).

In Florida *Bidens leucantha* is widely distributed, flowering all the year. The flower is about an inch in diameter with white rays and yellow center. The author has found the bees working it freely in January and February in widely separated localities. Probably it is seldom the source of surplus but of some value as a minor source of nectar.

SPARKLE-BERRY, **see Farkle-Berry.**

SPEARMINT, **see Mint.**

SPEEDWELL (Veronica).

There are a number of species of speedwell, several of which are commonly cultivated in gardens for ornamental purposes. They have a rather long blooming period and are very attractive to the bees. Several of them have been naturalized from Europe and are now found in widely scattered localities.

Because of limited abundance, they may be listed among minor sources of nectar. See also Culver's Root.

SPICEWOOD, **see Dogwood.**

SPIDER PLANT (Cleome spinosa).

The spider plant is a close relative of the Rocky Mountain bee plant and very similar in habit. The spider plant had quite a boom among beekeepers of a few years ago. The seed was sold quite gen-

The yellow spider flower reaches a height of six feet or more in our garden with as many as 300 flower clusters on a single plant.

erally and planted in gardens, but as it is of no value except for honey and as an ornamental its popularity soon declined and it is seldom mentioned of late. It secretes nectar abundantly and the bees work upon it freely just at nightfall, and again in early morning. It is a native of the tropics and has escaped and run wild in many places from North Carolina to Arkansas and Louisiana.

Much has been written about the remarkable secretion by this plant and the excitement it causes among the bees. Under favorable conditions it is one of the very best of honey plants and, if sufficiently common, would no doubt be the source of large quantities of surplus honey. There are reports to the effect that spider plant grows abundantly in neglected fields in some localities in southeastern Missouri and is there important. It is said to require a rich soil for best results in nectar secretion.

The yellow spider flower or golden cleome, (*cleome lutea*) is a vigorous annual found from southern California north to Washington and east to Colorado and western Nebraska. It reaches a height of six or seven feet and in late summer has as high as 300 flower clusters on a single plant. Bloom continues until killed by frost.

A plot of about one tenth of an acre in our test gardens in 1938 and 39 attracted the bees in large numbers in late summer and brood rearing in the experimental apiary continued much later than in neighboring hives beyond the reach of cleome. A test of forty-four samples by Dr. O. W. Park showed the nectar concentration to range from a low of 15% sugar content to a high of 35% and an average of 23%.

Vansell has indicated that the sugar content of this species is low in the far west.

Flea beetles have proved so destructive to young plants that we find it hard to maintain although the plant reseeds very freely.

SPIDERWORT (Tradescantia).

The spiderwort is a common prairie flower common to moist meadows. It attracts large numbers of bees to its bloom in early morning for pollen. They begin visiting the flowers almost as soon as it is light. By mid-morning the supply is exhausted and they pay no more attention to this plant until the following morning.

SPIKEWEED (Centromadia pungens).

Spikeweed is a common plant in California, where its range is given by Jepson as follows:

> "Abundant on the plains of the lower San Joaquin southward to southern California and westward to Walnut Creek and Alameda. On the alkaline plains of the upper San Joaquin this species covers tens of thousands of acres and often forms thickets four or five feet high. It is abundant in the low, more or less alkaline plains of Solano County and forms extensive colonies in summer fields. Extermination is often accomplished by means of

bands of sheep, which leave the fields perfectly clean and destitute of this spikeweed pest."—Flora of Western Middle California.

According to Richter, carloads of honey from spikeweed are shipped annually from Fresno County, the honey being of amber color, good quality and quick to granulate. He states that other plants are replac-

The bees seek the spring beauty in early spring.

ing spikeweed to such an extent that it is no longer of the importance which it was in the past.

SPOON-WOOD, see Laurel.

SPRING BEAUTY (Claytonia Virginica).

The spring beauty is one of the early spring flowers common in open woodlands from Nova Scotia and New Brunswick to Ontario, Saskatchewan and Alaska and from New England west to Minnesota and Nebraska and south to Georgia and Arkansas. Appearing so early in spring it is much sought by the bees at a time when there is little of either nectar or pollen available. At a later season it would be of such slight value as to be hardly worthy of notice on the part of the beekeeper.

SPRUCE (Picea abies).

For years past there has been an occasional mention in our beekeeping literature of spruce honey, or of bees working on spruce. As far as the writer has been able to ascertain this is the Norway spruce.

The Norway spruce is the source of honeydew.

Since the Norway spruce is not a native of this country, it is seldom found in considerable numbers except in the vicinity of cities, where it is planted freely for ornament.

It was at the Ontario Agricultural College at Guelph that the writer first saw the bees working on spruce to any extent. It was about June 12, and the bees were humming through these trees in large numbers. There are hundreds of these trees about the college grounds, and considerable honeydew seemed to be coming to the college apiary from this source. Honeydew is seldom desirable, as it is usually of poor quality and only serves to spoil the quality of good honey. However, this spruce honeydew seemed to be of rather better quality than is generally the case with honeydew, and, as it came ahead of the clover flow, was probably nearly all consumed for brood rearing.

For a time the writer was puzzled to know whether the bees were getting an exudation of sap from the tree, or were in fact getting honeydew. They were working on what appeared at first sight to be buds at the base of the new growth, but which under the microscope proved to be insects identified as *Physokermes picea.*

SPURGE (Euphorbia esula).

Asberry Singleton in a letter to the author wrote that leafy spurge, a weed pest of Manitoba, is a favorite with the bees and that they seem to prefer it to any other bloom. (See snow-on-the-mountain.)

SQUASH (Cucurbita maxim), also (C. moschata).

Squashes are widely cultivated for food. There are numerous varieties, but all are valuable sources of pollen and nectar. They secrete nectar freely, and where sufficiently abundant, a considerable quantity of honey is stored.

SQUAWBUSH or SQUAWBERRY.

In the vicinity of Phoenix, Arizona, a species of Desert Matrimony, *Lycium fremonti,* is commonly known as squawbush or squawberry. It blooms at any time from December until March, depending upon the rainfall, sometimes more than once. Beekeepers generally in that region regard it highly as a source of honey for winter and spring brood rearing. (See Lycium.)

In the Rocky Mountain region and also in California the same name is generally applied to the skunkbush, *Rhus trilobata,* which is a sumac found from Saskatchewan to Texas and westward. Western beekeepers report it as attractive to the bees, though probably not important. (See sumac.)

The redberry, *Rhamnus crocea,* is also known as squawbush in some California localities. (See Buckthorn.)

SQUAW CARPETS, see Mahala Mats.

STAR THISTLE (Centaurea). BARNABY'S THISTLE.

There are several species of star thistles widely distributed. The yellow star thistle (*Centaurea solstitialis*) is an introduced species from Europe that occurs from Massachusetts and Ontario, west to Iowa. It is also common in parts of California. It is only from the latter State that it is reported as an important honey plant. According to C. D. Stuart (American Bee Journal, page 340, October, 1918), it furnishes from one-third to one-half of Butte County, California's honey crop, an average of about sixty tons. Concerning the plant he writes as follows:

Yellow star thistle.

"Star thistle begins to bloom about the first of July and continues till frost, which usually comes between October 1 and November 1. The yield of nectar is slow but continuous. If it is stopped by drought it will start yielding nectar again after a rain. The plant has the faculty of existing in arid soils for long periods of drought, and when apparently dried up, it will start to grow and blossom after a rain. Some cattle growers find that thistle hay can be fed profitably when cut and dried like other hay, if it is moistened before feeding. The dampening of the fodder takes the sting out of the leaves and blossoms.

"Star-thistle honey is heavy-bodied, white, almost cloying in its sweetness as orange, and has a greenish yellow tinge, like olive oil. It is considered by large buyers equal in quality to any white honey in the State, and with the price at two cents a pound more than light amber of the alfalfa type, and still rising, beemen in northern California should worry."

Richter lists Napa-thistle or tocalote (*Centaurea melitensis*) as yielding light amber honey of good flavor 'nd fair body in Sacramento County, from May 15 to June 15. It is regarded as a bad weed and is abundant everywhere in fields and pastures. Like the foregoing, it has been disseminated with seed grain and grass seed. According to Richter, it does not yield nectar in southern California. Scholl lists American knapweed (*Centaurea Americana*) as not important in Texas.

Russian Knapweed, (*Centaurea repens*) is a serious pest in many localities but yields much honey. The honey is reported as light in color with a slightly bitter tinge. It is mentioned as of some importance in western Montana.

STICKLEAF, see Mentzelia.
STICK-TIGHT, see Spanish Needle.
STINKING CLOVER, see Rocky Mountain Bee Plant.
STINKWEED, see Jackass Clover.

STONECROP (Sedum pulchellum).

Stonecrop is common from Virginia to Arkansas and south to Georgia. It is abundant in many localities and is reported to be a valuable honey plant in the South.

Prof. Floyd Brailler in a letter to the author says: "Stonecrop does not receive the notice it deserves. It is our best honey plant (Madison, Tennessee) and gives a finer grade of honey even than clover.

STONE MINT (Cunila origanoides) Maryland Dittany.

The stone mint is a low growing perennial with small white or purplish flowers in terminal or axillary clusters. The plant is found on dry hills from New York to Georgia and west to Missouri and Arkansas. W. E. Bumgarner of Marshfield, Missouri reported it as important in his locality blooming for about a month in September and October. The bees filled the upper chamber of a two-story hive and stored twenty to forty pounds surplus per colony.

Stone mint is sometimes grown in perennial gardens but does not appear to be widely known.

A sample of honey sent by Eldon Martin of Goodwater, Missouri, which he regarded as a pure sample of typical stone mint honey is very light, almost water white in color, very mild flavor with slight spicy taste. He reports the hills for miles around covered with the plant and that it is a dependable source of winter stores with surplus every two or three years.

STORK'S BILL, see Pin Clover.

STRAWBERRY (Fragaria).

For some unexplained reason the strawberry seldom attracts the bees in large numbers. Since many varieties of strawberry require cross pollination the services of the honeybee would appear to be useful. There are numerous reports to the effect that the bees work upon strawberry, but the writer has been unable to verify them by personal observation. A chance bee can be seen flitting uncertainly from flower to flower, but the times when the blossoms are eagerly visited as though the bees were getting an abundance of nectar appear to be rare.

In 1895 an extended discussion of the relation of bees to the strawberry plant ran through a number of issues of American Bee Journal. Most of the reports were to the effect that the bees visited the strawberry plants freely only on rather infrequent seasons. The following extracted from page 408 are among the few showing favorable activity:

> "I am sure the strawberry is one of the important honey plants. It blooms just at the time to bridge the gap between the general crop of fruit bloom and the raspberry and alfalfa. The bees not only work on the bloom eagerly but they get enough nectar to keep brood rearing up to a vigorous rate. Go into the strawberry fields at any time when the bees are flying during the blooming season, and the hum of busy wings sounds like a swarm might be on the wing."—L. J. Templin, Colorado.
>
> "Bees do work on strawberries very strong some seasons; other years not at all. My bees this spring, have just swarmed over the blossoms, and gathered both pollen and honey."—Frank P. Stowe, Connecticut.
>
> "Some years they work on them but little—presumably because there is something else better; but this year in particular, the strawberries were one mass of bloom, and they were alive with bees."—S. H. Herrick, Illinois.

STRAWBERRY CLOVER (Trifolium fragiferum).

Strawberry clover is a low growing perennial native to the Mediterranean region. It has been grown as a pasture plant in Australia and New Zealand, and recently in this country.

Its name is derived from its fruiting head which is about the size and shape of a strawberry. It appears to be very promising for wet alkali or saline soils and according to E. A. Hollowell of the U.S.D.A. is the most salt tolerant legume that we have. It is thus likely to increase rapidly in the irrigated regions where accumulations of salt are present. In a publication of the USDA, (Agr. leaflet 176) Mr. Hollowell states that the bees apparently obtain considerable nectar from the blossoms.

STRINGY BARK, see Eucalyptus.
SUGAR GUM, see Eucalyptus.

SUMAC (RHUS).

There are about one hundred and twenty species of Rhus found in Asia, South America and North America. There are fourteen species common to North America. The red or scarlet sumac, (*Rhus glabra*) is most common, being found from New England west to Saskatchewan, Colorado and Arizona, and south to Florida and Louisiana. This species is a well-known source of nectar and is especially important in New England.

In Texas, Scholl lists the dwarf sumac (*Rhus copallina*) as yielding surplus throughout eastern and south Texas. He also lists green sumac (*Rhus virens*) as attracting the bees in west Texas.

In California, Richter lists poison oak (*Rhus diversiloba*) as yielding a superior grade of white honey which granulates readily. He also reports laurel sumac (*Rhus laurina*) as yielding amber honey of marked odor but fine flavor.

Leaves and blossoms of red sumac.

The poison ivy (*Rhus Toxicodendron*), common to the Eastern States from Nova Scotia to Wisconsin and south to Louisiana and Florida, is a vigorous vine, climbing by means of aerial rootlets. The flowers are inconspicuous, but secrete nectar freely. Where sufficiently abundant, surplus may be expected from this source.

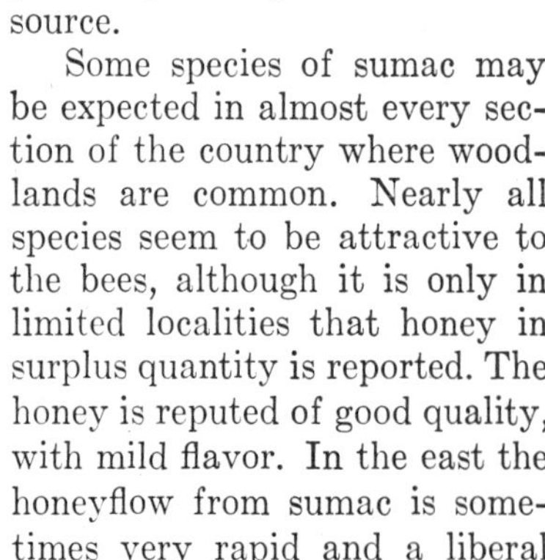

Some species of sumac may be expected in almost every section of the country where woodlands are common. Nearly all species seem to be attractive to the bees, although it is only in limited localities that honey in surplus quantity is reported. The honey is reputed of good quality, with mild flavor. In the east the honeyflow from sumac is sometimes very rapid and a liberal quantity of surplus secured, at times as high as 100 pounds per colony. New honey from this source is somewhat bitter to the taste, but this characteristic soon disappears. The blooming period usually comes in July and lasts for two or three weeks.

The dwarf sumac is a widely distributed species, known in some localities as mountain sumac. It is common in the Gulf States and is the source of honey in favored localities.

Coleman adds mahogany sumac (*Rhus integrifolia*), squaw bush (*R. trilobata*) and sugar bush (*R. ovata*) to Richter's list of species which yield nectar in California. The latter species, he states, is the source of considerable surplus in many southern California apiaries.

The mahogany sumac, (*Rhus integrifolia*), commonly called mountain mahogany by California beekeepers, was found by the author to be highly regarded as a source of nectar by many California beekeepers, but no reports of surplus of importance were found.

The skunkbush, often called "squawbush" (*R. trilobata*) was likewise valued by beekeepers of the mountain region of Colorado and Utah, who reported to the author that the bees work it freely during a short blooming period.

The mahogany sumac or mountain mahogany.

SUMMER FAREWELL (Kuhnistera pinnati).

The summer farewell is so named because of its late blooming. It is also known as pine-barren prairie clover. It is found on dry, sandy pine lands from North Carolina to Florida and west to Mississippi. It is reported as important to the beekeepers in the south portion of Georgia and throughout Florida. Frank Stirling, of the Florida College of Agriculture, wrote that it is the last important source of nectar in late summer, producing light honey which granulates readily.

The blossoms are white and first appear in July, but the blooming period continues until late autumn. See also aster.

SUNFLOWER (Helianthus).

There are many species of the sunflowers, some of which may be found from the Atlantic Coast to California, and from Canada to the Gulf. They are tall, coarse weeds, with yellow flowers. Large numbers of insects of many species may be found on the sunflower blossoms in search of the nectar. Whenever these plants are sufficiently abundant they are the source of large quantities of honey. M. H. Mendleson, of Ventura, California, reports that one year, following a wet winter, he secured a carload of surplus honey from sunflowers, although such yields are extremely rare in his locality.

The cultivated sunflowers are of little if any value for honey, but produce seed in large quantity, which is valued as poultry food. The Jerusalem artichoke is a variety of sunflower grown for the tubers. A variety of this plant grows wild in the upper Mississippi Valley States, where it is regarded as a weed. It is frequently reported as a good honey plant.

Many of the wild sunflowers are perennials, persisting for many years when once established. They are commonly to be found along wagon roads, railroads and other waste places. The honey is amber in color and strong in flavor.

In American Bee Journal for December 1868, page 112 appears the folowing by H. Faul:

> "My bees did nothing all season till late in the fall when sunflowers were in full bloom. Then they commenced storing honey in great quantities. They gathered from fifty to eighty pounds, per stand, in two weeks. Some of them filled all their surplus honey boxes, and when I took some of the honey for eating, I found it tasted as the sunflowers smell. It is scarcely fit for table use. There are hundreds of acres of the flowers growing here. (Council Bluffs, Iowa.)

SUPPLE JACK, see Rattan Vine.

SWAMP BASIL, see purple flowered mint.

SWAMP LAUREL, see Magnolia.

SWAMP-LILY, see Lizard's Tail.

SWAMP LOOSESTRIFE (Decodon verticillatus).

The swamp loosestrife, also known as water-willow, grass-poly, wild oleander, peat-weed, stinkweed, willow-herb and perhaps other common names, is common to swampy places from New England and Ontario west to Minnesota and South to Louisiana and Florida. It is an aquatic perennial blooming in mid-summer and yielding some nectar. Reports of surplus honey from this source come from Michigan and it may be of more importance in swampy regions than has been recognized.

SWEET BAY, see Magnolia.

SWEET CLOVER (Melilotus).

There are probably twenty species of melilotus native to the temperate regions of Europe and western Asia. Several have been introduced into America. Of these, two species, the white sweet clover (*Melilotus alba*) and the yellow sweet clover (*Melilotus officinalis*) are valuable plants and are widely distributed. The yellow variety blooms about two weeks earlier than the white and where both are present a long honey flow may be expected.

Sweet clover reaches the highest development in the secretion of nectar in the hot, dry summer climate of the plains region between the Mississippi River and the Rocky Mountains. In the East, the surplus secured from this plant has been disappointing, and eastern men insist that sweet clover is overrated as a honey plant. However, those who have seen the big flows that are frequent along the Missouri River and westward are enthusiastic in its praise. In the region about Sioux City, Iowa, it is grown extensively as a farm crop. In this section an average of 200 pounds surplus per colony from sweet clover is not uncommon. On the limestone soils of Alabama and Mississippi it also yields freely and fair yields are reported. In the irrigated regions of the West it is of great importance and beekeepers who ship sweet clover honey in carlots are not uncommon.

The quality of the honey is excellent. It is light in color and mild in flavor, although slightly peppery to the taste. It granulates more readily than white clover, but is regarded as of number one quality in the principal markets.

Selections have been made from plants which show special traits and have been distributed under special names.

The common white flowered sweet clover is a biennial, a well known plant which came from the old world many years ago. It is widely known as Bokara clover, Bee Clover, Melilot and sometimes by its scientific name "Melilotus." It is a native of Central Asia and is known to have grown for more than 2000 years in the region about the Mediterranean Sea. It is now scattered over most of the civilized world but has only in recent years been fully appreciated as a valuable forage plant.

Grundy County White Sweet Clover

About 1916, it was discovered that the strain of white sweet clover commonly grown in Grundy County, Illinois, was of finer stem and

Blossoms of white sweet clover (*Melilotus alba*).

earlier bloom than the common kind. This strain has since then been widely distributed under the name of its place of origin. Nothing is known of it, prior to its recognition in the Illinois neighborhood and it appears to have been the result of natural selection. Beekeepers are interested especially because its habit of early bloom prolongs the season of honeyflow when planted with other varieties.

Spanish

Spanish is a white flowered variety developed from seed secured from the Botanic Gardens in Madrid, Spain by the U. S. Department of Agriculture. It is said to be especially valuable for its ability to overcome competition with weeds in its first year of growth. It exceeds the common white in amount of growth for first season but has no special advantage in second year. It reaches maturity at about the same time as the common white but is rather distinctive in appearance.

Evergreen

Evergreen is a strain of the biennial white sweet clover selected from roadside plants by the Ohio Experiment Station. It is a late maturing variety of rapid growth and abundant forage yield. It is a good variety for weedy locations because of its quick germination and rapid first season growth. It is said to continue green in its second season for two or three weeks longer than the common white.

Madrid

Madrid is a yellow flowered variety from the Madrid Botanic Garden introduced by the U. S. Department of Agriculture. The foliage is very dark green with a heavy first season yield. It remains green late in the first season and starts growth early in spring of second year. It matures somewhat later than the common variety of *M. officinalis* and produces seed abundantly. It is meeting with considerable popularity and demand appears to be increasing.

Arctic Clover

This variety has been developed from a hardy strain of white sweet clover brought from Siberia by Professor N. E. Hansen, of Brookings, South Dakota. It differs but little from the common white sweet clover except in hardiness. It stands the most severe winters where others kill out. It has proved of special value in the western Canadian provinces and has been distributed under name of "Arctic Sweet Clover" by the College of Agriculture of Saskatchewan.

Yellow Sweet Clover

The large yellow sweet clover, (*Melilotus officinalis*), comes from the same region as the white form and is now distributed over the world in much the same manner. Like the other, it is a biennial. It is more spreading in habit, does not grow as coarse or tall and blooms earlier. Generally it can be depended upon to begin blooming about two weeks ahead of the common white sweet clover while but little ahead of the Grundy County White clover.

Redfield Sweet Clover (Melilotus suaveolens)

The Redfield sweet clover is a distinct yellow flowered variety which came from Manchuria. It was developed at Redfield, South Dakota and distributed from there, hence the name. It is especially adapted to the Plains Region from Kansas north and is said to make the best hay of any variety of sweet clover commonly grown. It blooms late and greatly extends the season of bloom by offering a honeyflow in most localities after the others have stopped blooming.

Yellow Annual

There is another yellow flowered variety of little value, (*Melilotus indica*), commonly called yellow annual or sour clover. It makes but little growth and is regarded of but little value as a field crop. Accordingly, it is but little cultivated in regions where the better kinds do well. It is well thought of in some sections of California where it does much better than in the cold regions. It belongs in a warm climate.

Alpha Clover

Alpha Clover is a white flowered mutation found in the fields of the Saskatchewan Experiment Station. It varies greatly from ordinary varieties of sweet clover in that it has many stems from each root, averaging about thirty. It thus provides many fine stems instead of one coarse one and so makes much better forage.

Hubam or Annual Sweet Clover

In the spring of 1916, Professor H. D. Hughes, of Iowa State College of Agriculture, discovered that certain sweet clover plants bloomed the first year. Professor Hughes saved the seed and distributed in small quantities to a number of interested persons. Later investigation revealed that annual white sweet clover was probably more widely distributed and more common than had been recognized. For a time great interest was manifested in the plant which was named "Hubam" as a combination of the name of the discoverer and Alabama, the state from which it came. For a few years the demand for seed approached the proportions of a small boom but it soon subsided. The fact that it matures seed the first year offers an advantage for certain uses in the farm rotation.

Melana Sweet Clover

Melana is an annual form of the Alpha sweet clover which was found at the Saskatchewan Experiment station in 1932. The crown of the plant forms from seven to eighteen stems and these in turn are much branched. It makes good hay and the bees seek the blossoms eagerly. In our test garden it offered great promise but in the field test it was unable to compete with the weeds and for this reason, its value will probably be limited.

There are annual forms of the yellow flowered, *M. officinalis,* also as well as numerous other strains selected for some special quality. New names for selected strains are constantly appearing and it is impossible to include them all.

The balsam clover is often called blue flowered sweet clover. Although sometimes classed as a melilotus it is usually known as Trigonella. (See balsam clover.)

Sweet Clover as a Farm Crop

When our older readers were beginners in the beekeeping business it was a popular thing for the beekeepers to buy sweet clover seed and stealthily sow it along the roadsides. So general was this practiced that whenever the plant appeared in a new locality it was generally charged up to the beekeepers living nearby. So great was the prejudice against the plant that much ill-feeling developed in some places because of it. It even went so far that in some States it was placed on the list of noxious weeds and its eradication required by law. When Frank Coverdale, well-known Iowa farmer who did so much to popularize sweet clover, first sowed it in his own fields, neighbors called on the county attorney to inquire whether he could not be prosecuted for sowing weed seed. For a generation the beekeepers kept up the fight, and constantly preached that sweet clover was not a weed, but a valuable forage plant. It remained for men like Coverdale, who were both beekeepers and farmers, to prove the assertion and convince the unwilling public, by making as much profit from sweet clover pasture for forage as the neighbors could make from other farm crops.

It was on poor lands which had been worn out by bad tillage, that the plant made the best showing. When lands which had been lying idle, because no other crop could be raised profitably, were made to produce good yields of milk, butter and beef from sweet clover, the neighbors were inclined to give it a trial on their own poor lands. The change in sentiment has been very marked during the past few years. There are large areas where sweet clover is grown generally as a farm crop, in Kentucky, Nebraska, Kansas and the Dakotas. In the early years of his experience, Coverdale kept bees in several outapiaries, so that much travel back and forth was necessary. Since sweet clover has become so generally grown in his locality, he was able to keep three hundred colonies in one yard in his orchard, where they were under his immediate care at all times. On visiting Falmouth, Ky., I was amazed at the stories they told of what sweet clover had done for that region. One of the pioneer growers was E. E. Barton, and his experience with it sounded like a fairy tale. Mr. Barton said that following the civil war, most of Pendleton County was given over to tobacco growing, with little live stock, and not much rotation of crops. It was a hill country, and although it had a fertile soil over a

Hubam clover.

clay subsoil, the heavy rains soon washed away the shallow surface soil, and one farm after another was abandoned. Hundreds of farms were abandoned, and many of them were sold for taxes, because no buyers could be found. More than a third of the population left the county, and the farmers who remained had hard lines to make ends meet. Sweet clover was stealthily sowed, probably by beekeepers intent on increasing the bee pasturage. At first it was regarded with disfavor and fought as a dangerous weed.

Mr. Barton came into possession of a farm, somewhat against his will, because the owner could not pay the mortgage. He tried renting it, and the tenant was unable to make a living, much less pay the rent. After it had been abandoned, he went to great trouble to keep down the weeds, especially sweet clover. Then came a year of drought, when there was very little feed for the cattle, and they were turned into the roads to graze. Even there there was but little except the sweet clover, which was by this time rather common along the roadsides. It was soon noticed that the cows were eating the sweet clover with relish and doing well. Then somebody tried an experiment by sowing it in a field. It thrived, the cows liked it, and the milk flow was increased. Mr. Barton by this time was quite ready to profit by the experience, and within five years the farm which would not grow grass was producing good crops. He bought more abandoned farms and sowed them to sweet clover, and his neighbors began to do likewise. One by one the farmers came back to their abandoned farms, new settlers came in, and everybody began to grow sweet clover. By 1917 there were fifty thousand acres of it in that county. Ask any farmer you meet on the streets of Falmouth what he thinks of sweet clover and he will tell you such tales of rebuilt fortunes from a combination of dairy cows and sweet clover as you never expect to hear. There were shipped from the county about half a million pounds of seed yearly, besides thousands of dollars' worth of dairy products every week. They find that an average of 300 to 600 pounds of hulled seed per acre can be secured from the white variety and 500 to 700 pounds of the yellow. An average yield of from $40 to $100 per acre is the return from the sweet clover, according to local reports picked up on the streets. Now one finds evidences of prosperity on every hand. The farmers have fine homes, automobiles, and money in the bank. (1917.)

Soil Requirements

There is no forage plant that will succeed on such a wide range of soil conditions as will sweet clover. It will succeed under unfavorable conditions on the heaviest clays and on light sand. It will grow on hardpan and on gravelly and stony land unsuited for general cultivation. It does well on soils too wet for either alfalfa or red clover and on soils so dry that neither of these will succeed. It will grow on land so poor and devoid of humus that no other clover or grass will grow. It is the greatest soil builder known, and now that the public has finally accepted the fact that it is not a noxious weed, it will be used to redeem untold thousands of acres of otherwise waste land. It grows all the way from sea level to the mountain sides, and is spreading in the semi-arid sections of Colorado and other Western States, where the annual rainfall is very light.

The growth of the plant, however, is no longer confined to the roadsides and worn-out fields, but farmers are growing it successfully and profitably on lands worth $100 per acre in Iowa and Illinois, because it pays them to do so. In some cases the railroad companies have discovered that sweet clover growing along the right of way is the best possible insurance against erosion of the roadbed. A heavy growth of sweet clover protects the banks from the washing of heavy rains, as no other plant will do. In places, one can see a continuous strip of sweet clover for miles along the railroads. It would seem the part of wisdom for the beekeepers' associations to bring this fact to the attention of the men in charge of keeping the lines in repair wherever possible. Once established along the railroads, it is bound to spread more or less along the byroads and into the fields, thus increasing the supply of forage within reach of the bees.

One of the most useful purposes which sweet clover serves is to smother out obnoxious weeds. So persistent is the plant where sowed in waste places that there are few weeds which can compete with it. Where bad weeds are present in old lots, along roadsides, etc., the easiest way to eradicate them is by sowing sweet clover freely. Within a few years the sweet clover will generally crowd them out. In spite of this fact, sweet clover itself is one of the easiest plants to destroy. Since it only lives two years and must come again from the seed after that time, all that is necessary to clear the ground of sweet clover is to cut it low when in blossom and before the seeds are formed.

Where there is difficulty in establishing alfalfa, sweet clover is often grown in advance to establish the nitrogen-gathering bacteria, which are peculiar to the leguminous plants. Following sweet clover, there is usually little difficulty in getting the alfalfa to grow, if the seed bed is carefully prepared.

There is no pasture crop which will support as many cattle or other live stock as will sweet clover during the second season of its growth. A small experimental plot of little more than an acre yielded two big loads of hay. The plants were permitted to get a good start after the hay was cut before pasturing, then two cows and a horse were turned in for the rest of the season. In addition to furnishing an abundant pasture for three, more than twenty bags of seed were secured. Allowing $15 per ton for hay, $1 a month per head for pasture, and $3 a bag for the seed, all very conservative figures, the crop returned about $96 per acre. While this small plot was experimental, there are numerous farmers who have received more than $100 an acre for seed alone.

Cultural Requirements

It frequently happens that, having seen sweet clover growing along the roadsides, on gravelly banks and other unpromising situations, we are surprised to fail in getting a stand in a well prepared field. Sweet clover requires a firm seed bed, and will not succeed on land where the soil has been deeply stirred and left in a loose condition. It is well to scratch the surface with a tool that does not penetrate deeply, leaving the surface loose for an inch or so, and compact below. While it will succeed on a great variety of soils, it requires that they be in a well settled condition and not freshly plowed to a depth of several inches, such as best suits many forage plants. This condition probably accounts for more failures in getting a stand of sweet clover than any other cause.

Sowing the seed on top of the ground or on the snow in winter, will often secure a good stand with no cultivation at all. Good results often come from sowing it with small grain in spring, on land that has been cultivated the previous season. Some succeed by sowing after the last cultivation of corn, the seed germinating to some extent the same season, while some does not sprout until the following spring. The ideal condition is to cover the seed with finely pulverized soil, with a firm soil underneath.

Time of Sowing

Sweet clover may be sowed in the winter or early spring, as above stated, or at any time from March until August. It should not be seeded when it is likely to start so late that it will not have time to establish itself firmly before winter.

The time of sowing will depend much upon the manner in which the crop is to be handled. Where it is desired to sow the seed on old meadows or pastures without plowing, it will probably be best to scatter it in winter or early spring. The freezing and thawing have a tendency to soften the hard coat of the seed, as well as to cover it with earth. As a field crop, the writer's limited experience would indicate that spring sowing, with a nurse crop that can be cut early, will be best, though winter seeding on stubble should bring good results.

There is a great diversity of opinion as to the proper amount of seed to sow. Where it is used to thicken up meadows or pastures a smaller amount is needed than where sown as a field crop on newly prepared land. Some growers say that 4 pounds of good unhulled seed per acre is sufficient to sow on grass lands. As high as 20 pounds of hulled seed per acre is advocated by some for a field crop. The seed covering is very hard, and, unless treated, only about half of it will grow the first year. If the seed is scarified, the hard coat is scratched until it germinates readily, and much less seed is necessary to secure a stand than otherwise. Ten pounds of hulled and scarified seed per acre should be sufficient on good land.

It is often difficult to get a stand on old land which is deficient in lime, for lack of the nitrogen-gathering bacteria that thrive on the roots of clovers. It is sometimes necessary to treat a small area with a good coat of manure, and sometimes with crushed lime. After the sweet clover is growing well on this land the area can be gradually extended.

Utilizing the Crop

Probably there is no forage crop which will furnish as much pasture per acre as will sweet clover in its second year of growth. It should be allowed to get a good start in spring before stock is turned in, and the area should be sufficiently large for the animals thus kept. Cattle, hogs and horses all eat it with relish after they become familiar with it, and thrive equally on it. It is a common practice to pasture the crop during the first part of the second season and then to turn the stock off and harvest a seed crop. The writer has harvested a very good crop of seed from a limited area, which was pastured lightly through the entire summer until the crop was cut. Of course, it is not possible to pasture heavily after mid-summer, and still secure a good crop of seed.

SWEET FENNEL (Foeniculum vulgare).

The cultivated fennel from Europe has become naturalized in some places along the Atlantic Coast in Maryland and Virginia and other Eastern States. It is often cultivated in gardens in many localities. Lovell lists it as the source of a light amber honey.

SWEET PEPPERBUSH, see Pepperbush.

SYCAMORE (Platanus occidentalis).

The sycamore, also known as buttonwood or buttonball tree, is sometimes reported as important as a source of pollen. It is a large tree found from Maine and Ontario west to Nebraska and south to Florida and Texas.

The California sycamore (*P. racemosa*), found in the interior valleys and foothills of California, is reported as a source of pollen from many localities in that state.

SYMPHORICARPOS, see Indian Currant.

T

TALLOW TREE (Sapium sebiferum).

The Chinese tallow tree or vegetable tallow is a tropical tree with milky juice which has been established in the warmer parts of the United States because of its rapid growth. From Houston, Texas, come reports of yields of sixty to 100 pounds of surplus honey from this tree in early spring. The honey is reported as amber in color and of mild though excellent flavor. In that area local beekeepers report a continuous bloom for about six weeks beginning in early May. Individual trees are reported as flowering about two weeks.

TAMARISK, see Salt Cedar.
TANGLEFOOT, see Aster, also Wild Alfalfa.
TARAXACUM, see Dandelion.

TARWEED (Hemizonia).

There are about eight species of hemizonia or tarweeds in California. Of these Richter lists three as sources of honey and indicates that others may also be of importance. The tarweed (*Hemizonia fasciculata*) is reported as common over a large part of southern California, except on the desert, and north to San Francisco. The blooming period is given from June to August and the honey is said to be dark amber with strong aroma. He states that it is an excellent producer, especially along the coast from Santa Barbara to San Diego, and that the

Tarweed blossoms.

honey is largely used in the manufacture of chewing tobacco and shoe blacking.

The yellow tarweed (*Hemizonia virgata*) is said to be common in the interior valleys and to be a "heavy and consistent yielder, beginning in August and lasting for about twenty days, according to Mr. B. B. Hogaboom of Elk Grove." The honey is stated to be of light yellow color, good flavor and heavy body.

The coast tarweed (*Hemizonia corymbosa*) is reported as yielding some honey, but not nearly as much as the foregoing species.

In Fresno County tarweed, known as yellow-tops, is stated to bloom from April to June and to yield an occasional surplus. The name vinegar weed is sometimes applied to tarweed in the San Joaquin Valley.—Honey Plants of California.

TASSIJILLA, **see Prickly Pear.**

TEASEL (Dipsacus fullonum). FULLER'S TEASEL.

Fuller's teasel is a European plant which was for a time cultivated in this country for the heads of stiff chaff with hooked points, used for raising the nap upon woolen cloth. Although no longer mentioned as a cultivated plant, it has escaped from cultivation and persists as a weed in some places.

During the time when it was generally cultivated in some parts of New York, the late G. M. Doolittle secured more than $1000 per year from less than 100 colonies, credited with being largely from teasel. (American Bee Journal, July 21, 1886.) The high price of teasel in the market led to a great boom for a few years. A beekeeper writing in the American Bee Journal in 1878 stated that the honey-yielding qualities of teasel were equal to basswood. He described the flavor of the honey as excellent, the color white and transparent. Reports of carloads of teasel honey shipped to Thurber & Co., are referred to and the statement made that it is one of the greatest honey-producing plants in existence.

In the same Journal (Aug. 18, 1886), G. M. Doolittle writes a long article giving the history and cultivation of the plant. From this article the following extracts are taken:

> "Bees work on teasel all hours of the day, and no matter how well basswood may yield honey, a few bees will be found at work on teasel. A bee that works on teasel is readily distinguished from those that work on basswood, by the abdomen being covered with a white dust. Black and hybrid bees work on it in larger proportion than the Italians.
>
> "The honey from teasel is very thin and white, in fact the whitest honey I ever saw: but it is not of as good flavor as either clover or basswood. This thinness of the nectar, and its coming just when basswood does is the great drawback to it. From careful tests I should say that it would take about four bee-loads of it to be equal to one bee-load of nectar gathered from basswood. Coming as it does with basswood, makes it of no great advantage, except that it usually lasts six to ten days after basswood is gone.
>
> "Again, my bees have to fly two to ten miles to get this nectar, as I am on the southern edge of the teasel belt. According to those who believe bees fly only 1½ to 2 miles for honey, I should not get anything from teasel. I have repeatedly seen my bees flying to and from the teasel fields from our church door, which is 2½ miles from my apiary in line with the fields.
>
> "As to what proportion of my honey has come from teasel the past fifteen years, I should say about one-tenth; some years more and some years not a single pound. In 1877 I got the largest crop, while from 1878 to 1884 little if any was obtained."

TENDRIL-BEARING SMARTWEED, **see Brunnichia.**

TENNESSEE—Honey Sources of.

Tennessee is a long and narrow state, extending from the Mississippi River, its western boundary, eastward for more than 400 miles to the crest of Great Smoky Mountains. Along its southern boundary are the States of Mississippi, Alabama and Georgia.

Within its borders are found a great variety of plants, as would be expected in a range from an altitude of about 250 feet above sea level in the Mississippi bottoms to more than 6000 feet in the mountains near its eastern boundary.

Taken as a whole, the state of Tennessee has good soils and is well adapted to agriculture. Even in the mountains, grass grows to the highest elevations and forests cover all the slopes. In Middle Tennessee is found a highly developed agricultural region where live stock breeding is a prosperous enterprise.

The climate is mild with rather frequent weather changes, but these changes are far less severe and occur less often than in states farther north. The temperature drops to zero occasionally, but the cold snaps are likely to be short. High temperatures are common in summer. In the higher elevations the summers are less oppressive than in the lower altitudes. The growing season is long, usually from early April until late October. Rainfall is ample, averaging about 50 inches annually.

Because of its latitude, there are few localities in Tennessee where heavy honeyflows are dependable. It is too far south for clover to yield its best except in high altitudes. There are, however, a large variety of sources of nectar which yield heavily when climatic conditions are favorable. In the eastern part of the state sourwood is found in the mountains and in some places yields good crops of fine honey. Since much of the state is timbered, trees may be expected to be of special importance to the beekeeper. Basswood, black locust, tulip-poplar, persimmon, maple and willows, are all important. In limited areas yellow-wood yields well. Among the shrubs may be mentioned, sumac, indian currant, huckleberry and blueberry.

White and alsike clover are grown commonly in meadows and pastures and sweet clover is common in some neighborhoods. Blackberries grow in abundance and yield nectar. Asters and goldenrods are important sources of fall honey. Cowpeas, cotton, dandelions, bitterweed, phacelia and ironweed add to the list. The orchard fruits are valuable sources for spring stimulation.

While conditions are not generally favorable in Tennessee for commercial honey production, the bees find them favorable for breeding over a long season and small crops can be secured with a minimum of effort and small winter losses.

TEUCRIUM, see Germander.

TEXAN EBONY (Siderocarpos flexicaulis).

Texan ebony is a beautiful shrub or small tree common to the Rio Grande Valley in Texas and abundant in northern Mexico. In the vicinity of Brownsville it is regarded as a valuable source of nectar, blooming two or three times during the year. The tree only blooms two or three days, so the flow is short, usually not exceeding a week, as the difference in blooming time between different trees is not great. The honey is of fine quality and light in color. In some localities beekeepers report that the flow only lasts about two days, but is heavy for this short period. Rains bring it into bloom, and in seasons when rains are frequent, it blooms several times.

TEXAS BUCKTHORN, see **Lotibush.**

TEXAS.—Honey Sources of.

Texas is a very large State, with a great diversity of soil and climatic conditions. In order to appreciate its vast extent, one needs to study the map and note something of the variety of climate represented within her borders. Orange, Texas, is very nearly south of Des Moines, Iowa, while El Paso is further west than Denver, Colorado. The northern line is near to southern Kansas, while Brownsville is a

Map showing natural regions of Texas as seen by the author after a tour of the state. 1—Rio Grande Valley. 2—The mesquite region. 3—The east Texas region. 4—The cotton region. 5—The west or plains region.

long distance south of San Diego, California. One might describe almost any condition of soil and climate with which he is familiar in any part of the United States and say with truth that it is like Texas, for almost every condition of soil or climate of the rest of the country is to be found somewhere in Texas. The climate ranges from a winter temperature of 20 degrees below zero in the Panhandle, to an almost frostless condition in the Lower Rio Grande Valley. In east Texas there is a heavy rainfall, with a consequent luxuriant vegetation, while in parts of west Texas one finds a desert flora and little rain. At one point which the author once visited he was told there had been no rain for eighteen months and the very dry and dead condition of everything, even the cactus, indicated that it was true.

Texas seems to be divided into about five natural beekeeping divisions. Each of these has a flora and conditions peculiar to itself, though, of course, there is a gradual merging of these natural divisions. The lower Rio Grande Valley is the southernmost section of the United States, except the extreme tip of Florida. Here we find conditions unlike any other part of Texas. There is a great variety of honey plants, with a light flow almost continuously, but no heavy honey flows. This condition favors the continuous breeding of bees and the consumption of the honey gathered in brood rearing. It is the most favorable situation in America for the rearing of queens and bees, but a poor place for honey production, since the average surplus per colony seldom exceeds twenty-five pounds.

In this section bees swarm as late as December and gather sufficient nectar to carry them through the brief period when vegetation is dormant. The huisache and catsclaw bloom in February and March. Hackberry also begins to yield in February. From February till December there are but short periods without some nectar coming from the field. Brazilwood, horsemint, Texan ebony, blackbrush, privet, coma, mesquite, whitebrush, and many others, yield light flows of nectar.

The Arid Region

The escarpment running east and west between San Antonio and New Braunfels marks a very definite boundary to the belt where cotton is the principal source of surplus honey. South of the escarpment and extending to the valley of the Rio Grande, we have the mesquite region. A line drawn in a southeasterly direction from San Antonio, through Cuera and Victoria to the Gulf, would mark the approximate eastern boundary of this region. A characteristic group of honey plants of southwest Texas. Agarita (barberry), mesquite, hackberry, Mexican persimmon, brazilwood, anaqua, prickly pear and huisache, all valuable plants to the beekeeper, are growing together. They are very characteristic plants for this region. Most of the honey comes from thorny shrubs, including catsclaw, huajillo (pronounced wa-he-ya), etc. This belt extends westward for a long distance and includes the famous Uvalde County, from which hundreds of cars of fine honey from catsclaw and huajillo have been shipped.

The Cotton Region

North of the escarpment already mentioned we find the cotton belt extending north to Oklahoma and including the black, waxy lands and other heavy soils. Within the cotton district, we find the highest development of agriculture in Texas. The soil is rich and the climate mild. Cotton, corn, alfalfa, small grains and truck crops are all profitably grown. Mesquite is an important source of nectar in the cotton belt, as is also horsemint, and in the northern portion, sweet clover. Broomweed is also important, some seasons.

East Texas

In the timbered regions of east Texas there is some splendid beekeeping territory. There is a variety of flora to be equalled in few sections, and remarkable yields of honey are reported. On heavy soils

Salmon berry, or thimble berry.

cotton yields freely. Basswood, rattan, huckleberry, partridge pea, horsemint, bitterweed and cowpeas are all reported as sources of surplus honey in this section. Beekeepers of extensive experience report that one can depend upon an average of more than one hundred pounds per colony per year for a ten-year period.

Although the average yield is much higher and the flows are much more dependable than in other parts of Texas, the business of beekeeping is less highly specialized here than in other parts of the State. This is probably due to the fact that general agriculture is profitable and public attention has not been called to the possibilities of this section.

In American Bee Journal, Sept. 1921, H. B. Parks gives an extended account of honey areas of Texas with map.

THIMBLE BERRY (Rubus parviflorus).

The salmon berry or Thimble berry, also known as flowering raspberry, is a well-known honey plant on the Pacific Coast. It occurs along the coast from Alaska to Mexico. It also occurs in the mountains as far south as Colorado and New Mexico. It is a shrub with erect stem and red fruit. It is known to some extent in Ontario west to Dakota.

THORN APPLE, see Datura, also Hawthorn.
THOROUGHWORT, see Boneset.

THYME (Thymus).

Thyme is an introduced plant from Europe which is sometimes cultivated in gardens and occurs occasionally as an escape. There are two species listed in this country *Thymus vulgaris,* the source of the famed honey from Mt. Hymettus, and *Thymus serpyllum,* the wild or creeping thyme.

According to J. E. Crane, thyme is found in southwestern Vermont in sufficient abundance to make excellent pasture for the bees during August.

T. serpyllum has become established over several thousand acres in Delaware County, New York. J. B. Merwin, of Prattsville wrote that it completely carpeted the ground for miles along roadsides and in the pastures. He regarded it as an ideal honey plant, beginning to bloom about the 15th of July as basswood was finishing its flow, and continuing until killed by frost in the fall, sometimes as late as November. No matter what the season, he never failed to get honey from thyme for a period of more than twenty-five years. One year he secured 4,000 pounds of comb honey from sixty colonies, and increased to 85 colonies. The honey was of the finest quality, and once a customer tried it he continued to want it permanently, or as Mr. Merwin stated it, "A customer once, always a customer." (Gleanings, Aug. 1,

1914). In a letter to the author Mr. Merwin states that the honey is a very light amber, of good body and of very good quality, but the bees do not winter on it as well as on clover honey. As high as 125 pounds per colony average has been secured from thyme in that portion of New York.

There are localities in Massachusetts and nearby states where the

Wild thyme.

plant has become well established, where honey from this source is secured in considerable quantity.

TIEVINE, see Morning Glory.
TISSWOOD, see Red Bay.

TI-TI.

The leatherwood or black ti-ti (*Cyrilla racemiflora*) is a shrub with shining leaves and large numbers of small flowers in clusters. It occurs in swamps from southern Virginia to Florida and west to Louisiana and Texas. In Alabama and Mississippi it occurs from the Central Pine Belt to the Coast Plain, in the edges of swamps and along streams. It is also known as white ti-ti, ironwood and red ti-ti. It sometimes reaches the height of 20 to 30 feet, becoming a small tree, though usually it is found as a shrub. The thin bark breaks up into large scales.

The bloom opens in June and July and is not regarded as a very dependable source of nectar. The honey is dark, with a mild flavor.

In Gleanings, (Dec. 1935), C. E. Burnside indicates that the so-called disease known as "Purple Brood" is caused by poisoning from leatherwood. He states:—

> "Brood starts dieing of purple brood about the time leatherwood is in full bloom. Most of the cases are found during June. Soon after bees stop working flowers of leatherwood the high death rate of brood ceases and the colonies recover.
>
> "The occurrence of purple brood in other states and reports of beekeepers whose colonies were affected support the assumption that the cause of purple brood is poisoning by southern leatherwood. * * No samples or reports of purple brood from states outside the range of leatherwood have been received at the Bee Culture Laboratory."

The ti-ti (*Cliftonia monophylla*), sometimes called buckwheat-tree or ironwood, also known as black ti-ti, occurs in wet, sandy soil and swamps from South Carolina to Florida and west to Louisiana. It is an attractive tree, often reaching a height of 30 to 40 feet. This shrub or tree, as the case may be, often grows in dense thickets called ti-ti swamps. The flowers are white and fragrant and appear in late February to April. There are large areas of ti-ti swamp in Georgia, Alabama and Mississippi, which furnish abundant bee pasture. According to Baldwin, it yields surplus honey in the extreme northwest portion of west Florida. He describes the honey as red, strong, and suitable mostly for baking purposes. (Gleanings, March 15, 1911.)

Ti-ti is of more or less value as a source of nectar in the swampy districts of all the Southeastern States from South Carolina to Louisiana. The honey is not of the best quality, nor is the yield generally heavy.

In the swamps and along streams in Florida and Louisiana is found *Cyrilla parvifolia*, a leathery-leaved evergreen shrub which blooms in spring. Frank Stirling, of the Florida College of Agriculture, mentions this species as one among the most valuable honey plants in Florida. He states that it is found in abundance in Calhoun, Liberty and Johnson counties. The principal drawback to this region lies in the few nectar-bearing plants which are available following the close of bloom from ti-ti and tupelo.

TOBACCO (Nicotiana tabacum).

The tobacco plant is a coarse annual. It is grown as a field crop in the South and also in a few northern localities, especially in Wisconsin and Connecticut. As a honey plant it is probably seldom important.

The fact that the plants are usually cut in advance of the time when the bloom is at its best, would make it unavailable, for the most part, as a source of nectar.

> "We are in a location where hundreds of acres of tobacco are raised every year. I have taken bees and placed them near the fields and they will store some honey from the plant some years. It is very dark, much like buckwheat in color, strong and very heavy body. Buckwheat is not my favorite honey, but I can eat it. Tobacco honey I cannot. It is very slow to

Tobacco blossoms about natural size. **(Photo by Lovell).**

granulate, and I have never seen it harden as other honey will, even when well ripened and two years old."—W. K. Rockwell, Bloomfield, Conn. American Bee Journal, page 63, 1919.

"Tobacco yields a heavy flow under certain conditions in Porto Rico."—Henry Brenner, in American Bee Journal, page 381. 1916.

"Tobacco in this section has always been raised in the open field and, when about 4 feet high, each plant has been 'topped' and not allowed to go to seed. * * * It is now being picked in the field instead of being cut by the old method. The plant is allowed to grow from 7 to 10 feet high and goes to seed. The leaves are saved by picking, this work commencing at the bottom, one row of leaves being gathered at a time, and the top leaves picked last. The plants are thus allowed to blossom, each one bearing hundreds of flowers, and they continue to bloom from August 1 till frost. Thus we have thrown open to our bees hundreds of acres of tobacco, containing myriads of flowers. The bees swarm on it, some days more than others, and the honey comes in as fast as during the earlier flows."—E. H. Shattuck, Granby, Conn. Gleanings, page 268. 1911.

TOCALOTE, see Star Thistle.
TOLLON BERRY, see Christmas Berry.
TOOTHACHE-TREE, see Prickly Ash.
TORNILLO, see Mesquite also Screw-bean.
TOY-ON BERRY, see Christmas Berry.
TRAILING ARBUTUS, see Arbutus.

TREE CLOVER, TREE ALFALFA or TAGASASTE (Cytisus proliferus alba).

The tree alfalfa, or white broom, is grown to some extent in California. In Australia it is regarded as important, as the following extract from Rayment (Money in Bees in Australasia), will show:

> "This rapid-growing hedge plant is now widely known as a grand honey yielder. As a wind break for the apiary it is unrivaled. The white blossoms burst the sheaves early in spring, almost before the winter has departed. Bees work upon it during a shower, as the drooping habit of the flowers prevents the nectar washing out with the rain. The pollen is cadmium in color and the honey very pale and clear, rather thin and of mild flavor. To make a close hedge it should be severely cut back. Unfortunately, the sheep and cattle, also the kangaroos are fond of it and keep it eaten back."

It is native to the Canary Islands and adjacent regions. As it thrives in California, it is promising for trial in the warmer parts of America.

TREE HUCKLEBERRY, see Farkle-Berry.
TREE OF HEAVEN, see Varnish Tree.

TRILLIUM.

One of the first spring flowers, the trillium, of which there are several species, is attractive to the bees. It is commonly known by the name of wake robin, or birthroot. As it blooms so early, it is seldom that the bees find many good days for flying, and such early spring flowers can never be regarded as important except for the purpose of stimulating brood rearing.

TROPICAL LILAC (Duranta plumiere). GOLDEN DEWDROP.

The tropical lilac is a hedge plant cultivated in Florida and California and other southern localities. W. K. Morrison lists it in Gleanings (Aug. 1, 1905), as extremely attractive to the bees and as blooming for some time. Probably not sufficiently abundant in America to be important anywhere.

TRUMPET CREEPER, **see Cow-itch.**
TRUMPET WEED, **see Boneset.**
TULE or TULE-POTATO, **see Arrow-head, also Pickerel-weed.**
TULEMINT, **see Mint.**

TULIP-POPLAR or TULIP TREE (Liriodendron tulipifera).

The tulip-tree, also known as yellow poplar, is a very large tree, often growing to a height of 100 to 140 feet, and a diameter of 6 to 9 feet. It is found from southern New England west to southern Michigan and south to the Gulf States. It is also found to a limited extent in southeastern Missouri and eastern Arkansas. It blooms in April and May and produces a dark amber honey of strong flavor.

According to Buchanan, the honey yield from this source is heavy and the tree is an important addition to the nectar-secreting flora of Tennessee and nearby States.

The possibilities of this source of nectar are not properly appreciated. Since it blooms so early in spring, few colonies of bees are sufficiently strong to gather the crop possible from tulip-poplar. The skilled beekeeper, who can bring his colonies through the winter in good condition, gets large yields of honey from this source. In the vicinity of Washington, D. C., it is the principal source of surplus, and strong colonies often store an average of 100 pounds. In many cases the bees build up on tulip-poplar only to become strong after the flow is over. In locations where this tree is common, too much care cannot be taken to get strong colonies early in spring to take advantage of this flow.

Farmer's Bulletin 1222 by E. F. Phillips and George S. Demuth is devoted to "Beekeeping in the Tulip-Tree Region." The authors state that tulip-tree is one of the most dependable early sources of nectar throughout a wide area.

G. E. Marvin while in the Bureau of Entomology made a study of the nectar flow of the tulip-tree. The average weight of nectar secreted by one flower was 1.64 grams. The average sugar content of nectar from freshly opened flowers was 16.7 per cent, increasing with the open flower to 35.9 per cent. The tree under observation was calculated to yield enough nectar to produce 2.16 pounds of honey.

Marvin observed that to one standing near a tulip-tree at the

Blossoms of tulip-poplar.

height of the blooming period with a gentle breeze, the falling drops of nectar give the impression of a light rain. During a favorable season nectar in tulip-tree blossoms is secreted so abundantly that honeybees and wild insects cannot carry it away as fast as it appears. (Am. Bee Journal April 1933.)

From an article on tulip-tree by Edgar Abernathy of North Carolina in American Bee Journal June, 1940 the following is quoted:

> "I have known it to secrete so profusely that great drops dripped out of the blossoms and were noticed on the ground beneath. So far as I have been able to learn, it always yields, regardless of weather. * * This nectar contains a very high percentage of sugar, and could not possibly be more accessible to the bee, which can get a full load on almost any petal it chances to choose.
>
> The quality of the honey is not of the best. It is so dark that it is often referred to as 'black poplar honey,' although it is really dark red. The flavor is not equal to the best either. Of course it is usually mixed with honeys from other sources, which improve both color and flavor."

TUNG-OIL TREE (Aleurites Fordii).

The Tung tree has been widely planted in Florida and along the Gulf Coast in recent years. It is the source of tung-oil or China-wood oil extracted from the seed. This oil is used in large quantity in the paint trade and millions of pounds are annually imported from China.

R. C. Daniels of Picayune, Miss., who is located near large tung orchards reports that the bloom yields enough nectar to attract the bees in large numbers but not enough to be of much value to the beekeeper.

The tree is an early bloomer and a source of pollen.

TUPELO (Nyssa).

The tupelo trees yield large quantities of honey in the swamps of the southeastern states. There has been a great deal of confusion regarding the different species, due to the use of common names by beekeepers when writing about them in the bee magazines. Dr. Everett Oertel of the Southern States Bee Culture Laboratory, has done much to correct misunderstandings in an article in the American Bee Journal, July 1934.

According to his conclusion, the white tupelo, *Nyssa ogeche,* is the most important of the group and is the source of the heavy honey crops along the Apalachicola River in Florida. This species is found from the South Carolina coastal district, through Southeastern Georgia and Western Florida where it blooms in May. This tree is also known as Ogeche-gum or Ogeche-plum and sometimes as wild lime tree. The honey is of good quality and when unmixed with other sources, will not granulate. For this reason, it is in much demand by bottlers.

The black tupelo, *Nyssa biflora,* is also common in this area but extends west to Louisiana and eastern Texas. This species blooms in April and is the source of a less desirable honey that is darker in color and sells at a lower price. This species is also known as black gum, water gum and water tupelo.

Blossoms of the tupelo gum.

Tupelo gum or cotton gum, *Nyssa aquatica,* is a large tree which is found from Virginia to Florida and Texas. It has long been regarded as the most important to the beekeeper and it does yield heavily in certain localities. Oertel records a gain as high as 17 pounds per day for experimental colonies at Port Vincent Livingstone Parish, Louisiana. This tree blooms in early April. Oertel says from April 1st to the 20th in South Louisiana. He found also that the honey from this species will granulate when subject to low temperatures.

Another species, the black gum or sour gum, *Nyssa sylvatica,* is apparently of less importance. It is widely distributed throughout the Gulf States and is also found northward to New England and Ontario. It grows to some extent, on the uplands as well as in marshy areas and is sometimes called highland black gum or pepperidge. It is regarded as of some value as a source of honey by some beekeepers inland from the Gulf Coast but is not recognized as important by Northern beemen as far as the writer can ascertain.

Apparently all four species yield honey in abundance and all are of major importance in some localities. It appears, however, that it is the white tupelo, *Nyssa ogeche,* which is the source of the famous crops of non-granulating honey which is so well known in the markets.

In favorable seasons a thousand barrels or more of this fine honey are produced in the Apalachicola area. A. B. Marchant reported an average of 82 pounds per colony of surplus honey from tupelo for a seventeen year period on the Apalachicola River, in Florida. During part of the time he kept as many as 500 colonies in a yard. In 1904, he took 250, thirty gallon barrels from 750 colonies, an average of about 120 pounds per colony.

Tupelo is the source of much honey in South Arkansas, east Texas and other southern states, although it is most often reported from the swamps of Alabama and Florida. Beekeepers complain that in many good tupelo locations there is a shortage of summer pollen so that it is necessary to move the bees away when the flow is over.

Because of the good body and mild flavor combined with non-granulation, tupelo honey brings a higher price than most southern honey brings.

TURKEY MULLEIN (Eremocarpus setigerus).

The turkey mullein is also known as coyote weed, woolly white drought weed, and verba del pescado. The latter, according to Jepson, is the Spanish-California name, derived from the custom of the Indians to use the heavy-scented herbage of this plant to stupefy fish in small streams, to catch them by hand. He gives the range as dry open areas from the plains of the Sacramento and San Joaquin, Sierra Nevada foothills to the low hills and valley fields of the Coast Ranges. It grows mostly in stubble fields following the grain harvests, contesting for survival with the blue curls.

From Western Honeybee we quote as follows:

"Turkey mullein is a different appearing plant (from blue curls), but grows under exactly similar conditions. Its insignificant pale green or white flowers yield a thick amber honey, of biting, astringent flavor, often in considerable quantities.

"I have known a beekeeper in San Diego County to market eight tons of this honey. It is fit only for manufacturing purposes. The foliage of turkey mullein is white, wooly, and exhales a pungent odor."—October, 1914.

TURNIP (Brassica).

Where turnips are grown for seed in large acreage, surplus honey may be expected from this source. The late J. S. Harbison said of them:

> "Turnip blossoms are eagerly sought by the bees, and afford so rich pasturage during March and April as to make it a profitable crop, if but for this purpose alone."—Beekeeper's Directory.

TURNIP WEED, see Boneset.
TURPENTINE WEED, see Blue Curls.
TWIN-BERRY, see Honeysuckle.

U

ULEX, see Gorse.
UMBRELLA TREE, see Wild China.

UTAH—Honey Sources of.

Utah produces large quantities of fine quality honey, mostly in the irrigated valleys. With her high ranges of mountains there is a great variety of conditions to be found within her borders. The greater part of the surplus honey reaching the markets from Utah is from alfalfa and sweet clover. The Uintah Basin, although having many points which are distant from railroad facilities is a famous honey-producing district.

Utah has a great variety of natural resources and many of the honey plants common to both east and west are found there. There are areas where commercial orcharding is an important industry and apple trees furnish large quantities of nectar for the bees. In other valleys canteloupes, watermelons and cucumbers are largely grown, as well as onions for seed, and all these are valuable sources of nectar.

Of the trees, box elders, maples and willows are to be found in abundance in many places, as well as wild plums, choke cherries and serviceberries. Among the weeds may be mentioned dandelion, mustard and gumweed. In the dry sections rabbit brush and greasewood are common. In limited areas in the southern part of the state mesquite and catsclaw are found also.

Clematis is common in the canyons and prickly pear on the plains. To the above list may be added snowberry, native currant, filaree or pin clover, and in some localities cleome. Ephedra is worked freely by the bees in the desert. White clover is reported as important in some sections.

V

VACCINIUM, see Blueberry.

VANILLA-PLANT or DEER-TONGUE (Trilisa odoratissima).

The vanilla-plant or deer-tongue is common to pine woods from Virginia to Florida and Louisiana, blooming in summer and fall. The blossoms are purplish, sometimes almost white. The heads contain several flowers, each arising from its own stem.

It is reported as a valuable honey plant only from Florida, but may be equally valuable elsewhere. C. H. Clute, of Sanford, Florida (Gleanings, Nov. 15, 1916), states that in August the deer-tongue yields well, giving a lot of nice honey. A report from southwest Florida credits it as a splendid nectar source with hundreds of acres available locally.

VARNISH TREE or TREE OF HEAVEN (Ailanthus).

This tree is met by a great variety of names, varnish tree, tree of heaven and Chinese sumac being common. The tree is a native of China and has been extensively cultivated as a shade tree in America. It spreads easily and is quite generally naturalized. Scholl lists it as a source of honey and pollen in Texas. Richter regards it as a wonderful yielder in California, where it produces an abundance of ill-tasting honey. It is reported as a honey producer in Georgia and is to be found in many States from east to west in greater or less abundance. The disagreeable odor of the staminate flowers is a well-known characteristic of the tree. In appearance it looks much like a very large sumac.

In the vicinity of Paris, France, the ailanthus, also called ailanthe, yields honey which must be removed before other honeys are harvested, as it spoils the taste of the honey of other plants.

Concerning the Ailanthus tree in Austria, A. Alfonsus writes in the August, 1921, Bee World, as follows:

> "The Ailanthus tree was introduced into Europe from China toward the end of the 18th century. Single specimens of it are met with in park and botanical gardens. The first was planted in Vienna in 1870. As it grows easily and rapidly from seeds, it has become so numerous in and near Vienna that the honey harvest from it exceeds in quantity that of all other plants taken together. The honey has a dirty green color; if a drop of it is spread on white paper, it looks as if it contained fine particles of dust. The flavor of the freshly extracted honey is too strong to be pleasant; but after a few weeks it becomes so fine that experts consider the honey from the Ailanthus tree to have the finest flavor of all the numerous varieties of honey produced in Central Europe. It crystallizes with a fine grain."

VELVET BEAN (Mucuna utilis).

The velvet bean is an important field crop in the southeastern states. It is an introduced plant which has found especial favor on the sandy soils of the Gulf Coast region. It is a luxuriant grower, single plants often attaining a length of forty to sixty feet in a season. The plant takes its name from the dark, velvety hairs covering the seed pods. Velvet beans are usually planted in rows and given field cultivation for a time. Later the plants cover the ground and smother out weeds and other growth. A large amount of forage is produced but it is not readily harvested, hence is often utilized by turning live stock into the fields to pasture it down. In Florida yields of twenty to thirty bushels of beans per acre are secured. As a soiling crop it is very valuable.

In localities where velvet beans are extensively grown considerable honey is secured from them. They begin blooming about August 15 and continue for about six weeks. Strong colonies are often reported to yield 100 pounds of surplus from this source.

The honey is regarded as of inferior quality, with a pronounced flavor. It has good body and is of a golden or amber color.

R. B. Willson says (Gleanings, December, 1922) that while it produces nectar freely, he has not found the bees working on it in Mississippi. J. Clay Dickman, in the same magazine, says that bees work it freely immediately after a rain in southern Alabama. W. C. Barnard states that velvet beans do well on the Tifton and Norfolk loam soils of south Georgia and on these soils can be counted on for a surplus year after year. Planted with corn, these beans are usually found in sufficient area to be of considerable value to the beekeeper in that region. He further states that he can rely on an average of one shallow super of chunk honey from this source. He regards it as of little value for honey on sandy soils.

Because of the poor quality it is principally used for feeding back to the bees.

VERBENA, see Vervain.

VERMONT—Honey Sources of.

Vermont appears to be better adapted to beekeeping than neighboring states. The western part of the state bordering on Lake Champlain offers perhaps the best territory. Through the center, the Green Mountains are largely covered with forest and offer but little bee pasture, while east of the mountains apiaries are smaller. In the Champlain Valley the limestone soils are favorable for farming and the clovers are abundant. Willows, maples, dandelions and fruit bloom furnish early nectar, while the principal crop comes from clover. On some soils buckwheat yields surplus. Sumac is important in some localities. Raspberry, goldenrod, asters and basswood may be counted on where conditions are favorable.

There is a long list of minor sources of nectar and pollen available to Vermont bees.

VERNONIA, see Ironweed.

VERVAIN (Verbena).

There are about 16 species of verbenas in North America, and several of them widely distributed. The figure shows the blue vervain (*Verbena hastata*), which is found from Nova Scotia to Quebec and Manitoba, south to Arkansas, New Mexico and California, and on the east, south to Georgia. This particular species is usually found in low lands, along streams, etc. Richter, in his "Honey Plants of California," mentions another species, *Verbena prostrata,* as yielding considerable honey in some localities in that State. In Iowa the hoary vervain (*Verbena stricta*), is very common in upland pastures, especially over the north half of the State, and reports of surplus honey from this source are not uncommon.

Bloom of the blue vervain.

Mr. Scholl reports *Verbena xutha* as yielding sparingly in Texas. While in the main the vervains can hardly be regarded as important honey plants, in limited localities some species are very valuable sources of nectar. Mr. S. W. Snyder, former Secretary of the Iowa Beekeepers' Association, reports the blue vervain as quite valuable in his locality, some years furnishing a surplus.

Vervain is the principal source of surplus honey over an area of considerable extent from Baton Rouge to near New Orleans in Louisiana. Three species are reported as yielding, *V. bonariensis V. rigida* and *V. littoralis.*

Verbena bonariensis which appears to be most important in many places is native to South America and came to this region as an immigrant from Brazil or Argentina. Apparently it was first noticed in this country in Mobile County, Alabama, in July 1893 and since that time has become common along roadsides and in waste places near the Gulf Coast much to the advantage of the beekeepers.

The honey is dark in color with pronounced though pleasing flavor.

In a letter to the author George W. Bohne of Luling, Louisiana wrote as follows:

> "This is a honey which often gives trouble to inexperienced operators as it is often gassy—not actually fermented—when freshly gathered we find large patches blown loose from the comb which form 'white blisters' and show frothy bubbles within the cells. When put into storage tanks such honey granulates very quickly. Moderate heating before storage overcomes this condition. I figure that vervain honey amounts to about forty percent of our total yield."

VERVENIA, see Phacelia.

VETCH (Vicia).

There are about a dozen species of vetch native to America and two or three varieties widely cultivated for forage. They yield some honey, and there are reports from beekeepers to the effect that their bees work upon vetch freely. Richter lists spring vetch as a source of honey, but does not indicate that it yields surplus in California, where it is grown as a cover crop. Some species have extra-floral nectaries.

Some of the group are regarded as good honey plants in Europe.

Concerning hairy vetch, which is widely cultivated, W. J. Sheppard states that "the bees are particularly fond of this plant and in some of the fruit growing districts of British Columbia, where the plant is largely grown in the orchards as a cover crop, it is a never failing source of nectar. The honey is a pale amber, of nice flavor, and large yields are obtained during June and July."

Edwin Trinder, of Simco, Ontario (Gleanings, Nov. 15, 1916), writes:

> "Honey from the sand vetch is mild in flavor, but it has a dark amber color. As the vetch blossoms at the same time as the white clover and alsike, the bees mix the two kinds of honey together, thus, of course, spoiling the color of the white honey. * * *
>
> "The honey, however, is so mild that many of our customers think that it improves the clover flavor."

Prof. A. L. Lovett writes to the author that "in western Oregon vetch is an excellent yielder of a white honey of very heavy body, in fact so heavy that it is rather difficult to draw it from the cells in the extractor without loosening the wax as well. This is yielded from super nectaries or is extra nuptial nectar. Frequently these glands are present at practically every place on the stem where it branches off. I have seen nectar produced in vetch at least 10 days before the blossoms open."

W. L. Arant, Forest Grove, Oregon, writes, "the honey is often as white as fireweed but of heavier body and richer flavor. Some years the flow is exceedingly heavy filling a fifty pound super in three days."

George H. Rea formerly of Cornell University, tells of a wild vetch growing on sandy hills in Northern New York. It thrives on land so poor and dry that it will support very little other vegetation. The bees work it freely and make a white honey of fine quality.

Concerning vetch in North Carolina, Edgar Abernathy writes, (Am. Bee Journal May 1940):

> "In my location it ranks second only to tulip-poplar as a source of surplus; in one very wet season it took first place. The honey is of average quality; light amber in color, and of good flavor, although just a shade strong for some palates."

VINEGAR-WEED, see Blue Curls, also Tarweed.

VINE MAPLE (Acer circinatum).

Vine maple is common on the Pacific Coast from Humboldt County, California, north to Washington. There are numerous reports

Virginia creeper, or American ivy.

of vine maple as an abundant source of early nectar in Oregon and Washington. The honey is amber and of excellent quality, but it blooms so early (April) that the bees are usually not in condition to store much surplus. (See Maple). In exceptional seasons as high as 50 pounds of surplus have been gathered by a single colony from vine maple.

VIPER'S BUGLOSS, see Blueweed.

VIRGINIA—Honey Sources of.

Conditions in Virginia are comparable to those of Maryland on the north and North Carolina on the south, both of which are described somewhat in detail elsewhere. Reference to the descriptions of these states will supply information which will apply generally to Virginia.

Like them, Virginia has a tidewater region near the coast, an intermediate area and a mountain section westward. In the vicinity of the swamps in the lowland country, the bees gather honey from maples, tupelo, gallberry and numerous fall flowers. In the Piedmont region, tulip-tree and sourwood, sumac, black locust, the clovers, persimmon, goldenrod and asters yield surplus.

In the higher elevations fruit growing is extensively followed and surplus is sometimes gathered in the orchards. Sweet clover is coming in and basswood is locally important. Honeydew is stored in large quantities some seasons.

Commercial honey production is but little developed in Virginia, although a few beemen have demonstrated the possibilities of that region. Few states would offer a greater total variety of honey plants, but the average yields per colony are smaller than prevail in most regions where beekeeping is extensively followed.

Hundreds of different plants yield something to the bees.

VIRGINIA CREEPER (Parthenocissus quinquefolia).

The Virginia creeper, also known as American ivy or woodbine, is a common climbing vine in thickets and woods from New England to Quebec and Manitoba, Dakota and Colorado and south to the Gulf from Florida to western Texas.

While the bees seek it eagerly at times and the vines fairly hum with them, it can hardly be regarded as of importance to the beekeeper.

This plant is often confused with poison ivy, but the two plants can easily be distinguished by the difference in habit of growth, and by the five leaflets in the creeper, while the poison ivy has only three leaflets to each leaf.

In Texas the seven-leaved ivy (*P. heptaphylla*) is often called by the name "cow-itch" and is regarded as a good honey plant.

Blossoms and leaf of Virginia waterleaf.

Like the foregoing species its value is probably limited because not often sufficiently common. The nectar yield is abundant. For related species see Cow-itch and Ampelopsis.

VIRGINIA WATERLEAF (Hydrophyllum virginicum).

The Virginia waterleaf does not bloom until after the fruit blossoms are gone, and so has less competition for attention than some other plants that come into bloom during the same period. It blooms abundantly and grows luxuriantly in moist woods. The bees have been so eager for the blossoms of this plant in the writer's wild garden and in the surrounding woods for several years past, that he has come to regard it as quite a valuable honey plant. Apiaries in the vicinity of woodlands should find this plant of considerable value.

Virginia waterleaf, massed in the author's wild garden at Atlantic, Iowa.

Some surplus has been reported from this source in Wisconsin.

VIRGIN'S BOWER, see Clematis, also Wisteria.

VITEX NEGUNDO INCISA.

A plant introduced by the Bureau of Plant Industry. Described as follows by Frank N. Meyer, explorer, who found it at Shantung, China:

"A sage which may prove to be a good plant for the arid Southwestern States. It is able to resist alkali remarkably well. The Chinese use it for basketry manufacture, taking the annual shoots for this purpose. It has pretty blue flowers and is diligently visited by all kinds of bees, and as such it might be grown in gardens as a semi-ornamental shrub. It grows, when left alone, up to 20 feet tall."

Specimens tested by the author in Iowa, which were mere whips about 20 inches tall, began blooming the same season, the last week in July, and had not entirely faded on September 10. The bees sought it eagerly and apparently, if abundant, it would be a valuable honey plant.

H. B. Parks writes to the author that this species is now becoming very common in the San Antonio district of Texas. He states that it makes a wonderful growth and that it is a most remarkable honey plant.

Vitex agnus-castus, according to Small, has become naturalized from the Old World on sandy soils from North Carolina to Florida and Texas. H. B. Parks states that this species is not of much value to the bees, although it blooms profusely from June till October. It is commonly known as alhuzama or Mexican lavender.

W

WAGNER PEA (Lathyrus silvestris wagneri).

In 1894 the bee magazines told the story of a German plant breeder named Wagner who had devoted thirty years to an attempt to improve a wild pea common to the mountains of Europe. He was search-

Wagner pea produces a heavy yield of forage.

ing for a good forage plant which was also the source of honey. When he offered it to the world there was much interest and numerous trials were made in this country.

When seed was sought for the American Bee Journal test garden (1938) it was so little known that several months passed before we were able to find seed from any source. After a small amount had been secured from Germany a few small fields were discovered in Washington which remained from the original planting.

The plant is slow in starting but once established will last for many years. It is deep rooting, drouth resistant and produces heavy yields of forage.

When bees visit the flowers they seem to find it necessary to make but few stops to secure a load and spend much time sucking a single blossom. Reports from western beekeepers indicate that it yields very satisfactory crops of honey and indications are that it is worthy of far more attention than it has so far received in this country. An extended history of the plant appeared in the December, 1941 issue of American Bee Journal.

WA-HE-YA (Huajillo), see Acacia.
WAHOO, see Hop-Tree.
WAKEROBIN, see Trillium.

WALNUT (Juglans).

The black walnut is a well-known forest tree in the eastern United States. Its usual range is from Ontario and New England west to Nebraska, and south to Florida and Texas. The wood is very valuable for the manufacture of gunstocks, furniture, etc., and is becoming somewhat scarce. The tree leaves out later than most forest trees, not developing its foliage until May or June. The blossoms are long catkins borne on the wood of the preceding year. The blossoms appear before the leaves. Quantities of pollen are produced, and, at times, the bees seek the trees in such numbers as to make a continuous roar. The walnut blooms after the maples and willows, and is not as valuable as earlier blooming trees, because it comes at about the same time that the dandelions are in bloom. May is the month of blossoming in most northern localities.

The white walnuts or butternuts of the Eastern States, and the English walnuts, Japanese walnuts and California walnuts grown in the warmer parts of the country, especially in California, are relatives of the black walnut, and probably equally valuable for pollen.

There are reports of occasional crops of honeydew stored by the bees in surplus quantity from the walnut groves in California. The product is of inferior quality and sells at a low price for baking or manufacturing purposes.

A. C. Burrill of the Missouri Resources Museum reports a flow of honeydew of unusual intensity from black walnut in Jefferson City. Fallen leaves and twigs were sticky with the honeydew and bees were humming in the walnut trees as loudly as when basswood is in flower. The walnut leaf hopper, (*Oncopsis distinctus*), appeared to be the source of the honeydew.

Pollen-bearing bloom of black walnut.

WAMPEE, see Pickerel-weed.

WASHINGTON—Honey Sources of.

Washington, the state which lies farthest to the north and west has a land area of slightly more than 69,000 square miles. It is divided into two natural divisions with very different climates. The area west of the summit of the Cascade Mountains has an abundant precipitation which is unequally distributed through the year. About eighty per cent of the rainfall is during the wet season beginning in October and ending in May. In the coast Counties there is a heavy rainfall averaging from 60 to 138 inches annually in different sections. In the Puget Sound Basin the rainfall is considerably less, ranging from 21 to 55 inches annually. On the western slope of the Cascades the rainfall is often above 100 inches. In the high elevations there are heavy snows which move off slowly.

In this region the temperature is not subject to great change. The winters are mild, the summers cool and the growing season is long.

East of the Cascade Mountains lies the Great Plain of the Columbia River. It is a high rolling plateau with few trees. There are also low mountain ranges in the northern and southern counties which are covered with forest.

Rainfall is scant in this region and so-called dry farming is practiced where irrigation is impossible. Great extremes of temperature prevail, ranging from 20 or more degrees below zero in winter to more than 100 degrees above zero in summer.

Beekeeping is important principally in the irrigated regions where alfalfa and sweet clover are abundant. The Yakima, Columbia, Okanagon and other Valleys offer some good locations.

In the west part of the state good yields are secured from fireweed, white clover, snowberry and other plants. In this part of Washington we find the greatest variety of honey sources and the largest number of apiaries. (H. A. Scullen, American Bee Journal, March, 1921).

The principal honey sources are sweet clover, alfalfa, white clover, alsike clover, fireweed, the maples, dandelion, locust, Cascara sagrada, Cat's ear, snowberry, milkweed, goldenrod, madrona, dogbane, wild sunflower, phacelia, horehound, mustard and huckleberry.

WATER CHINQUAPIN (Nelumbo lutea). Duck Acorn.

The water chinquapin, also called American lotus and duck acorn, as well as water-nut and yankapin, is distributed over a wide area. It is found in ponds and streams locally from Massachusetts to Minnesota and from Florida to Louisiana and Texas.

In a personal letter to the author, H. B. Parks writes as follows concerning this plant:

> "*Nelumbo lutea* occurs in vast areas in the waters of east Texas. Nothing can equal the sight of this immense plant with its leaves and flowers 6 to 8 feet above the water and so thick a boat cannot be pushed through the stand. The flowers are often as large as a cabbage head and the nectaries are large. At the Elkhart Lake and at Long Lake, near Palestine, in the spring of 1919 bees secured thirty pounds of surplus from this plant."

Miss Hasslbaur also wrote the author about visiting at Elkhart, near Palestine and being particularly interested in a wonderful honeyflow from the water lilies. She stated that a beekeeper who had a number of colonies near the lake had a good surplus from this source.

WATER ELM (Planera aquatica). PLANER TREE.

The water elm is a small elm-like tree common to the southern states from southern Illinois to western Florida and Texas. It is found principally in densely shaded lowland areas where it blooms in early spring. Everett Oertel of Baton Rouge, Louisiana has found it worked heavily by bees for both nectar and pollen. It is reported as a small tree of 20 to 30 feet in height and seldom more than a foot in diameter.

WATER-GUM, see Tupelo.

WATER HOREHOUND (Lycopus). Bugle-Weed.

There are several species which are found in wet places in the states east of the Missouri River from New England to the Gulf Region. They are also found to some extent on the Pacific Coast. They are similar to the mints, but less aromatic. The flowers are small and closely clustered in the axils of the leaves. In the vicinity of marsh lands the beekeeper is likely to find that the water hoarhound contributes nectar in important quantity. F. L. Wright, of Plainfield, Michigan, writing for the Beekeeper's Exchange, Nov. 1881, credits water hoarhound as follows:

> "During the severe drouth two years ago we were surprised to find that our bees not only gathered honey enough to keep the brood-rearing going on finely, but some of the most industrious were actually storing a little. They seemed to gather it quite slowly, but kept at it all day long. Since then we have watched it very closely and find bees at work on water hoarhound more or less from June to October, and even yesterday (October 30) we saw bees busy on the blossoms.
>
> "The honey is of good color and pleasant flavor. We think that where this plant abounds in sufficient quantities, it will be found to be a valuable one in long-continued drouths, as it seems capable of secreting honey when everything else fails. It likes a low, mucky soil, reclaimed marsh, etc."

WATER LILY (Castalia).

There are several species of *castalia* common to the eastern and southern states. H. B. Parks states that *Castalia elegans,* which is common in ponds in southern Texas and adjacent Mexico, is the source of honey.

Lovell states, (Gleanings p. 20, Jan., 1931) that the water lily yields pollen only and that it is a mistake to list it among honey plants.

WATERMELON (Citrullus citrullus).

Where grown on a commercial scale the watermelon is the source of some honey. The bees visit the blossoms eagerly for both pollen and nectar. Important only in a few localities.

WATER PEPPER, see Heartsease.

WATER PLANTAIN (Alisma).

The water plantain is common in swamps, ponds and ditches in nearly all parts of America from New England to Washington and from Florida to California. It is a perennial herb growing in shallow water or mud.

According to H. B. Parks, it is much visited by bees and probably adds something to the crop in the swampy regions of Texas. It is well known that the bees gather nectar from a great variety of sources in swampy regions where but little information is available as to the relative importance of particular plants.

WATERSHIELD (Cabomba caroliniana).

The cabomba or water shield is found in ponds and sluggish streams in the southeastern states from southern Missouri and Illinois to the Carolinas and south to Texas and Florida. H. B. Parks writes to the author that the bees work it heavily and that since it grows with the water chinquapin it is probable that much of the honey credited to that plant comes from the cabomba. Parks states that most of the water plants are nectar-bearing and that bees store much nectar from them during the long blooming period from May to October.

WATERWEED (Jussiaea californica). PRIMROSE WILLOW.

The California waterweed is common to wet lands in the lower Sacramento and San Joaquin Valleys, where, according to Richter, it is the source of considerable ill-tasting honey.

Dr. Oertel reports that *J. decurrens* yields both nectar and pollen in Louisiana.

WATER-WILLOW (Dianthera americana).

The water-willow is found in water or moist places from New England and Ontario to Wisconsin and south to Georgia and Texas. It is a perennial herb with entire leaves and purple or pale flowers in dense spikes.

In Texas this species is common wherever there is clear fresh water and is reported as a valuable source of nectar. H. B. Parks writes that it yields from June to October and that honey from it is sometimes secured unmixed with that from other sources. He states that it resembles the white clover honey of the north, and where running water abounds produces more honey than clover. Growing at the water's edge the hot dry weather does not affect the yield. It is visited by the bees early in the morning and the yield continues throughout the day.

WATTLE, see Acacia.
WAXBERRY, see Snowberry.

WAX MYRTLE (Myrica).

The California wax myrtle (*M. californica*) is common on sand dunes and forest slopes near the ocean for nearly the whole length of the California coast. It is an evergreen shrub or small tree 8 to 20 feet in height. Coleman, in Western Honeybee (Feb. 1921), lists it as a source of both pollen and honey. He also lists the sweet bay-berry *M. Hartwegii,* which is a deciduous shrub of the Sierra Nevada mountains.

The wax myrtle of the east (*M. conifera*) is found in sandy swamps from Maryland and New Jersey south to Florida and west to Arkansas and Texas. It is also found in the Bahamas and the West Indies. It is also known as bayberry, waxberry, candleberry and puckerbush.

The odorless myrtle (*M. inodora*) is a rather rare evergreen shrub found in wet lands near the coast in Florida, Alabama and Mississippi.

The waxberry (*M. carolinensis*) is found on sandy soil along the Atlantic Coast from Nova Scotia to Florida. It is also found in New Jersey and Pennsylvania. It was formerly utilized as a source of wax by the early settlers of the eastern states.

There is still another species, *M. pumila,* found in the sandy barrens of Georgia and Florida, where it blooms in winter and early spring.

Frank Stirling lists three species, *M. cerifera, M. pumila,* and *M. carolinensis,* as of some value to Florida beekeepers. He places them with andromeda among the minor sources.

WEATHER AND HONEY PRODUCTION.

It is a well-known fact that nectar secretion is very sensitive to weather conditions. However, the same conditions are not most favorable for all plants.

J. L. Strong, a beekeeper of Clarinda, Iowa, kept a careful record of weather conditions in connection with the daily gain or loss of a colony on scales for 29 years, from 1885 to 1914. In his locality, in southwestern Iowa, white clover was the principal source of honey. While the bees begin the season with the willows, maples, fruit bloom and dandelion, there is little surplus stored until the blooming of clover. Leslie A. Kenoyer, of the Iowa College of Agriculture, made a careful study of the Strong records and prepared a bulletin outlining his conclusions. While the records kept by Mr. Strong for a long period give material for fairly accurate conclusions regarding the effect of the weather on nectar secretion in white clover, it can hardly be expected that the same conclusions will apply to all other plants. (Bulletin 169, Agricultural Experiment Station, Ames, Iowa.) From the foregoing the following conclusions are extracted:

Rain

Abundant rain seems essential to stimulate plants to the vigor necessary to nectar production and to furnish the water contained in the secretion.

Some honey years are poor because of excessive rain. In June, 1902, there was 11.64 inches of rainfall. June, 1911 represents the opposite extreme, with .76 inch, and was also poor. Abundant rain in advance of the honeyflow is of great importance, particularly in the month of May. When there is 5 inches or more of rain in May an abundant honey harvest seldom fails.

It is also shown that a heavy fall of snow during the previous winter tends to favor a good yield of honey. The abundance of snow provides both available moisture and protection to the clover plants during the cold months.

A rainy day is, of course, unequal to a clear one for honey production, and the fact is shown that best yields are gathered on the days just preceding and following the rains. A gradual increase in honey stored is apparent each day following a rain, until the fourth day, when it begins gradually to decline.

Wind

An average of results for 200 days shows that south winds are slightly more favorable, probably due to the warmer and clearer weather. East winds are shown to be somewhat unfavorable, probably due to the clouds and rains which frequently accompany them.

Temperature

It is generally recognized among beekeepers that hot weather is most favorable for nectar secretion. The above records show that the hot months do yield better than the cooler ones, this being especially noticeable in the months of May and September, which are often too cool in Iowa for best production. It was found that little honey was stored at a temperature below 70 degrees and that 90 per cent of the entire amount gathered was at temperatures between 80 and 100 degrees. Days attaining a maximum between 80 and 90 degrees are the best yielding days, being slightly better than those with higher temperatures. A variation in the temperature seems to favor nectar secretion. Mr. Kenoyer was able to show experimentally that low temperatures favor the accumulation of nectar. Days, then, with a wide range of temperature favored the bees.

A cold winter previous to the honeyflow cannot be said to be beneficial, yet it is shown that it is not detrimental. A warm March, however, often favors a good season. This is explained as due to the favorable conditions for building up the colonies of bees in spring as well as to the increased vigor of the clover plants through lack of severe weather at the start of the growing season.

Kenoyer's Conclusions

"Mr. Kenoyer concludes that, 1. June yields 56 per cent of the annual hive increase in honey, with about half the remainder gathered in July. 2. A large June increase indicates a good year. 3. There is an evident alternation between good and poor years. 4. A good year has a rainfall slightly above the average, the honey season being preceded by an autumn, winter and spring with more than the average precipitation. 5. A rainy May scarcely fails to precede a good honey season. 6. South wind seems favorable and east wind unfavorable. 7. The yield shows a gradual depression preceding and a gradual increase until about the fourth day following a rainy day, after which it remains fairly constant until about the fourteenth day following a rain. 8. Good honey months average slightly higher in temperature than poor, this being especially true of the spring and fall months. 9. Clear days are favorable to the production of honey. 10. Yield is best on days having maximum temperature of 80 to 90 degrees. 11. A wide daily range of temperature is favorable for a good yield. 12. A low barometer is favorable for a good yield. 13. The fluctuations in the yield for a producing period seem to be closely correlated to the temperature range and the barometric pressure, acting jointly. 14. A cold winter has no detrimental effect on the yield of the succeeding season, but a cold March reduces it. 15. A winter of heavy snowfall is, in the great majority of cases, followed by a larger honey yield."

In the American Bee Journal for March, 1927, Prof. A. V. Mitchener had an extended article on "Influence of Temperature on the Honeyflow." This dealt particularly with conditions on the northern plains of Manitoba where sweet clover (*Melilotus*) is the principal source of surplus. His conclusions are based on careful record of several observers in different parts of the province with hives of bees on scales during the honeyflow.

His conclusions follow:

"A study of the data seems to prove that, other things being practically the same, the more hours of sunshine there are daily, the more nectar the bees bring into the hive. Evidence is also produced to show that the greater the spread between the night and the day temperature, the greater the increase in the weight of the hive. Further, it seems that the higher the spread is, namely, the hotter it is during the day, the greater the gains made by the colony."

WEST VIRGINIA—Honey Sources of.

The greater part of West Virginia is rugged and mountainous with forest clad hills and fertile valleys. The streams are swift, furnishing abundant water power for industrial purposes. Abundant coal and other minerals offer especially favorable conditions for industrial development. The weather is variable with precipitation ample and well distributed throughout the year.

Beekeeping is not important in the state, although in a few counties considerable honey is produced. Willows, maples, dandelion and fruit bloom furnish abundant spring forage. Tupelo and black gum yield surplus in some localities. Tulip-tree, black locust, sourwood, basswood, sumac and huckleberry are important. Alsike and white clover are the principal sources of surplus in some sections. Buckwheat, viper's bugloss, goldenrod and asters are also sources of surplus. Fireweed or willow herb is valuable in forest regions.

In Bulletin 33, West Virginia Department of Agriculture, Charles A. Reese listed the following additional sources of bee pasture: redbud, Juneberry, hawthorn, azalea, holly, chestnut and sourwood.

WHEAT.

The wheat plant, the source of most of the world's bread, is nectarless and hence yields no honey. However, when the plant is cut before the stalk is fully ripened, the bees often work on the stubble, storing the sap that flows from the open stems. They are usually reported as working in this manner for a short period of two or three days. In most cases the results do not show in the hives, although there are cases on record where honey, or rather syrup, from this source has been extracted. (Gleanings 1883, page 9.)

It is probable that in few cases do the bees get enough sap to amount to anything nor is it likely to be of much value when they do.

WHISK-BROOM PARSLEY, see Cogswellia.
WHITE ALDER, see Pepperbush.

WHITE BRUSH (Lippia ligustrina).

White brush is a very common shrub throughout south Texas. It has some resemblance to the Indian currant of the Northern States, except that it is larger in size. It has long sprays of fragrant white flowers which appear several times during the year, following rains. According to H. B. Parks, it yields little honey during the regular blooming period of spring, but during the rain-induced bloom, in late fall, it yields heavily.

In the lower Rio Grande Valley, beekeepers reported to the author that the blooming period is short, but that it usually yields well, while the bloom lasts. In many localities similar reports are received. Most beekeepers regard the honey as of good quality, but the plant not dependable. In seasons when there are frequent heavy rains considerable honey is harvested, as it blooms after every heavy rain.

WHITE CLOVER (Trifolium repens).

White clover once held first place as a honey plant in America. It is important as a source of nectar from Maine to Nebraska and south to Kentucky and Missouri. In all the Northeastern States it is one of the principal sources of nectar and, in many localities, it stands alone as the source of marketable surplus. Remove white clover, and beekeeping would be a poor dependence in a large portion of this great area. Alsike is similar in yield and quality of its honey, but it is not so widely spread. The white clover plant is a perennial and establishes itself in pastures, along roadsides and in waste places everywhere. It is a good lawn plant and holds its own with bluegrass in a way that few plants will do.

It yields more heavily in the northern part of its range. One Minnesota beekeeper reports that a yield of less than 200 pounds per colony from white clover is uncommon. The author kept bees in southern

Iowa for many years, and it was a rare season when the yield in that locality totaled 200 pounds per colony. In northern Iowa the average yield is much better than in the southern part of the State.

Best yields come in seasons following a year of excessive rainfall. In wet years the conditions favor the rooting of thousands of new plants, which are ready to produce a crop of nectar the following summer. Most readers of beekeeping literature are familiar with Dr. C. C. Miller's phenomenal crop harvested in 1913. From 72 colonies of bees

White clover.

he harvested more than 19,000 finished sections of honey, or more than an average of 266 sections per colony. His best colony produced 402 sections of white clover honey. The crop was due to a favorable season, combined with expert management. The flow was unusually long, lasting from early June to late August.

White clover yields best when the weather is hot, with plenty of moisture in the soil.

Honey from white clover is of the best quality. It is mild in flavor, light in color and commands the highest prices in most markets. It is the one honey of high quality produced in sufficient quantity to fill a distinct demand for long periods of time.

Most people prefer white clover honey to the somewhat more spicy flavors of alfalfa or sweet clover, though in color they are lighter.

WHITE IRON BARK, **see Eucalyptus.**

WHITE MANGROVE or BUTTONWOOD (Laguncularia racemosa).

The white mangrove is a tree, sometimes reaching a height of 60 feet, which is common to the seashore of southern Florida, the West Indies, Mexico and Central America. In Florida, however, it seldom attains large size, and usually occurs as a shrub.

It has thick and leathery leaves and short, stout branches. The blooming period is usually in June and, according to Frank Stirling, it is an important source of light amber honey. He states that it does not yield as freely as the black mangrove.

WHITE SNAKE-ROOT, see Boneset.
WHITE-WEED, see Ox-eye Daisy.
WHITEWOOD, see Basswood, also Tulip-Poplar.
WICKY, see Andromeda, also Laurel.

WILD ALFALFA or DEER CLOVER (Lotus glaber).

Wild alfalfa is also known as wild broom, deerweed and tanglefoot. In California it is regarded as an important source of honey over a large part of the State. It is a plant growing two to three feet high, with a woody stem at the base. The flowers are yellow, later turning red. Some years the plant is very abundant, then it dies out for a time, so that it varies greatly from year to year. The blooming season is from June to September.

We quote Richter as follows:

> "A very erratic honey producer. Some years, in some sections, yielding twice as much as the sages; this is true for either the coast or the valley side of the coast ranges, yet a good wild alfalfa flow on the east coast does not necessarily mean such is the case on the west side. Beekeepers report wild alfalfa honey as being white, light amber and at times with a characteristic greenish tinge. This is one of the main honey plants of the Coalinga district. This plant, according to Mr. Z. Quincy, of Ramona, upon reaching its second year of growth, after a mountain fire, is said to give us a great amount of nectar."

Coleman states that the honey from deer-clover is water white and that it candies in a few months, when pure, but that it is usually mixed with honey from the sages or other source, which prevents granulation.

WILD BALSAM APPLE, see Wild Cucumber.
WILD BERGAMOT, see Horsemint.

WILD BUCKWHEAT or FLAT TOP (Eriogonum).

A group of low annual or perennial plants common to the Rocky Mountain and plains States. It is a large group and numerous species are to be found from Nebraska to California. In New Mexico the Eriogonums are among the commonest plants, there being something like forty species recorded from that State. The best known is the wild buckwheat of southern California (*Eriogonum fasciculatum*). This is an important source of nectar in that region. The honey is said to be light amber and of good flavor. It granulates readily.

Richter lists it as the most important honey plant in many southern California localities.

> "Wild buckwheat (*Eriogonum fasciculatum*) is one of the principal honey plants of southern California. In the Acton, Antelope Valley and Elsinore districts it is the main reliance. Although quite abundant at lower levels, it does not yield much nectar below 1,500 feet nor above 5,000 feet. The honey is a deep light amber, but of fine flavor, and the comb is very white. The honey has such a heavy body that it is seldom extracted."—J. D. Bixby, Western Honeybee, May, 1917, page 116.

Mrs. May Lovett reports that *E. fasciculatum* is found in a rather restricted area in the mountains near Phoenix Arizona. There it yields a very good quality of light amber table honey which she states granulates readily. The flow lasts until killing frosts, sometimes until near Christmas.

Blossoms of wild buckwheat.

In Colorado *Eriogonum effusum* is sometimes known as "heather." According to Herman Rauchfuss it yields nectar nearly every year when there are late rains. It blooms in August and September and is usually in bloom for some time before the bees are attracted to it. He sometimes gets a super of comb honey per colony from this source. The honey he reports to be amber, of fine quality, and strong aroma, of which one does not easily tire.

It grows abundantly on the prairie about Denver. The plant is small and inconspicuous, with minute flowers, and is not generally recognized as a honey plant among beekeepers of Colorado.

There are several species of *Eriogonum* common to the northwest and it is probable that some honey is secured from this source in various localities in British

Columbia, Washington, Oregon and eastward. In northwestern Oregon some species are locally called sage and are reported as important about one year in four. Most years the flow is said to be light. J. Skovbo reports that in 1920 he secured 100 pounds from Eriogonum in Umatilla County. He states that the honey is strong, though not unpleasant, and of amber color. It granulates quickly, according to his report. The flow lasts four to six weeks with him, coming in late summer. In 1916 it continued to yield until late in October.

Scullen and Vansell record *E compositum* as a source of nectar and pollen near Medford, Oregon. Also *E. niveum* as yielding heavily of light colored honey, some years in Northeastern Oregon. (Bul. 412 Oregon Ex. Station.) See also bindweed.

WILD CABBAGE (Caulanthus crassicaulis).

Wild cabbage is a succulent biennial with flowers in long racemes. It is found on rocky soils and hillsides in the northwest. It is reported from Oregon, California, Utah and Idaho.

From Hermiston, Oregon, J. Scovbo reports that although not very common it seems to produce nectar very plentifully. He states that usually several bees are to be found on the same plant and that they seem to try to get a load from a single flower rather than by flying from one to another. The blooming period is May and June.

James A. Green of Grand Junction, Colorado, reports that in that vicinity *Thelypodium elegans,* a plant belonging to the mustard family is known as wild cabbage. In a letter to the author he wrote as follows concerning it:—

> "It is now (April 24) in full bloom on the desert hills near our apiaries, where it is quite abundant. It is visited very freely by the bees and we consider it of great value as a source of early honey to encourage brood rearing. As dandelions are also in bloom and abundant, I do not know from which source the honey coming in very freely now, yellow with a decided green tinge, is most to be ascribed."

WILD CARROT, see Carrot.

WILD CHERRY (Prunus serotina).

The wild cherries are widely distributed over the North American Continent, and beekeepers who live in timbered sections may expect to find one or more species within reach. The wild black cherry, is a large tree with reddish-brown branches and oblong taper-pointed leaves. This tree is common in the woods of Newfoundland, Ontario and Manitoba, south to Florida and Arizona. There is a smaller tree with very similar flowers, the choke cherry (*P. virginiana*) to be found over much of the same territory, while the western choke cherry, or western wild cherry (*P. demissa*), ranges from Dakota, Kansas and New Mexico west to California and British Columbia.

The larger tree *P. serotina,* is also said to occur in Mexico, Peru and Columbia. There is also a varietal form known as the mountain black cherry, found in southwestern Virginia, Georgia and Alabama. It is found on the open rocky summits of the higher altitudes. This form is a tree 25 to 35 feet high, with very rough bark and drooping branches. The wild red cherry, or pigeon cherry (*P. pennsylvanica*) is common in the Northwestern States, and secretes nectar freely.

Both leaves and seeds of all these forms are poisonous, although the fruit is edible. There seem to be well authenticated cases of poisoning of cattle from eating the leaves, and of children dying from swallowing

Blossoms and leaves of wild cherry.

the seeds. Pammel, in his book of poisonous plants, gives an extended description of the chemical action in such cases. The poisonous property of all species of cherry leaves, according to authorities quoted there, is due to prussic acid. The poison does not exist as such in the growing plant, but by the action of moisture and a vegetable ferment which exists in the plant, a complicated chemical reaction takes place when the leaves are separated from the stem. Wild cherry bark is used to some extent in medicine.

Wild cherries are not often reported as valuable sources of nectar. Richter lists the western choke cherry as a source of honey in California, and Lovell mentions the wild red cherry in the Eastern States.

The writer has a sample of wild cherry honey sent to him from the apiary of W. S. Pangburn, of Jones County, Iowa, having a distinct cherry taste and bright yellow color. After two years it shows no trace of granulation, although subject to all changes of temperature of Iowa climate, both summer and winter. All but few of the samples of honey in the collection have candied under similar conditions.

Since in the Northern States it blooms after the domestic fruits and just before the opening of white clover, it should prove of considerable value where present in quantity.

The author found the bees storing honey which apparently came from wild cherry near Mexico City, Mexico, in February, 1945.

R. E. Newell of Holliston, Mass., reports that wild cherries are more valuable for honey than most people realize. He says that the bloom follows that of cultivated fruits and yields a mild, light colored honey in considerable quantity so that some surplus is taken.

Blossom, fruit and leaf of wild cucumber.

WILD CHINA (Sapindus drummondi).

The wild China tree, also known as chinaberry, or soapberry, is a common shade tree in the Southeastern States. It is also found in the Southwestern States to some extent.

It is frequently mentioned as a honey plant in the Southern States, but is probably not sufficiently common in many places to be important. It is sometimes confused with the China tree (*Melia azedarach*), which see.

WILD CRAB APPLE, see Crab Apple.

WILD CUCUMBER (Echinocystis lobata).

The wild cucumber, or wild balsam apple, is a climbing vine common along streams from New England to Texas. It is also commonly cultivated as a shade for arbors and porches. The plant is an annual and comes from the seed each year. There are few localities where it is sufficiently abundant to be of value to the beekeeper, and it is seldom mentioned among honey plants. However, in a few localities along the Mississippi River it is reported as quite an important source of nectar in mid-summer. On river bottoms it is sometimes to be found in great abundance. The honey is reputed to be white and of good flavor.

The one-seeded bur cucumber (*Sicyos angulatus*) is also common on rich soils along the river banks from Maine and Quebec to Minnesota and south to Florida and Texas. It yields a white honey which is stored as surplus in some localities. It is reported as especially valuable in southern Indiana, where it is found with bluevine or vining milkweed.

WILD CURRANT, see Currant, also Barberry.

WILD HOLLYHOCKS (Sidalcea malvaeflora). CHECKER-BLOOM.

The wild hollyhock is common in the valleys and plains of California, where it is of some importance as a honey plant. It is reported as of special importance in the Imperial Valley. Related species occur in New Mexico and north to Utah.

WILD HYACINTH (Camassia esculenta).

Specimens of wild hyacinth have been sent to the author by beekeepers who have found the bees working upon them freely. The scientific name of the plant is an adaptation of the Indian name quamash and the plant is commonly known as camass also.

Charles Robinson in Botanical Gazette March, 1896, mentions finding the honeybees abundant on this plant which is common at Carlinville, Illinois.

WILD LIME-TREE, see Tupelo.

WILD OLIVE, see Oleaster.

WILD PEACH (Prunus Caroliniana).

The wild peach of the south is an evergreen tree common to the coast region from the Carolinas to Florida and west to Texas. It is known by a variety of names as wild orange, mock orange, laurel cherry, evergreen cherry and Laury-Mundy. It reaches a height of thirty or more feet and blooms in spring from February to April. The fruit is black and hangs on the tree for long periods.

Many south Texas beekeepers report is as a source of nectar. The bees evidently secure both nectar and pollen and the honey is said to be dark in color and of mild flavor.

WILD PENNYROYAL (Satureia rigidi).

Wild pennyroyal is a square-stemmed plant of the mint family that grows abundantly on the sandy pine lands of the south half of Florida.

Wild pennyroyal.

It begins blooming in December in the southern part of its range, and blooms till early in March. Weather conditions are too uncertain during the winter months to favor storing much surplus honey. However, according to Poppleton (Review, Jan., 1893), it is the source of some surplus, and from it the bees are stimulated to begin heavy brood-rearing about Christmas. In an occasional season a fair amount of surplus was secured, sometimes as much as 50 pounds per colony. The honey is said to be light in color, good flavor and heavy body—a first-class article.

Blooming as it does in the winter months, it is invaluable to the beekeeper whose bees have access to it. If no surplus is secured, it serves to fill the hives with bees and honey at an important season and to prepare for the later crops to follow.

WILD RADISH, see Radish.
WILD SUCCORY, see Chicory.
WILD SUNFLOWER, see Sunflower.
WILD SWEET POTATO VINE, see Bluevine.
WILD TEA, see False Indigo.
WILD THYME, see Thyme.

WILLOW (Salix).

In the Northern States the blooming of the pussy willow (*Salix discolor*) is among the first signs of spring. It is a small tree, growing along streams and on wet lands. Furnishing as it does about the first honey of the season, as well as pollen in abundance, it is highly regarded by the beekeepers.

The pussy willow is one of the first trees to bloom in the North.

There are about 160 species of willows, mostly confined to the cooler and temperate regions of North America. Some species extend their ranges into the Arctic regions, where the vegetation is sparse. While the number of varieties is not so great in the Southern States, it is regarded as valuable in the Gulf States and in California. As an example of the comparative abundance of willows North and South, it may be mentioned that four species are recorded for Alabama and eighteen for Connecticut. The willows bloom too early in the spring in the Northern States for the bees to store surplus from this source, but both nectar and pollen are supplied for early brood-rearing.

In Richter's "Honey Plants of California" I find reference to numerous localities where surplus has been secured from the willows. It is said to be "a dark amber and bitter honey." In a few other southern locations surplus yields from willow are reported. The flowers on one tree will be staminate and on another pistillate.

In British Columbia Prof. J. Davidson ("Native Flowers for Bees") mentions the willows as probably the most valuable plants to the apiarist. Although furnishing the first nectar of the season, the bees sometimes store as high as 8 to 15 pounds of honey per hive from this source. He states that the honey has a pleasant aromatic taste not unlike that obtained from fruit blossoms. He also mentions the fact that no early blooming flowers furnish such an abundance of pollen as willows.

Newman I. Lyle of Sheldon, reports one good honeyflow from willow in Iowa in a period of 23 years. It came in April and some of the strong colonies were hanging out with every cell in the brood combs filled with honey. Strong colonies stored about thirty pounds of white honey with mild flavor. The weather was dry with several days of still warm weather.

Percy H. Wright in American Bee Journal, May 1942 writes concerning Bebb's Willow, (*Salix Bebbiana*) as follows:—

> "In Saskatchewan there is one willow, widespread and fairly common, which blooms late, at a time when the bees are almost sure to have a chance to work it. It grows freely in low spots throughout the province and far into the wooded north. In the Plains area the bloom comes during the second week of May, a little before dandelion bloom. The bees flock to it in preference to dandelion, or to fruit bloom. Examination of the combs indicates that the plant is an exceedingly generous producer of nectar. It is also a most abundant producer of pollen."

WILLOW HERB, see Fireweed.
WINTER-BERRY, see Holly.

WINTER CRESS (Barbarea vulgaris). YELLOW ROCKET.

Winter Cress is a widely distributed weed, native to Europe. It was long ago introduced into this country in the region around Lake Superior, and has gradually extended its range.

Reports come to the author of bees working it freely but there is little available information as to the extent of its yield or the quality of the honey. Von Mueller, (Select Extra-Tropical Plants), states that it is a valuable honey plant, especially for cold regions.

WINTER HUCKLEBERRY, see Farkle-Berry.
WINTER SAVORY, see Savory.

WISCONSIN—Honey Sources of.

The greater part of Wisconsin's surplus honey comes from white and alsike clover. Basswood is still important in the northern part of the state and to some extent along the western border and in numerous interior localities where the basswood forests have not yet been cut.

Fireweed is an important source of surplus honey in the burned over forest areas and buckwheat yields well on the sandy soils. Wild raspberry yields surplus also in the cut-over country. Goldenrods and asters are to be found over the state and in many places give a good fall flow along with heartsease.

The spring forage is similar to that of other states in this region, elms, maples, willows, dandelion, fruit blossoms and early wild flowers furnishing an abundance of pollen and nectar for spring brood rearing.

Sweet clover is gradually coming into favor among the farmers and bids fair to become the most important honey plant of Wisconsin, as of other states in the central part of the continent.

In the May 1923, issue of the American Bee Journal, Prof. H. F. Wilson gave an extended account of the natural regions of that state. He divides the area into eight more or less arbitrary divisions, which are not clearly defined, some of which overlap two or three different soil areas. He describes the region bordering on Lake Michigan about central north and south, as the best beekeeping territory in Wisconsin. Alsike clover is largely grown for seed in this section. To the south of this area sweet clover yields a dependable surplus. Goldenrod may be expected to give a crop of honey in the central part of the state.

WISTERIA.

The Chinese wisteria, the "Fugi" of China and Japan, is widely grown in America as an ornamental climber. It is a very hardy and long-lived vine and attains great size where properly supported.

There are two native species in the southeastern states. *Wisteria frutescens,* known as virgin's bower or kidney-bean tree, is found in low lands from Virginia to Arkansas and south to Florida and Texas. *W. macrostachys* is found in swamps in Arkansas, Louisiana, Missouri and southern Illinois. Both are climbing vines.

The wisterias are very attractive to the bees and some references indicate that where sufficiently abundant they may be very important. The cultivated sorts may be of some value in cities and towns, while the wild ones yield nectar in the swampy regions of the southern states.

WOLFBERRY (Symphoricarpos occidentalis).

Wolfberry is a relative of the snowberry and Indian currant and is probably of very similar value as a source of nectar. Much honey is gathered from all these plants when conditions are favorable. All three are commonly known as "buckbrush."

Root gives wolfberry as one of the most important honey plants of northern Idaho, stating that an average of 25 pounds per colony is secured from this source. ("A. B. C. of Bee Culture," 1919 edition).

It is frequently mentioned as a source of nectar in British Columbia, although usually snowberry is reported as of more importance. Wolfberry is widely distributed, being found from Michigan and Illinois to Kansas, Colorado and British Columbia. (See also Indian Currant and Snowberry).

In the Agricultural Journal for February, 1925, W. J. Sheppard writes that A. H. Smith at Natal in the Fernie district of British Columbia where no bees had previously been kept, harvested 3,800 pounds from sixteen 3-pound packages or an average of 238 pounds per hive from this source.

Wolfberry is called badgerbrush in some sections of Western Canada.

WOLF WILLOW (Eleagnus argentea) Silverberry.

The wolf willow or silverberry is a shrub without thorns but with silvery leaves and fragrant flowers, which are also silvery without and yellow within, which is found from Quebec to Alberta and southward to Minnesota and Utah. It is also common to portions of British Columbia. It is reported as of special importance to the beekeepers of the prairie provinces of western Canada. Numerous reports of bees working freely on wolf willow are available.

WONDER HONEY PLANT, see Pentstemon.
WOODBINE, see Virginia Creeper.

WOOD MINT (Blephilia ciliata).

The wood mint is native to dry open places from New England west to Minnesota and south to Missouri and Georgia. It is perennial with small bluish-purple flowers in mid-summer. It is rarely mentioned either as an ornamental or a bee plant but is valuable for either. In the garden the plant is very attractive and is swarming with bees

Woodmint is attractive both as an ornamental and as a source of nectar.

during its period of bloom. Wood mint appears to be worthy of far more attention than it has thus far received.

WOOD SAGE, see Germander.
WOOLLY WHITE DROUGHT WEED, see Turkey Mullein.

WYOMING—Honey Sources of.

Wyoming is situated between Montana and Colorado and the conditions are so similar that the general description of those states should serve for Wyoming also. The state lies at a high altitude and the rainfall is light. While commercial honey production is highly developed and large quantities of honey are shipped to market, it is nearly all produced from alfalfa and sweet clover.

The beekeepers are to be found in the irrigated valleys and little forage is available for the bees outside the irrigated territory. Occasionally a small amount of surplus is harvested from Cleome, commonly called Rocky Mountain Bee Plant and from the gumweed. Cottonwood and willow trees along the water-courses furnish pollen for spring brood rearing. This is supplemented by dandelion, which is spreading over the cultivated areas.

XYZ

YAUPON, see Holly.
YELLOW INDIGO, see Indigo-weed.

YELLOW JESSAMINE (Gelsemium sempervirens).

The yellow jessamine is a well-known poisonous climbing vine common to the Southern States from Virginia to Florida and west to Mexico. Its yellow flowers, in short axillary clusters, appear in early spring (February and March) and are very fragrant. The vine climbs over trees to a great height, often 30 feet or more. It yields pollen and probably some nectar. It is reported as poisonous to the bees.

> "For the past nine years I have observed, commencing with the opening of the yellow jessamine flowers, a very fatal disease attacking the young bees and continuing until the cessation of the bloom. The malady would then cease as quickly as it came. The symptoms of the poisoning are: The abdomen becomes very much distended, and the bees act as though intoxicated. There is great loss of muscular power. The bee, unless too far gone, slowly crawls out of the hive and very soon expires. The deaths in twenty-four hours, in strong stocks with much hatching brood, may amount to one-half pint, often much more. My observations have been verified by dozens of intelligent beekeepers breeding pure Italians where *Gelsemium* abounds."
> —Dr. J. P. H. Brown, American Bee Journal, Nov., 1879.

With reference to the condition described by Dr. Brown, the matter was referred to T. W. Livingston, of Leslie, Georgia, who writes as follows:

"I have for many years noticed the disease described. I have seen the same disease where there was no yellow jessamine, that I knew of, but much more of it where that plant was plentiful. It may be caused by it. It was told several years ago by the Florida State Chemist, who had analyzed a sample of honey reported as poisoning some people, that the honey contained pollen grains from yellow jasmine."—March 13, 1919.

As to the effect on animals poisoned by the plant we quote Pammel as follows:

"Dr. Winslow gives the toxicological effect on animals as follows: Muscular weakness, especially in the forelegs, staggering gait and falling. These symptoms are followed by convulsive movements of the head, forelegs and sometimes of the hindlegs. The respiration is slow and feeble, temperature reduced, and there is sweating. Death occurs because of respiratory failure." —Manual of Poisonous Plants.

YELLOW POPLAR, see Tulip Tree.
YELLOW STAR THISTLE, see Star Thistle.

YELLOW-TOP (Verbesina encelioides).

The yellow-top is a common annual plant from Kansas and western Missouri to Colorado, Texas and Arizona. Mohr describes it as a frequent and persistent ballast weed along railroads in Alabama. The golden yellow flowers appear from July to September. It is reported as important to the beekeeper principally from Texas, where it is common to low grounds throughout the state. H. B. Parks writes that it is common all over southwest Texas, where it is very valuable at times. He describes the honey as dark in color and strong in flavor.

Yellow-top is said to be one of the most dependable sources of midsummer honey in Texas, being little affected by heat or drouth.

Specimens of this plant have been received by the author from H. E. Weisner of Tucson, Arizona, who writes as follows:—

"*Verbesina enceloides,* which Doctor Vorhies gave the common name of 'sore-eye,' is a rather important plant despite its atrocious odor, for it grows with comparatively little moisture at any time from April until heavy frosts and is much visited by bees for both honey and pollen. It is of quite general distribution in this part of the state."

YELLOW-TOPS, see Tarweed.

YELLOW WOOD (Cladrastis lutea).

The yellow wood is a tree confined to a limited range. It is found principally in Kentucky, Tennessee and North Carolina. While it may occur to some extent in the States adjoining the three mentioned, it is rare, except in very limited areas. It is recorded as occurring on shaded bluffs in the Tennessee Valley in Alabama, and may be looked for in similar situations in Mississippi, Georgia or South Carolina. The flowers are white, and appear in April and May. The panicles are

Blossoms of the yellow-wood.

sometimes a foot long. According to the notes furnished by J. M. Buchanan, the honey has a strong, distinctive flavor and is light amber in color.

The wood is heavy and hard and yields a yellow dye. It is known also as Kentucky yellow wood and gopher wood.

YERBA BUENA, (Micromeria chamissonis).

Yerba buena is a trailing perennial herb with slender stem and small, white, solitary flowers. Jepson gives the range as common in woods near the coast: Humboldt County, Marin County, Berkeley, San Francisco, Belmont, Monterey and southward to southern California.

Richter states that it is considered a fair honey plant in places.

Yerba santa. (*Eriodictyon crassifolium*).

YERBA DULCE, see Baccharis.
YERBA DEL PESCADO, see Turkey Mullein.

YERBA SANTA (Eriodictyon Californicum).

Yerba santa is a low shrub common to some parts of California. Richter lists it as important in Ventura County, where it frequently yields surplus, blooming in June and July.

Coleman says that it is common over extensive areas throughout the coast ranges and at middle altitudes in the Sierra Nevada Mountains. He reports the honey as of amber color and good quality, frequently secured in surplus quantity where the plants are plentiful. (Western Honeybee, Dec., 1921). The blooming period is June and

July. It is also known as Mountain Balm. E. crassifolium, the related species shown in picture is apparently less valuable to the bees.

YUCCA.

Over vast areas of the arid west there is little for the bees. A few plants stand the long continued periods of drought even where there is no irrigation, and add to the total production of the apiaries in the irrigated regions. Among the attractive plants may be mentioned the yucca, also called Spanish bayonet, Spanish dagger, Adam's needle, mountain queen and Roman candle. There are about a dozen species,

Yuccas are showy plants when in bloom.

mostly from Dakota west to the Pacific and southward. They are common in Mexico and Central America. They are also to be found in the sandy sections along the Atlantic Coast from North Carolina to Florida and Louisiana.

When in bloom the plant is very ornamental. A single tall flower stalk contains many large, white or cream-colored flowers. In many localities where the plant does not grow wild, it is grown for ornament.

In "Honey Plants of California," Richter lists *Yucca whipplei* as an important source of nectar, which, in localities where it is abundant, yields surplus. In that State its blooming period is June and July.

Trelease, in 13th annual report of the Missouri Botanical Garden (page 124), says: "*Hesperaloe* secretes much nectar and appears to be adapted to birds, as are the Cape aloes, to which it bears no inconsiderable resemblance in its flowers. The other genera are sparingly if at all nectariferous, though all have septal glands, which are rather small in *Clistoyucca*, but very large in others."

It is a well known fact that most of the yuccas are dependent for pollination upon small Pronuba moths which lay their eggs in the developing seed cases where their young spend the early stages of life. It is doubtful whether nectar plays an important part in attracting these insects to the blossoms.

ZYGOPHYLLUM.

Zygophyllum fabago has been naturalized in the vicinity of Messilla Park, New Mexico. J. W. Powell, a beekeeper of that place, reports that the bees roar over it from the time the first bloom appears until the last one fades. He says that he has never known it to fail to produce nectar. This species, the Syrian bean-caper, had not been known elsewhere in the United States previous to 1925 when its presence here became known through publication in the American Bee Journal. It is probable that its range will be greatly spread through the southwest by beemen interested in increasing the available bee pasture.

Several native species of the same family occur in New Mexico and adjoining states where they are probably of some value to the bees.

INDEX